Particle Acceleration and Detection

The series “Particle Acceleration and Detection” is devoted to monograph texts dealing with all aspects of particle acceleration and detection research and advanced teaching. The scope also includes topics such as beam physics and instrumentation as well as applications. Presentations should strongly emphasize the underlying physical and engineering sciences. Of particular interest are

- contributions which relate fundamental research to new applications beyond the immediate realm of the original field of research
- contributions which connect fundamental research in the aforementioned fields to fundamental research in related physical or engineering sciences
- concise accounts of newly emerging important topics that are embedded in a broader framework in order to provide quick but readable access of very new material to a larger audience

The books forming this collection will be of importance to graduate students and active researchers alike.

More information about this series at http://www.springer.com/series/5267

Sören Möller

Accelerator Technology

Applications in Science, Medicine, and Industry

Springer

Sören Möller
Forschungszentrum Jülich GmbH
Jülich, Germany

ISSN 1611-1052 ISSN 2365-0877 (electronic)
Particle Acceleration and Detection
ISBN 978-3-030-62307-4 ISBN 978-3-030-62308-1 (eBook)
https://doi.org/10.1007/978-3-030-62308-1

This Springer imprint is published by the registered company Springer Nature Switzerland AG
The registered company address is: Gewerbestrasse 11, 6330 Cham, Switzerland

Preface

Or why you have to read this book:

Accelerators often seem like an abstract idea, a toy of scientists that grows bigger and bigger with every billion euros you invest, a technology as far away from our lives as quantum physics. This book shows you the secret players amongst the accelerator field that you might not have noticed so far. The working horses in science, industry, and medicine do not work in a distant fundamental science, but those are improving our lives on a daily basis.

The concept of particle accelerators is now about 120 years old. The beginning of this technology was made in 1897 with the cathode ray tubes, later being used in TV devices. The technical climax is currently the Large Hadron Collider (LHC) at CERN, a device of 26.7 km length. Within these 120 years, the maximum energy of the accelerated particles was increased from some keV to several TeV. The incredible increase of a factor of more than a billion compares well to the advances made in the computing power of supercomputers. In between these extremes of particle energy nowadays lays a manifold of devices being used in science, industry, and medical applications worth to learn more about.

Efficient and powerful accelerators have become a key technology for a multitude of applications, although often unnoticed. Silicon wafers, the basis for modern micro-chips, are usually prepared by accelerators called implanters in this context. Cancer can be treated more effectively, yet with less side effects than ever before by proton beams or accelerator produced isotopes. In the development of new combustion engines, accelerator-induced radioactivity visualizes wear processes. In material science, accelerators open a view to the atomic scale in the form of electron microscopes. Accelerator development and applications therefore became an import industrial and development factor, exploited by numerous companies worldwide. All of this connects in accelerator technology.

Accelerator technology is in many aspects substitutive for classical fission technology. For example in scientific applications of neutrons in Europe, accelerator-based neutron sources will soon be more common than fission reactor sources. In nuclear medicine, already today radioactive isotopes produced by accelerators are more frequently administered to patients than those produced in fission reactors. The advantages responsible for this success are in particular related to the low secondary costs of accelerators: less or even no side activity and nuclear waste are produced and also no fissile fuel is required. Therefore, the laboratories can be compact, close to application and de-centralized, e.g. directly in a hospital. In conclusion, accelerators have the potential to grant nuclear industry and nuclear applications new acceptance after the extensive discussions on their safety, especially in Europe.

The knowledge of accelerator technology is exclusively taught in the context of physics related studies, while the knowledge on accelerator applications is part of mechanical, electrical, material, and nuclear engineering, medicine, and biology studies. This book intends to grant an overview over all of these aspects to students, scientists, engineers, and users. In contrast to a physics book focusing on a specific topic, this book will discuss accelerator technology in a broad view, allowing seeing the connections between the different aspects and applications. It combines the technological and physical basics of accelerator applications amongst the accelerator and beam-matter physics and application sides. This is going to be a book for physics interested people with a strong tendency for hands-on work, device layout, and application layout. Of course it is still a book, so expect some equations and fundamental considerations, but even when working hands-on you have to know what you are doing. Technical layouts, technological limits, economic aspects, and radiation protection will be even more intensely covered than physics. This book will present traditional and emerging fields of accelerator applications and, hopefully, inspire the reader to totally new developments.

The basics of accelerator and elementary particle physics will be briefly discussed, but we will not go deeply into collider physics or the standard model. There are enough dedicated books on the market for these topics to which the reader is sincerely referred. Instead of focusing on the development of accelerators for fundamental research with ever growing beam energy and luminosity, we will take a look at robust and industrially established technology. These devices cover ranges from some keV to about 250 MeV with beam powers of μW to MW with alternating current (AC) and direct current (DC) driving for accelerating ions or electrons and combine with clever end-stations forming valuable applications. The terms capabilities, efficiency, productivity, and costs dominate these applications, and in contrast to other accelerator-related literature they will be discussed in this book and identified as positive drivers of innovation and progress.

By far most of these applications work on the basis of stopping and attenuation of charged particle beams in matter, their nuclear reaction processes with matter and nuclear inventories. By combining these physical approaches with the chain of technologies involved in accelerator applications up to the final products, the book can provide the reader a unique insight one cannot get from reading a dozen specific books on these topics. The introduction of modern computer codes for describing the individual processes amongst application examples will enable developing own ideas. Basics on radiation protection help to keep these ideas safe. Finally, the overview of applications in science, industry, and medicine allows integrating the knowledge to see the big picture beyond the individual fields.

So if you want to become an accelerator engineer or you are just curious on understanding how your latest PET scan was performed and what it took to apply it, continue and explore this book!

Jülich, Germany

Sören Möller

Acknowledgements

I want to thank some of my colleagues for the interesting and inspiring discussions on the different topics. I want to thank Rahul Rayaprolu for his knowledge of nuclear reactions and applications and his support in literature research. Thanks to Tobias Wegener for his expertise in nuclear and chemistry matters. Thanks to Marcin Rasinski for giving me a deeper insight into electron microscopy and providing numerous excellent images. Thanks to Timo Dittmar and Christian Scholtysik for our time in the ion-beam analysis laboratory. Daniel Höschen for his inspiring engineering qualities. Ulrich Rücker and Johannes Baggemann for our joint developments towards neutron sources and their input regarding neutron science. Tom Wauters for sharing his knowledge on RF technology. Thanks to Antje Vollmer for the interesting and fruitful discussion of synchrotron light sources. Special thanks to Christian Linsmeier for his extensive support of my scientific career and accelerator competences. Without more words required: thanks to my oldest and best friends Giovanni Nocera, Florian Schmitz, and Simon Vogt. Thanks to my soccer group around Bernd Düppmann and Markus Towara for maintaining a healthy body for my brain and for our good time outside the soccer court. I thank my mother for her general and unconditional support. Thanks to my father for inheriting his talents and making me strive for even more. Finally yet most importantly, I thank Kathrin Pytel and our families for bearing the weight of my world during the exhaustive write-up work and giving me Max.

Jülich, Germany Sören Möller

Contents

Chapter 1
Introduction

Abstract The atomic nucleus and the electrons around it form the basis of matter. Acceleration of particles in natural processes such as supernovae and the hot plasmas of stars represent the origin of all elements except for hydrogen. Here on earth, particle accelerators enable a controlled reproduction of these natural processes in order to control and modify elementary particles and atomic nuclei. The introduction starts asking central questions on the physics of these processes, the required technology, and the use of it.

We work to better ourselves and the rest of humanity (Picard, Star Trek).

The first steps towards our world were made shortly after the big bang when the hot and dense matter cooled down to a level where the hydrogen nucleus can survive. All elements heavier than helium were afterwards generated in stars. About half of the elements heavier than iron are nowadays believed to be synthesized in supernovae. These giant explosions accelerate several solar masses of particles first to their inside where electrons and protons fuse to neutrons forming an incompressible object. This object reflects the impinging matter, leading to the giant supernova-explosion. Neutrons follow the explosion front and react with the elements bred by nuclear fusion forming all the elements and isotopes we have in our nuclide chart displayed in Fig. 1.1. From this extensive set of isotopes, the most decay according to their nuclear properties. Only the stable nuclei survive in the end in a mixture characteristic for this chain of fusion, nuclear reaction, and decay. This nucleosynthesis produced all the atomic nuclei we know. After many years of decay, we end up with the isotopes we find in our earth, but still the numerous particles and radiations emitted from the explosion allow obtaining a thorough understanding of the supernova process from extreme distance.

Supernovae work as giant gravitational accelerators, which are capable of producing basically any isotope and elementary particle, but we are left with the thin black line of boxes in the centre of the known nuclide chart in Fig. 1.1. What if we want to make use of the isotopes aside this line or use their information properties here on earth (in safe distance to a supernova)? So far, we have mastered the manipulation of macroscopic matter for example by metallurgy and machining. We were able to separate the elements from ores and other compounds, leading to the

S. Möller, *Accelerator Technology*, Particle Acceleration and Detection,
https://doi.org/10.1007/978-3-030-62308-1_1

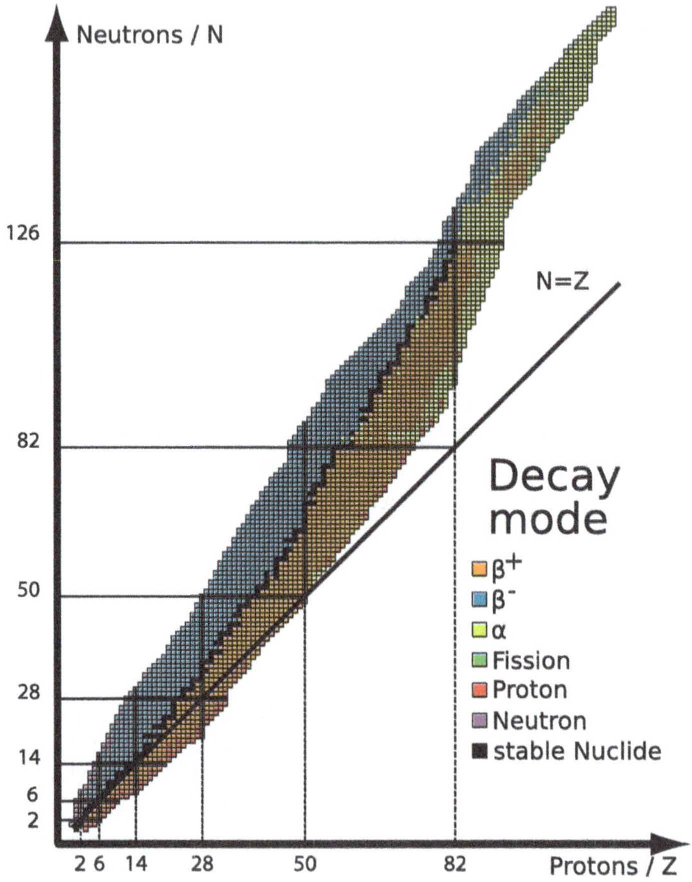

Fig. 1.1 The nuclide chart displays all known nuclear isotopes in the space of protons and neutrons in the nucleus. The highlighted numbers are so-called magical numbers representing particularly stable configurations. Modified version of Table_isotopes.svg: Napy1kenobi, CC BY-SA 3.0, via Wikimedia Commons

formulation of the periodic table of elements. Chemistry allowed us to manipulate the atomic bindings and rearrange them. All of these controls over matter already became part of our everyday life. Now we want to take control of the elementary particles and nuclei by analysing them, controlling their internal state, and finally synthesising any particle in any state to make use of its specific properties. Here on earth we cannot exploit the power of gravity as the supernova does, but we need to apply technology. Accelerators open this door by breaking down the extreme conditions present in supernovae using electro-magnetic technology.

Our modern society is based on technologies to make our lives sustainable, long, and prosperous. As such, we have a strong demand for new possibilities in medicine, industry, and science. Accelerators provide the additional degree of freedom of

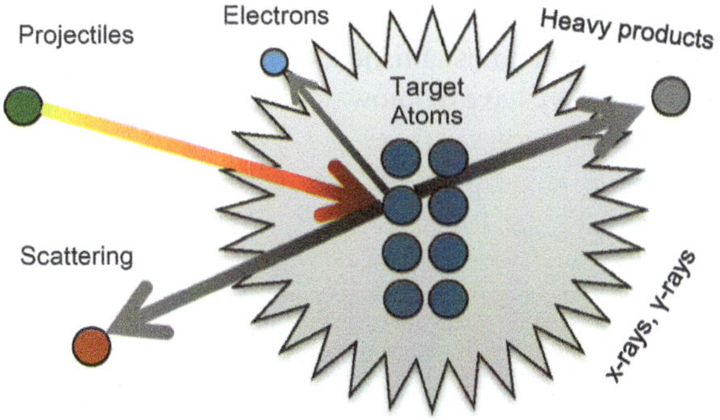

Fig. 1.2 The basic situation considered in applications of accelerator technology. Accelerated projectiles hit atoms in a target leading to possible emission of photons, electrons, and heavy particles depending on projectile energy and species. The application defines the focus

controlling elementary particles and nuclei and interacting with them, allowing finding better and more specific technologies. Accelerators, or more precisely particle accelerators, exploit electric and magnetic fields for separating, accelerating and controlling atomic nuclei and electrons. How exactly does this work and how does it allow us accessing elementary particles and the nuclide chart in an industrial fashion?

Physical interactions facilitate the control over the constituents of matter. Many different interactions are possible, as depicted in Fig. 1.2, but their number is finite. A large fraction of these interactions was already unravelled by science. We learned most of them follow, at least partially, deterministic rules or at least statistical distributions we can exploit in accelerator technologies. The theoretical part of the book will demonstrate and discuss these physics, teaching how to exploit and combine the physics for mobilizing the full potential of accelerator technologies.

What can we do with the new degree of control over matter and the particles? The main focus of this book will be on practical aspects and apparatuses. Hands-on and physics may appear contradictory, but there are fields where they become two sides of a coin. Accelerators exhibit this two-sided coin aspect. The complex apparatuses feature many technical aspects, starting with vacuum, electrical engineering, and safety discussed in Chap. 2. The second part of this book, starting from Chap. 5, forges theory and apparatuses to useful methods. Applications were developed all over our modern society. This book's subtitle *applications in science, industry and medicine* describes this in three buzzwords. The three disciplines will be roughly separated, but remember: One of the main messages of this book is to visualize the strong connections between the applications in these three fields through physical and technological overlaps.

Every good product starts with an idea, but requires lots of qualified feedback in its growth process. Without feedback, may it be positive or negative, the product will never reach an outstanding quality, whether it is in science, medicine, industry, or personal life. Therefore, feel free to write down your own ideas, criticism and corrections into this book and submit them to the author, but also accept feedback yourself. The final sentence of this introduction might be one of the very reasons accelerator scientists found and established such a vast range of applications. This search ever continues and we have to know our individual motivation to become part of this continuing success story.

Chapter 2
Technology

Abstract Technology is the engineering concept of theoretical knowledge. Technical implementations of a technology allow for optimizing a technology towards certain application directions. Think of a house heated by gas or oil, both combustion technologies with different technical realisations. Accelerator applications require combining multiple complex technologies into a new one. This chapter discusses these supporting technologies, their limits, and safety aspects. The understanding of the concept of technology enables the reader to select suitable technologies and identify the point where the technical possibilities of a certain technology ends and new ways have to be found.

A man and a donkey are always smarter than a man.

Technology describes the fundamental way to solve a problem. Let us assume, for example, that a quantity E of electrical energy is to be generated. For this purpose, a solar cell with the output power P could be exposed for t hours to the sun. A combustion engine consuming an amount P of gasoline energy per time with the efficiency μ could also be operated with a generator for $t = E/(\mu * P)$ hours. Furthermore, a nuclear reactor with a steam generator could be run for $t = E/P$ hours. We see that all of these different technologies enable reaching the same goal in completely different ways. Each technology has its own unique features and problems.

In order to clarify the reasoning behind this, let us look at the three examples in detail: The solar cell has an output power P, but it requires the input of the solar hours. At the same time, this power P is limited to a value between 0 and the power of the solar radiation (in practice even less). A combustion engine also achieves an output power P, but there is in principle no upper limit. In contrast to the solar cell, which requires solar hours, the internal combustion engine requires fuel and air inputs, and only achieves an efficiency $\mu < 1$ with these inputs due to its thermal process. The third and final example in this list is the nuclear reactor. Its advantage: It only requires nuclear fuel as input allowing it to be operated in any surrounding. However, due to physical properties, its power has a certain minimum value $P > 0$. Its fuel requirements are small so efficiency plays a negligible role, but it creates nuclear waste as a disadvantage to the other two technologies.

S. Möller, *Accelerator Technology*, Particle Acceleration and Detection,
https://doi.org/10.1007/978-3-030-62308-1_2

The example of power generation shows that every technology has its justification and its technical limitations. On earth we will never be able to build a solar cell that generates the power density of an internal combustion engine, and at the same time we will never be able to develop an internal combustion engine that can be used in space like a nuclear reactor or a solar cell. We will, however, be able to develop a combustion engine with $P = 1$ kW and one with $P = 1$ MW, but presumably we will be able to reach $P = 1$ GW only with a nuclear reactor. Furthermore, we will be able to develop a solar cell based on crystalline and one based on amorphous silicon or a combustion engine that burns diesel and one burning gasoline. These are the *technical* features that are possible within the framework given by the *technology*. The technical solution describes the unique approach within the technology. When solving a problem, it is important to know where the technical and the technological *limits* are. An equivalent statement would be to know where we reach a solution by improving an existing approach (technically) or we require a completely new approach (technology).

This book will address the accelerator technology in terms of accelerator based technologies. The aside on power generation intended to clarify that this technology also has its limitations and unique features. As we shall see, these boundaries are exceptionally broad, so there is great flexibility for the use of accelerator technology. Moreover, in many cases the technology can be well connected to other technologies forming a large set of complex applications.

2.1 Vacuum

Accelerated particles will interact with any form of matter. Although this is the basis of all accelerator applications, it also disturbs the preparation of the particle beams in an accelerator like viscous engine oil. In this analogy, it is therefore necessary to dilute this oil in practically all accelerators, that is, to create a vacuum within the accelerator. A good reading for vacuum physics is actually a free book of the company Pfeiffer vacuum (Pfeiffer vacuum GmbH 2013), although the reader should be careful regarding its sales aspect. For the physical understanding and classification of the necessary technical equipment, we first define the concept of pressure p with the common and equal units of hecto-pascals (hPa) and milli-bar (mbar). This pressure p connects via the ideal gas law

$$p = \frac{N}{V} * k_B * T, \tag{2.1}$$

with the (chamber) volume V, the gas particle count N in the volume V, the Stefan-Boltzmann constant k_B, and the absolute temperature of the gas T. The viscous effect for accelerated particles connects to the gas particle count they pass as we will discuss later in more detail. Equation (2.1) does not refer to the nature of the particles, hence it is valid for individual components of a gas (for example, nitrogen and oxygen in air) as well as their partial pressures, but also for all components together and the total pressure. From the equation, we can see how the pressure is related to the amount of particles N in a volume: Linear!

Table 2.1 shows the relationship between pressure, particle density and other variables relevant for accelerator technology for various pressures from atmospheric pressure to the lowest pressures, such as those in space (extreme ultra-high vacuum). The relevant pressure range spans over 15 orders of magnitude, which are usually divided into 6 pressure ranges. The division of these ranges roughly follows the limits implied by current vacuum technologies. The mean free path, i.e. the average distance between collisions of two gas particles, is the main quantity for understanding the technical aspects of a vacuum. It increases with decreasing pressure and reaches a technically relevant dimension in the range of 10 mm in the high vacuum range ($\approx 10^{-3}$ mbar).

Starting from atmospheric pressure, gas has to be removed from a closed volume, the so-called vacuum vessel, by pumps in order to reach the high vacuum pressure range. The rate of this pumping is referred to as throughput. Practically, the fine vacuum range is always reached by compression pumps. Compression pumps enclose the gas in a sealed displacement capacity inside the pump. This capacity of volume V is alternately reduced (compression phase), expanded into an exhaust gas line, and increased (suction phase). This process chain can take place in various technical implementations for example in the form of reciprocating pistons, rotary vanes, or screws. The pressure changes result in vibrations in a wide spectral range with amplitudes of a few 10 μm up to a few mm. The amplitude depends strongly on the design type of the pump and on the pressure conditions. Often vibration transfer to the vacuum vessel needs to be mitigated by special damping elements or by shutting down the pumps in critical operational phases.

If the mean free path (Table 2.1) is greater than the dimension of the vacuum vessel, the compression process reaches a critical limit. If the pressure is below this limit, compression procedures are ineffective, as shown by the curve of the compressor throughput in Fig. 2.1. In this case, a pressure gradient in the gas no longer drives a flow, since a gas flow is induced by collisions between gas particles. However, by definition collisions between the gas particles and the wall dominate the gas behavior when the mean free path λ is significantly greater than the smallest dimension d of the vacuum vessel (for example a pipe diameter or the capacity diameter of a pump). Collisions between gas particles become decreasingly probable with decreasing pressure. This transition in the ratio of length scales is described by the Knudsen number $K_n = \lambda/d$, with a transition region, the Knudsen flow, around $K_n = 1$. In most accelerator applications, significantly lower pressures than those at $K_n \leq 1$ are required. The regime $K_n < 1$ is referred to as a viscous flow (right half of Fig. 2.1), the regime $K_n > 1$ as a molecular flow (left half of Fig. 2.1).

In molecular flow, pumps therefore require other technological approaches than in the viscous range. Hence, in order to reach molecular flow (HV and lower), starting from atmospheric pressures, a multi-stage pumping system consisting of viscous

Table 2.1 Relevant vacuum pressure ranges and their connection to vacuum technology and physical quantities

Pressure range	Pressure p in hPa or mbar	Particle density N/V in molecules/m^3	Mean free path λ	Energy-loss S for 15 MeV proton in eV/m	Vacuum pumping technology	Vacuum meters
Atmospheric pressure	1013	2.7×10^{25}	68 nm	3×10^6	Compressor	Membrane, heat loss
Rough vacuum	300–1	10^{25}–10^{22}	0.1–100 μm	10^6–10^3	Compressor	Membrane, heat loss
Fine vacuum	1–10^{-3}	10^{22}–10^{19}	0.1–100 mm	10^3–1	Compressor	Membrane heat loss
High vacuum (HV)	10^{-3}–10^{-7}	10^{19}–10^{15}	10 cm–1 km	1–10^{-3}	Turbo	Membrane, friction, ionisation
Ultra-high vacuum (UHV)	10^{-7}–10^{-12}	10^{15}–10^{10}	1 km–10^5 km	10^{-3}–10^{-6}	Turbo/Getter/Cryo	Ionisation
Extreme ultra-high vacuum (XHV)	$<10^{-12}$	$<10^{10}$	$>10^5$ km	$<10^{-6}$	Getter/Cryo	

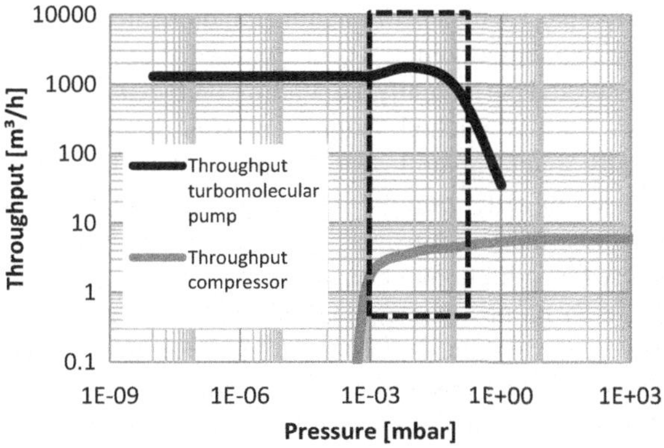

Fig. 2.1 Comparison of throughput of a rotary vane compression pump with 25 mm open diameter to a turbo molecular drag pump of 100 mm open diameter. In the transition region of molecular and viscous flow both devices reach technological limits. Fortunately an overlap exists (box), allowing to combine both in a 2-stage pumping system.

and molecular flow pumps is required. For molecular flow pumping technologies (for HV, UHV and XHV) the effective opening area $A = \pi d^2/4$ of the pump body to the vacuum vessel is the determining factor for the gas-type-dependent throughput S, since collisions with the vessel walls dominate transport in this regime. Because of the molecular flow, an ideal gas with the temperature T consisting of particles of mass m leads to a throughput of

$$S_N(p, m) = \alpha L * Ap * \sqrt{\frac{1}{12mk_BT}} \quad \left[\frac{\text{Particles}}{\text{Time}}\right]$$
$$S_V(m) = \alpha L * A * \sqrt{\frac{k_BT}{12m}} \quad \left[\frac{\text{Volume}}{\text{Time}}\right] \tag{2.2}$$

With the gas species dependent efficiency factor α given by the properties of the pumping technology and the conductivity of the piping between pump and vacuum vessel. The throughput can be stated in particles/time or gas volume/time, which is interchangeable by the ideal gas law, Eq. (2.1). The throughput in particles per time is proportional to the pressure p, accordingly at lower pressure fewer particles get pumped per time.

2.1.1 *Pumping Technologies for the UHV*

In practice, the three most common technologies are: getter pumps, cryopumps, and turbo-molecular pumps. These types differ in their gas-dependent efficiency factors α and their application characteristics.

Getter pumps bind gas by means of chemical and physical reactions with solids, liquids, or gases, the so-called getter. These getters fill up like a sponge and must be replaced or regenerated periodically, depending on the amount of gas throughput, when their capacity limit is reached. Either a plasma (ion getter pump) or a reactive metal (especially titanium sublimation pump) stimulates the binding process. Titanium is often used in getter pumps, since it forms extremely stable compounds with relevant elements, namely nitrogen (TiN), carbon (TiC), and oxygen (TiO_2) and correspondingly has a high getter capacity. The binding of noble gases is only possible with ion getter pumps. For noble gases and other gases which can only be physically bound, the throughput is, however, smaller than for chemically reactive gases. Without chemical reactions, the saturation limit of getter to gas component is by 2–3 orders of magnitude smaller compared to chemically reacting species.

In cryopumps, unlike getter pumps, a cold surface is brought into contact with the vacuum. Thereby, condensation and physical binding of gas on the cold surface, similar to an air dryer, induce the pumping effect. For the condensation of the gas components, their respective boiling point must be above the temperature of the cold surface. Nitrogen features the lowest boiling point of the main components of air (Table 2.2), hence it is generally used in liquid form (LN_2) to cool the cryopump. As shown in Table 2.2, hydrogen and helium are still gaseous at the boiling point of nitrogen and consequently cannot be effectively pumped by condensation in an LN_2 cooled cryopump. In special applications in the XHV, this is solved by cooling with liquid helium. Similar to the getter pump, the cryopump also has a limited capacity. Similar to those, this requires frequent regeneration by in-vacuo heating and pumping of the stored gas through a secondary pump. To further improve the cryopump properties, porous adsorbers such as activated carbon can be deposited on the cold surface. This increases pump capacity, reduces the vapor pressures of the condensed liquids, and also binds non-condensing gases in the pores. In order to maintain the pumping effect even when a cryopump is being regenerated, vacuum vessels must be equipped with several separate cryopumps which are alternated by valves. The technological approach of the cryopump requires only a coolant, so it is particularly fault tolerant, does not emit electromagnetic fields or vibrations and can be scaled up cost-effectively. For this reason in ion traps or fusion reactors, large

Table 2.2 Boiling points for the most relevant gases for cryo-pumps

Gas	Boiling point (K)
Nitrogen (N_2)	77
Oxygen (O_2)	90.2
Carbon dioxide (CO_2)	194.7
Water (H_2O)	373
Argon (Ar)	87.2
hydrogen (H_2)	21.2
Helium (He)	4.2

parts of the vacuum vessels up to the entire vacuum vessel are used as cryopumps to reach maximum throughput.

The third and currently most common pump technology is the turbomolecular pump. Its great advantage is quasi maintenance-free and continuous operation. The latest models with magnetic bearings are non-contact to run in principle over decades without user interaction, although this still needs practical proof. Due to these bearings and the electromotive drive technology, turbomolecular pumps generate local magnetic stray radiation in the range of some 100 μT potentially exerting detrimental effects on charged particle beams. The pumping effect is generated by inclined impellers running at up to 100,000 revolutions per minute. These impellers continuously hit gas molecules reaching the pump pushing them into the pumps exhaust. Cascades of typically 10 such impellers compress the gas within the pump to fine vacuum pressures at the exhaust. The flat and un-aerodynamic impellers experience strong friction forces under viscous flow, limiting their speed and thus the gas compression effect. In order to expand the compression into higher exhaust gas pressure regions, a so-called drag stage can be connected downstream. In the drag stage, a rotating helical channel of small open diameter induces a viscous flow, similar to a drill dragging out material from a hole in the wall. As a result, a robust overlap of the areas of high throughput of both turbomolecular and compression pumps is present in the fine vacuum range (Fig. 2.1). Despite the drag stage, the turbomolecular pumps are limited in their compression ratio, i.e. the maximum ratio of inlet to exhaust side gas pressure (p_{in}/p_{out}). This technological disadvantage, in comparison with the two aforementioned technologies, is particularly significant for hydrogen and helium with typical compression ratios of 10^3 to 10^7. Heavier and therefore slower gas components reach compression ratios up to 10^{12}. The compression limit is not a fixed number, but the result of a technical design of a pump and can be shifted e.g. by series of turbomolecular pumps.

When the compression ratio is reached for a certain gas component, the throughput of this component drops to zero. In this case, equal amounts of the gas component flow from the inlet to the exhaust side and vice versa. Taking the example of the situation depicted in Fig. 2.1. We have 2-stage pumping system with a turbomolecular pump delivering a compression of 10^3 for hydrogen and 10^{12} for nitrogen and a minimum pressure of approx. 10^{-3} mbar between the compressor and the turbomolecular pump. This system delivers a compression limited minimum vacuum pressure of 10^{-6} mbar with a hydrogen partial pressure of 10^{-6} mbar and a nitrogen partial pressure of 10^{-15} mbar. This type of gas species dependency typically produces residual gas compositions that are dominated by hydrogen and other light gases, as shown in Fig. 2.2. A careful design of a pumping system, considering the residual gases present in the vacuum vessel (e.g. the accelerator) and the gas species dependent pressure requirements is therefore essential for an accelerator requiring UHV pressures.

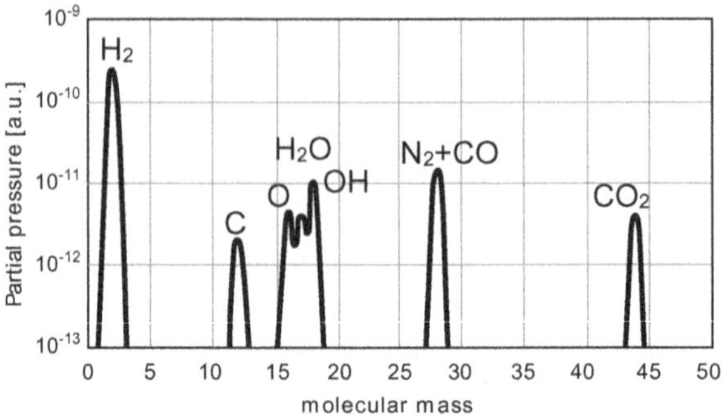

Fig. 2.2 Typical residual gas spectrum of a clean vacuum vessel pumped by a turbo-molecular pump. Due to the low compression of light gases, the residual pressure is dominated by hydrogen

2.1.2 Pumping Systems and Vacuum Vessels

Accelerators usually require UHV pressures to mitigate the energy-loss and beam energy broadening of particle beams, as indicated by the energy-loss column in Table 2.1. Precise analytical applications and accelerators with sub-keV particle energy can require even lower pressures. The interaction with the particle beams breaks up residual gas molecules, potentially leading especially to carbon condensation on analytical samples. Furthermore, even UHV pressures provide oxidizing conditions for a number of common materials, referring to the *Ellingham diagram*. Since this aspect is application-specific, the necessary details are provided in the corresponding sections.

Obtaining these pressures requires a two-stage pumping system. Such a system, consisting of a compressor pump and e.g. a turbomolecular pump, can be considered as an independent unit separate from the vacuum vessel of the accelerator. For reasons of maintenance and operation the pumping system is usually equipped with pressure monitoring and various separation and venting valves, in addition to the pumps. Installing elastic components after each pump suppresses vibration transfer and avoids mechanical stresses. Figure 2.3 shows such a structure in the schematic representation. In this system, a bypass is additionally installed in order to reduce the pump-down time for example in a sample exchange chamber. With this equipment a *Programmable Logic Controller* (PLC) can monitor and automate the system based on the pressure values and pump parameters to intercept operating risks and failures and to protect the nearby experiments and accelerator components.

For measuring the vacuum pressure different technologies are required for the different pressure ranges, see Table 2.1. Pressure, being a force per area, can be measured as such down to the HV region by the deflection of elastic elements. In the same range the heat conduction of the vacuum can be measured using hot wires, but the heat conduction is gas species dependent. Recalculation of heat-loss to vacuum pressure requires gas composition dependent correction factors. Below the HV range, however, the particle densities in vacuum are so low that the vacuum

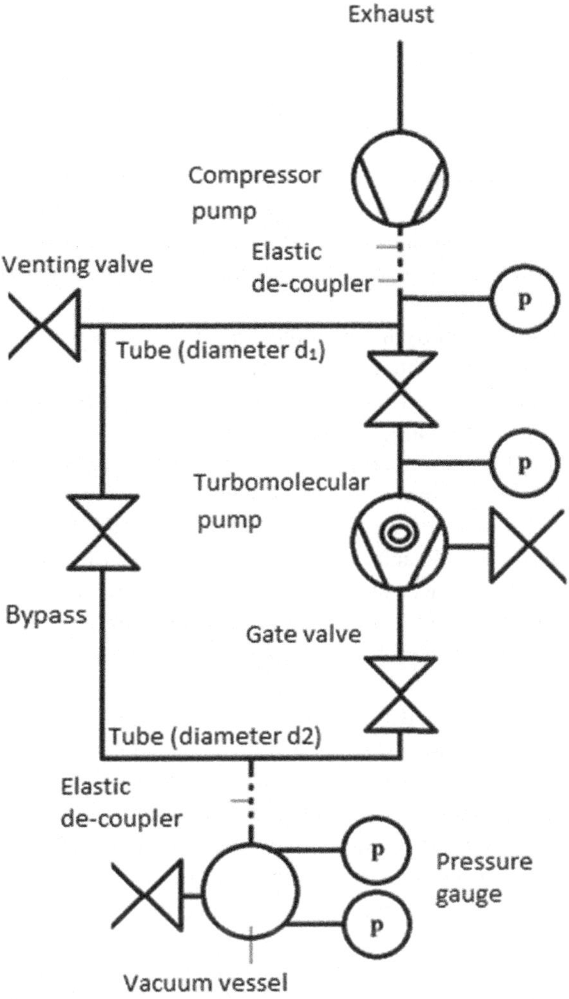

Fig. 2.3 Schematic of a typical pumping system connected to a vacuum vessel. The system contains 2 pumping stages with continuous pressure monitoring using 4 gauges, 3 vacuum valves/gates, 3 venting vales, elastic elements for de-coupling vibration and thermal expansion, and a bypass for quicker pump-down

contribution to forces and heat transfer becomes negligible compared to other effects as electronic noise and thermal radiation. Therefore, plasma discharges are used in the UHV. These discharges carry an ion current, which is dependent on the particle density N/V of the vacuum and hence provides a measure for the pressure (see 2.1). The ionisation efficiency in these plasma discharges is also gas species dependent, requiring calibration curves. In the XHV range, the particle density is so low that even plasmas are difficult to maintain. Only specially designed devices can provide means of pressure determination in this difficult regime.

The minimum achievable pressure p_{min} (final pressure) of a vacuum vessel is given by the equilibrium of pumping throughput S and gas input F into the vacuum vessel

$$p_{min} = F/S(p = p_{min}) \tag{2.3}$$

Here the throughput S is pressure-dependent (2.2) and the gas input F is pressure-independent. For a given S, (2.3) says the final pressure is defined by F. F results from the sum of the vacuum leaks, the outgassing, and the permeation through the vessel walls. Also intended gas inlet, for example hydrogen for a proton ion source, contributes to F.

Vacuum leaks are generally holes in the wall of the vacuum vessel. Don't think of holes a size sinking the Titanic, but only microscopic defects such as surface scratches. They can occur, in particular, on connections between components, the so-called flanges, and on welding seams. The flange systems ConFlat (CF), clamp flange (ISO-K) dominate the market for HV and UHV components and the small flange (KF) is the most common for the higher pressure ranges. All three flange systems produce a contact between two components in the form of a closed ring by means of a gasket. This is achieved either by gasket O-rings with a round cross-section and flat contact surfaces on the components (ISO-K, KF) or by cutting edges on the components with flat sealing rings (CF). Interruptions in this closed ring, caused by scratches, cracks or non-centrically seated seals, lead to vacuum leaks and a gas input $F > 0$.

Gases and liquids adsorbed on the surfaces within the vacuum vessel form an exhaustive reservoir by outgassing. If a drop of water or a fingerprint is left in a vacuum vessel, the vapour pressure of it continuously releases gas, until it is exhausted. In the roughness of surfaces and in porous materials, e.g. hydrophilic plastics, relevant amounts of humidity and other gases can adsorb. The quantities are negligible at normal pressure, but result in enormous volumes in the case of the reduced pressures of a vacuum (see Eq. 2.1). Since the vapour pressure of a liquid depends exponentially on the temperature, the depletion of the outgassing reservoirs can be accelerated by heating the chamber (so-called baking) to temperatures of 100–400 °C for hours to days. First, the gas input F increases sharply as temperature rises, but as soon as the reservoir is exhausted, significantly lower values of F are achieved.

In particular in the low UHV and the XHV, the permeation of gases through the vacuum vessel walls and gaskets becomes the dominant contribution to F. Small gas particles such as He, H_2, H_2O, CO and CO_2 can penetrate solids at relevant rates. The materials used and their individual and temperature-dependent permeation coefficients K determine the quantity of the permeation related gas input. Besides the materials, the gas input by permeation is given by the penetrated surface area A (e.g. vessel inner surface), the outer pressure p_0, and the material thickness d:

$$F = K * \frac{A p_0}{d} \tag{2.4}$$

The permeation coefficients K of polymers are too high for the UHV range. Metals, on the other hand, offer by orders of magnitude smaller permeation coefficients, allowing reaching XHV pressures. Even if the vessel itself is made from metal, the permeation through polymer gaskets will limit the final pressure to values $\geq 10^{-7}$ mbar. Lower pressures require gaskets made from soft metals such as copper or aluminium. For reaching XHV, the permeation through steels becomes relevant since these have particularly high K values for hydrogen. A further reduction in the hydrogen permeation can be achieved with vacuum components made of aluminium alloys or by applying special barrier coatings with low K. Additionally (2.4) indicates solutions by reducing p_0 by a vacuum outside the vacuum vessel or by multiple differential pumped concentric seals each having a small individual p_0.

In this chapter, the concept of vacuum was introduced and discussed on the basis of several criteria. On the basis of these criteria, technologies for the gradual reduction of the pressure in a vacuum vessel for achieving accelerator compatible values were presented. Two of these technologies were used for a technical implementation in the form of a two-stage pumping system case study. Independent of the pumping technology, the final pressure is given by the equilibrium of gas input and pumping throughput.

2.2 Accelerators

In this section we will implant the heart, an accelerator, into the body of the vacuum vessel discussed in the last section. In contrast to a heart, there are several technological options for an accelerator. Accelerators qualify themselves for an application by a set of performance parameters. The most important parameters are the beam particle kinetic energy (measured in eV), the beam current (in A) or ion flux density (ions/s/m^2), and the cost and size.

Particle physics research thinks of the beam-particle kinetic energy, or just beam energy, as the most important parameter. It defines the types of possible reactions the particles can undergo and their probability. Discovering new physical aspects of fundamental particles therefore requires a higher beam energy, with currently TeV (10^{12} eV) defining the technical limit of science, but what is an eV? Physically it is the kinetic or movement energy a singly charged particle, such as an electron, gains when passing an accelerating potential of 1 V, hence the name electron Volt (eV). Accelerating 1 A s (=1 C) of electrons corresponds to accelerating 6.25 * 10^{18} electrons (1/elemental charge e). These many electrons would bear 1 J (=1 W s) of kinetic energy at 1 eV beam energy, hence requiring to run a 1 V power supply for 1 s with 1 W output power. Now you can scale all values by kilo (10^3), mega (10^6), giga (10^9), and terra (10^{12}) to get an idea of the currents and powers involved in large accelerators.

In applications the envisaged reactions/interactions define the desired beam energy. Particle physics research can work with a low quantity of interactions between beam particles on the scale of quarks and gluons with energies in excess of GeV (rest

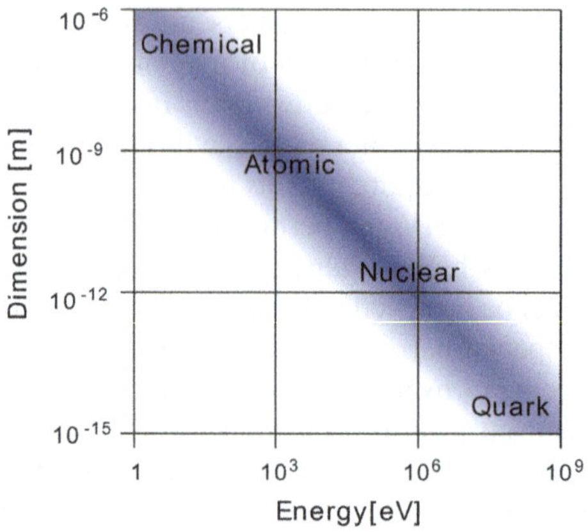

Fig. 2.4 The higher the energy of accelerated projectiles, the smaller the investigated and interacting structure. For example in the MeV (10^6 eV) region of energies, the nuclear structure becomes visible and nuclear reactions can be induced

mass energy of a proton = 938.272 MeV). In contrast, the applications discussed in this book require a substantial quantity of interactions of beam particles with large amounts of target particles in materials, atoms (>keV), and atomic nuclei (>MeV), see Fig. 2.4. Imagine the quantity of atoms in one gram of ^{18}F used for PET analysis (Sect. 6.1.2) of $3.35 * 10^{22}$ and the corresponding amounts of current (A), time (s), and power (W) required from an accelerator to generate these many particles! Consequently, lower beam energies with higher beam currents are demanded from these application accelerators.

Accelerators for applications accelerate ions and electrons to induce these interactions. Physically, these particle types are very different. The electron is an impartible object, with a low rest mass of 511 keV/c^2 and a fixed negative elementary charge $-e$. In contrast ions are a group of objects, consisting of protons and neutrons and therefore also of quarks and gluons as depicted in Fig. 2.5. All of the isotopes listed in Fig. 2.1 could be accelerated as ions. This divisibility allows mixing of their constituents with target nuclei generating other particles with the same basic constituents (the quarks and gluons), but in a different composition. Practically, we call this fusion and fission reactions, or in general nuclear reactions.

Ions feature a mass at least 1836 (for protons) times higher than that of electrons and charge states of several negative up to their individual amount of protons of positive elementary charges. Electrons being impartible always bear the exact same charge quantity and polarity. Ion charge on the other hand can vary due to the number of electrons attached to them. The close binding of electrons and the atomic nucleus makes them appear as a single object (the ion or the atom) with a summed charge q in most technical applications. From the outside, the object appears as uncharged if the amount of protons equals the amount of electrons, although in the nuclear zoom level it is not. Ion charges can change by removing or adding electrons. Due to these differences in particle structure, electron and ion beams find different applications.

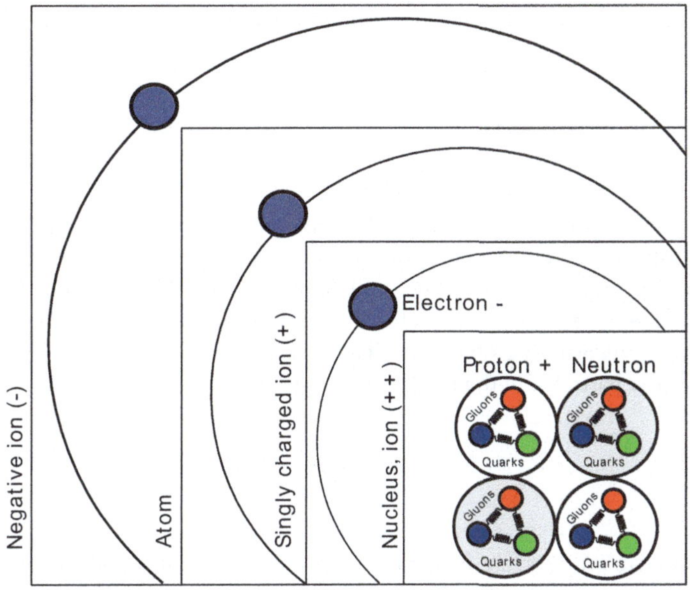

Fig. 2.5 The atom, ion, nucleus, and electrons of helium. Electrons are impartible, ions and nuclei are not. The atomic nucleus is an ion of maximum charge, but its charge can be changed by binding (orbits) of electrons to the nucleus, leading to different charge states. The picture shows a ^{4}Helium nucleus consisting of two neutrons and two protons with 0 (double charged ion), 1 (positive ion), 2 (neutral), or 3 (negative ion) electrons

Despite these differences, the acceleration and control of electron and ion beams is quite similar. Similar types of accelerators are used for both, yet taking into account mass, velocity and charge differences for the technical details. Hence the physics and technological aspects of ion- and electron accelerators are largely identical.

Strong differences arise when Einstein's relativity enters the game. Einstein defined the speed of light as the absolute maximum relative velocity between two physical objects. The constant $c = 299{,}792{,}458$ m/s quantifies this speed of light. Accelerators honour Einstein by easily accelerating particles to velocities close to this speed c. Due to their lower mass, electrons reach relativistic speeds quickly, e.g. 3 keV (typical electron energy in an electron microscope) is enough for 10% of c (=0.1c). For ions the relativistic physics remain negligible until about 10 MeV due to their higher mass. We will see in Sect. 2.2.2 how the relativistic physics affects the design and construction of accelerators, in particular of the accelerator types reaching the highest beam energies.

Throughout this book, the mass of particles will be given by m and their charge will be q. The ratio of particle speed v to the speed of light c plays an important role for the relevance of relativistic effects. The ratio v/c usually comes by the name β. From this, we can derive the Lorentz factor γ as depicted in (2.5). The Lorentz factor can be understood as the value of relativistic length contraction or time dilatation, but it also increases the particle mass from its rest value m to the relativistic mass m_{rel}.

$$m_{\mathrm{rel}}(v) = \frac{1}{\sqrt{1 - \left(\frac{v}{c}\right)^2}} * m = \gamma * m \tag{2.5}$$

This relativistic mass increase is the limiting mechanism for the particle velocity, since the faster a particle becomes the higher its mass and consequently the more difficult to accelerate it further. In theoretical physics often c is taken as 1 to shorten equations. This book will avoid such abbreviations and try to only use the regular and practically measurable quantities. This velocity ratio β connects to the kinetic energy E of the particles in the relativistic case via

$$E = \frac{mc^2}{\sqrt{1 - \left(\frac{v}{c}\right)^2}} - mc^2 \tag{2.6}$$

The first term on the right hand side represents the total particle energy and the second the energy equivalent of the particle rest mass according to Einstein's famous equation $E = mc^2$. The kinetic energy is the determining value for an accelerator, as it is one the one hand easily measured by the applied voltages and the energy unit electron volt (eV) and on the other hand the relevant parameter for physical interactions. The particle mass connects energy and velocity. The reference value for different particles is their rest mass, the value without any relativistic corrections according to (2.5). The main difference occurs between electrons and ions, with a minimum ratio of 1836 (electron to proton mass).

2.2.1 *Direct-Current Driven*

The era of accelerators started with the application of direct-current voltages (DC) on conductive plates. Every charged particle positioned in between two conductive plates will get accelerated by a DC potential towards one of these plates, depending on the voltage polarity and particle charge sign. Cathode ray tubes, as the first accelerators, work on this approach by applying a voltage between an electron emitter and a plate anode with an aperture hole (Fig. 2.6). This simple approach quickly reaches its limits in the achievable beam energy and hence new technologies were required.

The main technological challenge of electro-static DC accelerators is the generation of the high electro-static potential and its isolation from the surrounding structures. Remember, every eV of beam energy requires 1 V of potential difference in the device, with applications requiring keV to about 250 MeV. Free electrons, naturally present in all gases and on all surfaces due to the thermal distribution functions and

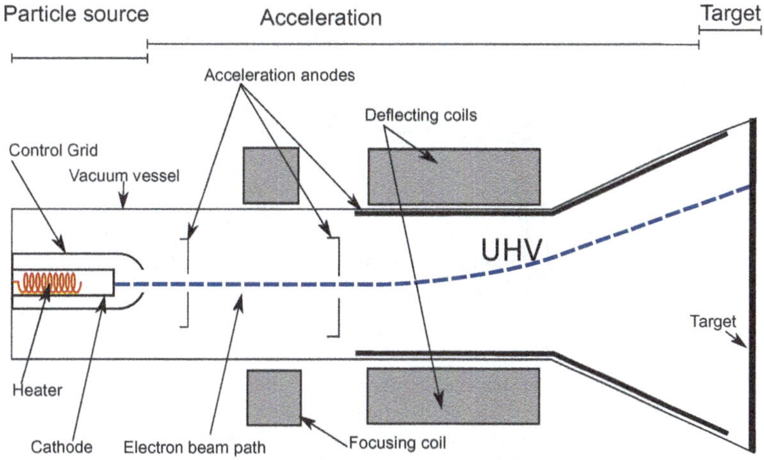

Fig. 2.6 Basic principle of a DC accelerator realised in the form of a cathode ray tube. The electron source is a hot filament, from there on the electrons are accelerated by a cathode-anode DC potential and directed by a magnetic deflection system onto a target. A similar setup, but with reduced beam diameters is the basis for electron microscopes

cosmic radiation, will be accelerated by these voltages unintendedly. These accelerated rogue electrons quickly gain enough energy to release new charged particles (requiring about 10 eV/particle) from gases and surfaces, an exponential avalanche effect can initiate. This effect depends on the voltage and the distance over which it is applied due to its connection to the mean free path. The plasma created by this effect has a low resistance over which the DC potential will discharge, preventing the efficient build-up of high voltages.

The characteristics and resistance of this plasma can vary over orders of magnitude, depending on the combination of voltage gradient and outer conditions, in particular pressure, as demonstrated in Fig. 2.7. The first visible discharge type, carrying also relevant amounts of current is the corona discharge. This type of discharge can be observed on high-voltage land lines and generates the crackling noise often audible around these lines. The resistance of these discharges is still high, hence inducing acceptable loads on the high voltage power supply, but it indicates the onset of a problem for DC accelerators. Exceeding the highest tolerable voltage results in the breakdown of an arc discharge. The resistance of the arc plasma discharges is in the order of metallic conductors, leading to significant power consumption and heat dissipation on the surrounding elements, limiting the achievable voltages of DC accelerators. The voltage gradient limit $V_{\text{Breakdown}}/d$ of the arc transition, with its sudden decrease in discharge resistance demonstrated in Fig. 2.7, is described by the Paschen-curve:

$$V_{\text{Breakdown}}(p, d) = \frac{Bpd}{\ln(Apd) - \ln\left[\ln\left(1 + \frac{1}{\gamma_{SE}}\right)\right]} \tag{2.7}$$

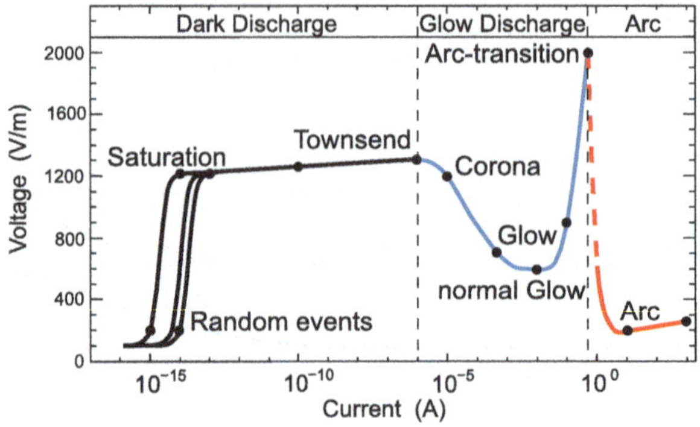

Fig. 2.7 The different regimes of plasma discharges (in 1 mbar Neon), limiting the acceleration potential. After the dark discharge region, two regions of negative differential resistance (higher current with lower voltage/gradient) follow. Several types of DC accelerators make use of this negative differential resistance to filter voltage ripple. Original work by Wikigan, CC-BY-SA-3.0, via Wikimedia Commons

With the gas species dependent parameters A and B, the gas pressure p, the distance between the two charged plates d and the number of secondary electrons (see Sect. 4.1) generated per electron and ion impact on the plate surfaces γ_{SE}. The arc-breakdown voltage-limit of (2.7) strongly depends on the involved transport medium (e.g. gas) between the voltage poles. Tabulated values of the breakdown voltages for different solid, liquid, and gaseous isolation materials exist in handbooks. Table 2.3 illustrates these values are in the order of 10 kV/mm for typical insulators. Plastics and ceramics reach similar values as a vacuum. The problem of vacuum-based isolation lies in the strong dependence of its breakdown limit on the surface conditions of the parts isolated against each other. Roughness results in strong local voltage gradients at the roughness peaks and surface absorbed gases potentially form a relevant gas pressure in the sense of (2.7) upon release. In applications these properties often remain hidden. Gases achieve slightly lower values at standard pressures, but the breakdown voltage in gases is proportional to the gas pressure. At a pressure of 10 bar, the special isolation gas sulphur-hexafluoride (SF_6) provides breakdown

Table 2.3 Breakdown voltages of selected materials

Material	PE	PTFE	Al_2O_3	Demineralised water	Oil	Air	SF_6 @1 bar	Vacuum
Breakdown-voltage (kV/mm)	20	24	17	65	<30	0.1	8	20–40

The breakdown in gases depends on the gas pressure and is given for 1 bar. In vacuum, the breakdown strongly depends on the considered surfaces, their roughness, cleanliness, composition, and temperature. In the literature, large spans of these values can be found, indicating a connection to these partially hidden parameters limits of 80 kV/mm, surpassing most solid isolators. Consequently, most DC accelerators above 1 MeV apply SF_6 for keeping the devices small, while lower energy accelerators mostly operate in air or vacuum.

For the generation of voltages up to $\approx$100 kV industrial stand-alone power supplies with standard polymer insulated cables and polymer or ceramic insulators are sufficient. These efficient and compact devices power mostly tabletop devices such as electron microscopes, X-ray tubes, or ion sputtering devices (see Sect. 5.3, Chaps. 6 and 7). Due to the low charged-particle energies, the effort for beam handling and control is significantly smaller than for beam energies above some MeV, with more details in Sect. 2.3. For this reason, these low beam-energy devices are highly integrated and compact.

For reaching higher acceleration voltages, the power supply separates into a high and a low voltage part, electrically connected to each other by a transformer isolated with oil or gases (e.g. SF_6). The low voltage part is already a high voltage part in the usual understanding, with voltages in the order of 100 kV. Electronic power supplies generate these voltages with high efficiency. For the high voltage part, two different systems have been established so far. In the Cockcroft-Walton accelerator type (Fig. 2.8), the low voltage side provides an alternating-current voltage, which is electrically decoupled by a transformer from the high voltage side. On the high voltage side, the AC is subsequently rectified by diodes and smoothed by an RC low-pass filter. These diodes guide the negative and positive parts of the AC to charge up two separate capacitors, resulting in a doubling of the voltage between these two capacitors compared to the AC peak to zero value. Stacks of these doublers allow reaching proportionally higher terminal voltages. Technical aspects limit the

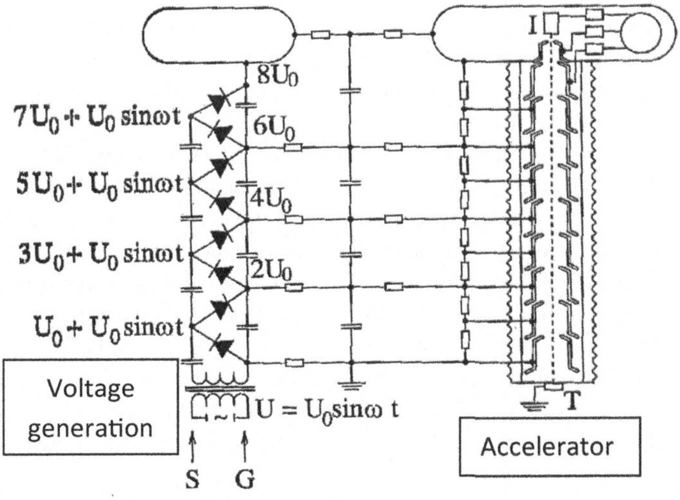

Fig. 2.8 High voltage generation in a Cockcroft-Walton accelerator. A high frequency voltage $U_0 * \sin(\omega t)$ couples into the accelerator via the coil at points S G. A capacitor-diode cascade transforms it to even multiples of the input voltage amplitude U_0 and rectifies it. The increasing voltage applies to a cascade of plates isolated against each other (right), accelerating the beam in steps of U_0. The resistors lead to a voltage drop (loss of accelerating potential) along the individual plates with increasing beam current. The magnitude of this voltage drop depends on the technical layout. Reproduced from Hinterberger (2008) with permission by Springer

number of stacks at some point due to the electrical resistance and the required isolation distances.

The van-de-Graeff generator system in contrast places DC charges on an isolating belt. This belt rotates in the device to mechanically transports the charges from the low voltage to the high voltage side. At the low voltage side the belt can be charged up via the triboelectric effect like a plastic belt charges up when drawn over a carpet. For accelerators, a design using sharp tips to induce charges via short distance corona discharges with high voltage gradients generated by the tip field offers practical advantages. In this design metallic balls are embedded in a rotating isolating belt as charge buckets. This mechanical charge generator results in a lower energy efficiency but also lower investment costs compared to the solid-state design of the Cockcroft-Walton type.

Both accelerator types can accelerate all types of charged particles regardless of mass or charge due to the linear and constant acceleration field. Furthermore, they are continuously adjustable in their acceleration voltage via the primary voltage or a voltage load, giving a high degree of freedom valuable especially for experimental and varying applications. The beam "travels" down the electrical potential to the grounded end for acceleration, like a river where the water travels from a high gravitational potential to a low potential. The accelerator applies the voltage between conducting parts, therefore either the charged particle source or the target side must be electrically in contact with the high voltage side of the accelerator with the other part grounded. Since these devices are often used in scientific applications, involving several different experiments attached to the accelerator, usually the target is grounded for practical considerations and the ion source is attached to the high voltage side inside the isolation gas. After the accelerator, a dipole switching-magnet directs the beam towards the serviced end-station.

For achieving even higher beam energies and avoiding the river problem, at least for ions, a clever trick can be exploited in the so-called tandem accelerator. The tandem accelerator principle requires the injection of negative ions which are converted to positive ions in the centre of the accelerator by charge stripping with a foil or gas. Changing the sign of the charges generates a virtual ground at the position of charge exchange, allowing the beam particles to accelerate twice with the same voltage. In contrast to gravity, the two different signs of electrical charge allow gaining energy when travelling the river up and down. This way, the same DC potential accelerates the ions twice, first the negative then the positive ion. This results in acceleration to 2 times the DC potential for 1+ ions, 3 times for 2+ ions and so on. The disadvantage is, the generation of negative ions is significantly less efficient in most ion source types compared to positive ions, see Sect. 2.4 for details. Furthermore, electrons cannot be accelerated. Due to this source side limitation, tandem accelerators typically provide lower beam currents than single acceleration devices.

Gas strippers typically use nitrogen for ion stripping which is injected into a separate volume with two small holes, which is placed in the accelerator tube centre. The holes result in a relatively low conduction, leading to a higher pressure inside the volume compared to the lower pressure in the beam tube. The more gas passed, the higher the charge exchange probability. Gas strippers are robust and allow adjusting the stripping ratio via the variation of gas pressure/injection rate. Thin foils of typically μm thick graphite provide a constant stripping, but require frequent replacement due to heat load and beam damage. A foil avoids injecting gases into the vacuum resulting in generally lower beam energy spreading after stripping, but foils also limit the beam current handling capability. In both cases, the charge exchange generates singly and multiply charged ions (except for hydrogen, which has only one nuclear charge) with efficiency up to 90%, with flux ratios depending on beam energy, species, and stripping. The remaining beam particles pass the accelerator as neutrals accelerated to the terminal voltage (no charge = no further acceleration) or hit the vessel walls, since negative ions cannot pass a tandem accelerator.

All the discussed technological variants of DC accelerators commonly share the problem of slight acceleration voltage variations, the ripple. They all have to bridge the DC voltage isolation, either by an AC transformer or by a belt with moving charge buckets. Both methods have periods where no charge is delivered, the zero-crossing of the AC or the distance between two charge buckets. The charged particle beam, being a DC beam, induces a constant load on the DC potential, creating the voltage ripple with charging frequency, voltage, and system capacity dependent amplitude. This DC voltage ripple is directly transferred to the charged particle beam, broadening the beam energy distribution. In modern devices, ripples down to 10^{-5} of the terminal voltage are possible, but for analytical applications, even these small ripples impose a relevant technical limitation as we will see in Sect. 2.3.

In addition to variation of the acceleration voltage, DC accelerators allow for control of the charged particle current from pA to A, typically from the injection or source side, respectively. A control of beam currents assists many applications by allowing for time-of-flight analysis in between the pulses, for probing with methods of different sensitivity, or for adjusting the power loading dynamically to the target conditions. On the injection side, the particles still have low energies in the keV range, easing beam control and power loading. The adjustment of the beam current by changing of the source parameters, allows for changes on the 100 ms time-scale. For an operation with reduced duty cycle and short pulses, a switching of the power supplies for the beam injection optics allows to achieve pulse-length down to μs. Special chopping systems are required to compensate the sudden change in beam load, depending on the high voltage capacity, see the many capacitors in Fig. 2.8. For even shorter pulses, bunchers have to be installed on the injection side. These devices accelerate and decelerate chopped beams consecutively to compress the pulses from μs to ns with a simultaneous increase in the maximum instantaneous beam current. Pulse length below ns require special source constructions working with a short pulse driver, e.g. a short pulse Laser.

Two main application regimes exist for DC accelerators. The first and by far largest are accelerators for energies in the keV up to some hundred keV range for electron beams. This comprises electron microscopes, electron based sources for bremsstrahlung (e.g. medical X-ray), and electron beams for material processing and modification. The usage of ion beams from DC accelerators is dominated by material processing on the nano-scale for example for micro-electronics by some 10 keV heavy ion beams. DC ion-beams on the MeV scale nowadays mostly occur in the scientific context of material analysis and modification.

Only few industrial manufacturers for MeV scale DC accelerators exist. The available standard products range from 100 kV to about 25 MV delivering currents between 10 μA and several mA (Quax et al. 2010). The technology is not fundamentally limited in its voltage or current, but there is also no scaling advantage above the standard values and furthermore the devices become fairly large because of the necessary voltage isolation. Van de Graeff devices typically provide higher acceleration voltages, Cockcroft-Walton devices, in contrast, provide higher beam current. Producing beams with high currents in DC accelerators yields higher electrical efficiency with increasing current, as typically some kW are required to maintain the DC potential. The ion acceleration itself reaches >90% electrical efficiency. In conclusion, DC accelerators are energy efficient accelerator types with high industrialisation and numerous products on the market.

2.2.2 *Alternating-Current Driven*

At this point we have to realize it will be technically very challenging to reach beam energies above a few 10 MeV using DC accelerators, therefore other concepts are required to surpass this technical barrier in a more cost efficient way. DC accelerators make use of the accelerating field only once, at maximum twice with the tandem's charge exchange. If the accelerating voltage could be used multiple times, much higher energies could be reached without extreme voltage levels, therefore circumventing this technological difficulty of DC accelerators, but how to make multiple use of the same structure? According to Coulomb's law, a particle cannot gain more energy by moving through an electro-static field several times. So how to circumvent fundamental physics? The tandem accelerator provides the basic idea, as it breaks the law by charge exchange. Alternating-current (AC) driven accelerators apply the same trick, but instead of changing the particle charge polarity, they work by changing the acceleration voltage sign/polarity. AC's change their voltage polarity infinitely often with their given frequency, avoiding the technological limit of voltage isolation. For passing the AC voltage multiple times, the AC accelerators grow larger making the highest beam energy AC accelerators the largest single technical devices build by humanity.

We start the conception of an AC accelerator by simply reusing the concept of DC accelerators with its acceleration chamber, the fundamental building block where any accelerator applies the accelerating potential to the charged particle beam. The chamber now converts an AC input power with typical frequencies of 10 MHz to 10 GHz and peak voltages in the order of 100 kV to a directed acceleration (see Fig. 2.9).

Additional technical degrees of freedom arise for AC in the method of coupling power to the chambers, compared to applying a DC voltage between two plates. Depending on the accelerator layout and in particular the AC frequency we can either apply the AC directly to plates or, for higher frequencies, drive resonant modes in a closed resonance cavity (Fig. 2.10) with an antenna or even by transferring energy from a resonant mode in another cavity or a waveguide. The resonant mode inside the cavity works like a vibration on a guitar string with a resonance frequency according to length and properties of the cavity/string. In contrast to acoustic waves, electro-magnetic waves can also propagate in vacuum. Usually the so-called transversal magnetic 010 (TM_{010}) mode is exploited as it has the lowest amount of nodes in the different axis. For understanding what this means imagine a straight conducting wire with an electric potential/voltage applied between both ends. A current will flow along the wire and induce a circular magnetic field around it. Placing the wire along the beam direction and removing it yields the field distribution of the TM_{010} mode. Physical language would describe it as a flat longitudinal electric field with a maximum in the centre and zero field at its outer radius (1 node) in combination with a transversal magnetic field, since Maxwell's laws of electro-magnetism dictates electric and magnetic fields to be always perpendicular to each other.

The beam draws the power required for acceleration from this cavity wave. Delivering this power into the cavity requires either a current flow in wires or, in particular for higher frequencies, the power coupling by waveguides. Figure 2.9 shows an example of a TM_{010} cavity powered by a magnetic antenna and a matching network. The current flowing through the antenna ring generates a magnetic field field with a vector directed through the ring. This matches to the local magnetic field vector of the

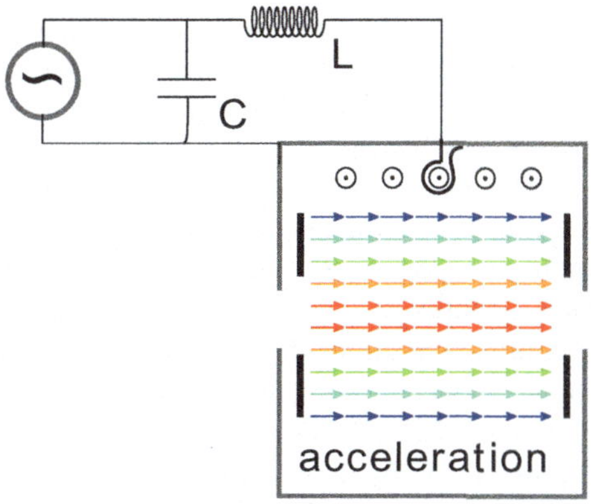

Fig. 2.9 Sketch of a TM_{010} mode with its longitudinal electric field (arrows) and a coupling antenna connected to a power source through an impedance matching network. The matching the power source impedance (e.g. 50 Ω) to the TM010 coupling antenna using a coil (*L*) and a capacitor (*C*) of variable inductivity and capacity, respectively. RF impedance matching calculators can be found on the web for determining *L* and *C* depending on source and load impedance. The field oscillates with time but can be considered constant for sufficiently fast (relativistic) particles passing through

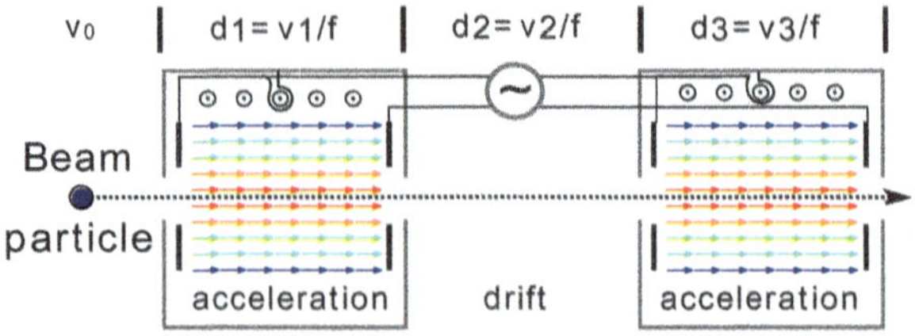

Fig. 2.10 Schematic of the 2π-mode AC accelerator design with a beam particle in phase with the accelerating voltage (arrows). This voltage originates either from electrostatic plates or from field coils magnetically driving e.g. the TM_{010} mode with its central voltage maximum. The beam particle first enters an accelerating structure, then passes a drift part until the voltage sign reaches its original polarity at the moment the particle enters the second structure. In the case of constant speed (e.g. close to speed of light) all v_i and hence all d_i are equal, otherwise the d_i have to increase with index. In the π mode both cavities have opposite polarity and $d_2 = 0$.

TM_{010} mode as discussed above, enabling coupling of the power to the cavity mode. The L-C network matches the impedance of the input lead to the cavity impedance.

High beam energies require frequencies in the GHz range. In this range, semiconductor power supplies reach their limits due to their inherent capacity and finite electron mobility, respectively, limiting their output power at higher frequencies. Vacuum electron resonators, mostly Klystrons, provide high output power up to several 100 GHz with conversion efficiencies of 50–75%. These electron resonators have their own cavity wave mode (not necessarily the TM_{010}) requiring a matching connection to transfer power to the acceleration cavity. The connection between the power source and the cavity depends on the modes and the power transmission method. A Klystron actually consists of a DC electron beam flowing through a series of chambers similar to the cavity depicted in Fig. 2.9. The electron beam couples to an input AC wave amplifying it since the electrons bunch according to the wave field. This electromagnetic wave travels through vacuum/air just like the power delivered in a household microwave. Instead of cables, hollow waveguide tubes allow for conducting the power to the accelerator. This conduction in not loss-free, but losses occur on the waveguide walls. A technical advantage lies in the coupling of the power and its impedance matching via geometrical modifications (tuners) in the waveguide instead of an *L-C* matching network.

A series of the acceleration structures depicted in Fig. 2.9 provides the repetitive use of the driving voltage we were aiming at. A series of acceleration cavities form a so-called linear accelerator (LinAC). Of course, DC accelerators are also linear accelerators, but the term LinAC by convention describes the AC type. The alternating voltages in the cavities constantly change amplitude and polarity, as they are literally alternating currents. In order to expose the beam particles to the same voltage-polarity, and hence acceleration direction, in each cavity, beam and AC voltage in every cavity have to be synchronised or in phase, respectively, with the propagating beam. In other words, each particle has to pass each cavity within equal half-periods of the AC.

The space in between two cavities has to equal the time required for the other half-period of the AC to finish, as depicted in Fig. 2.10. The higher the frequency, the shorter this time and the smaller the accelerator, making higher frequencies desirable for AC accelerator design as they allow for smaller structures and higher specific acceleration (MeV/m). The dimension d_i indicates this in Fig. 2.10.

Ultimately, the skin effect limits the conduction of AC power, since it restricts the current flow to a shallow surface layer, may it be over a wire or the walls of a waveguide. The skin effect reduces the effective conductor thickness with the square-root of the frequency, reaching values of about 10 μm at 50 MHz. Correspondingly the resistivity increases with frequency, representing a technological limitation. Transferring digital information over an USB3 cable might be possible with such thin conductors, but the power requirements of an accelerator require a different solution in order to keep power losses and voltage damping within tolerable limits. A solution would be the use of superconductors since they feature infinite conductivity. In reality this is only half-true, surface resistance R_{surf} from adsorbents (see Eq. 2.1) present at their cryogenic operating temperatures and the alternating nature of AC currents lead to small, yet relevant power losses P_{loss} per cavity area A_{cav} (2.8) on superconductors at high AC frequencies ω and magnetic fields B_{HF}. These different conduction physics in AC accelerating structures lead to a technological disadvantage and generally smaller energy efficiency of AC compared to DC accelerators.

$$\frac{P_{\text{loss}}}{A_{\text{cav}}} = \frac{R_{\text{surf}}(\omega)\, B_{\text{HF}}^2}{2} \tag{2.8}$$

Being in phase with the AC allows a single ideal particle to travel through multiple acceleration cavities. This particle defines our ideal situation, but a real beam consists of numerous particles spread across a certain volume given by beam size and spread around this central particle. Just like for a society it is important for the beam performance to not only take along the individuals who have the ideal starting conditions, but as many as possible by finding means to support the less privileged reaching the common goals. Due to the alternating nature of the accelerating voltage, the beam needs to adapt to a similar structure, see Fig. 2.11. The beam arranges in *bunches* which move in phase with the AC trying to catch the maximum benefit from every cavity passage. In contrast to a DC beam, particles can only survive the acceleration if they stay in a certain phase window of the sine wave of the given forward directed polarity. Intuitively, we would propose the peak of the wave in order to exploit the maximum accelerating field.

Unfortunately, some of the underprivileged particles will be slightly slower, have an angular deviation in their direction of movement (=longer path), or be just displaced compared to others, hence the beam will have a certain spatial and momentum extent. These beam aspects will be discussed in more detail in Sect. 2.3.1. The non-ideal underprivileged particles will reach the cavity slightly off timing and it is the task of the accelerator design to allow them to participate in the beam in spite of their deviations from the ideal. Slower particles reach the cavity after the AC peak and therefore see smaller amplitude. Smaller amplitude means less acceleration and

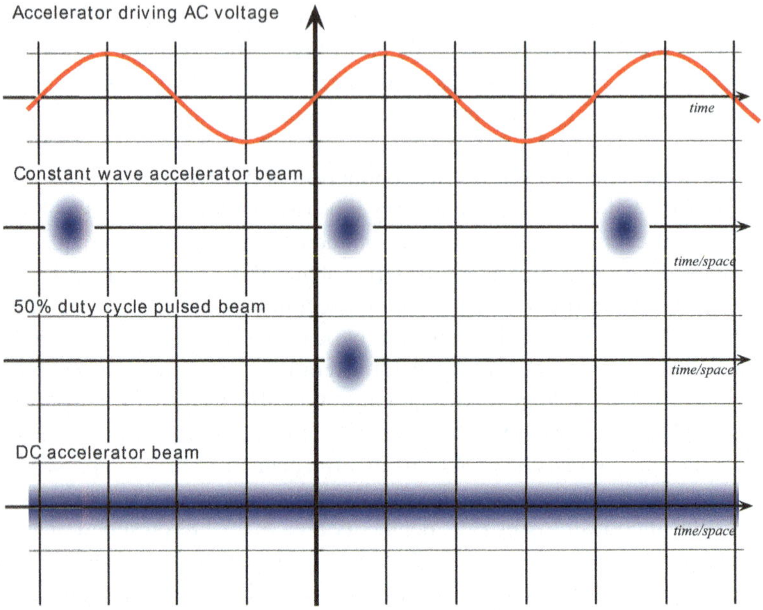

Fig. 2.11 The AC accelerator voltage is given by a sine wave (1st row). Correspondingly, the beam particles arrange in the wells of equal voltage sign (2nd row). A DC beam (bottom) in contrast remains unchanged over time and along the acceleration direction. The Y-axis represents voltage amplitude for the sine (top) and the transversal direction for the beams, respectively. Darker colours represent higher beam density

hence the particles become even slower compared to the rest of the bunch. Quite quickly, they will be left behind, the beam current reduces. Thinking how often the bunch will pass the cavities with acceleration voltages of $\approx$100 kV until it reaches some 100 MeV, it is easy to imagine that the whole bunch will disintegrate prematurely. Mathematically said, the bunch averaged phase-deviation from the ideal phase increases if we aim at the voltage peak. In order to avoid this issue and focus the phase, we have to choose a slightly earlier phase, see Fig. 2.12. On the positive slope of the amplitude faster particles receive less acceleration, since they reach the cavity early equal to a lower voltage, and slower particles receive more acceleration by reaching the cavity later. This focusses the bunch onto its central phase. The acceptance (maximum phase mismatch still focussed) of this phase focussing effect marks an important accelerator design parameter. It says how supportive the accelerator design is to underprivileged particles. The acceptance depends on the position of the ideal phase on the positive slope part, since significant deviations would bring particles into the bad phase regions. Besides this phase focussing, the AC beam also requires transversal focussing exactly the same way as in DC accelerators.

In order to synchronise the phase position of the individual bunch's cavity passages, the beam has several options (modes). The basic choice: It could drift/wait for another half-period through a field-free drift space in a setup where subsequent

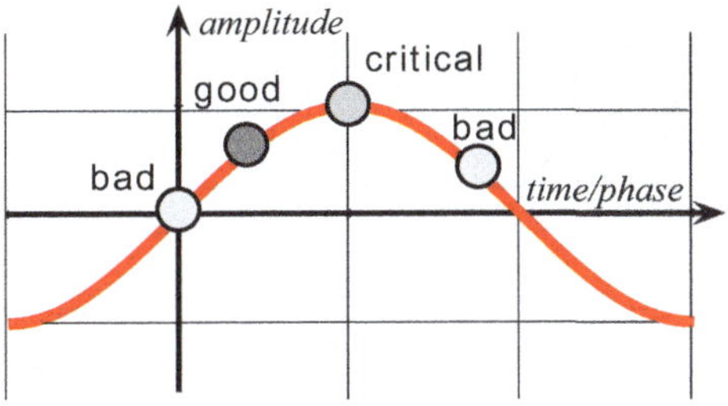

Fig. 2.12 Principle of phase focussing. The circles mark different options for the central phase position of the particle beam in the sine wave. A position at zero voltage amplitude allows no acceleration (bad). At the peak of the wave, slower particles defocus (critical phase limit). At the negative slope, all particles defocus (very bad), only at the positive slope phase focussing works to confine the beam in phase space (good)

cavities feature opposite polarity. Besides this so-called 2π-mode also π, $\pi/2$, or $2/3\pi$ resonance conditions come into application, depending on the technical layout of the accelerator. In these modes, length d_i of cavities and drift parts are adjusted to fit the resonance condition. As Fig. 2.10 shows, these dimensions d_i depend on the particle velocity, a quantity changing upon acceleration. Electrons quickly approach the speed of light, reaching e.g. 94.2% of light speed c at 1 MeV. Close to the speed of light, the particle mass increases, but the increase in velocity levels out, (2.5), hence the electron speed remains practically constant above a few MeV. Ions on the other hand require at least 1836 times (protons) higher kinetic energy (1836 MeV for 94.2% of c) due to their mass, making their velocity highly variable in our 250 MeV region of interest. Technical realisations of AC accelerators usually try to reduce complexity by using either the same frequency or similar or even the same acceleration structure throughout the accelerator. Consequently, the acceptable range of particle velocities becomes limited, restricting the dynamic range of serviceable beam energies to a window around the design value. At relatively high energies (see above), relativity limits the velocity gain close to c, expanding this beam energy window compared to the low energy range.

The particles increase their velocity upon passing a cavity and this has to be compensated to maintain the resonance condition when entering the next cavity. The solution is either to adapt the length of the drift parts in between the individual cavities or to use different frequencies (given a fixed phase relation between them). Several different types of these so-called high-frequency systems (sometimes also RF-systems) were developed with solutions based on waveguide or cable + antenna coupling and different solutions for compensation of variable particle velocities. When using a single frequency, the optimal spacing ensuring the resonance condition is called the Alvarez structure, see Fig. 2.13. In this structure drift and cavity length increase proportional to the particle velocity. LinACs for higher energies (see e.g. 4.3.3) apply numerous of these structures, while the high energy simplifies the design as soon as the velocity increase levels off in the highly relativistic regime.

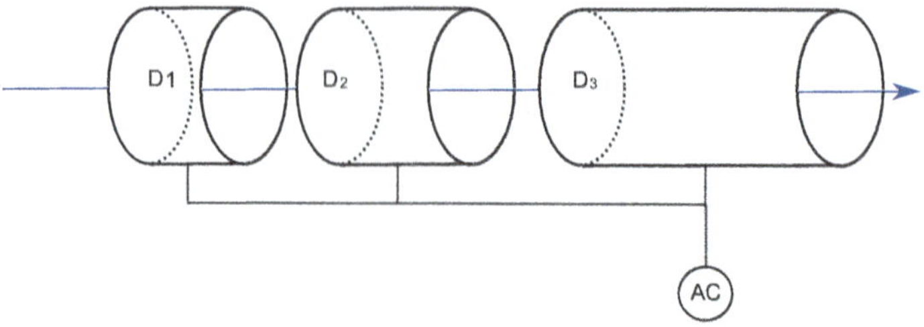

Fig. 2.13 The Alvarez LinAC structure works with a single AC frequency and variable drift and cavity length D_i to compensate for the increasing velocity

Besides the geometrical solution of the Alvarez structure, a LinAC construction can also exploit physical solutions. In all the discussion of this section, we discuss standing waves. It was never mentioned directly, but the hills and valleys of the AC are always assumed to rest inside the cavities. Jumping one frequency period further will always result in the same voltage at a given position. With the aspect of bunching and particle velocities close to the speed of light, in principle particles can also surf on the AC wave peak. These travelling wave acceleration structures require the phase velocity (the movement speed of the hills and valleys) to be equal to the beam particle velocity. The phase velocity is not equivalent to the photon velocity (=c), but it can be influenced/reduced by the propagation geometry. Figure 2.14 shows a schematic example of such an acceleration structure with four cavities.

We started by making multiple use of a single voltage, but also a single acceleration cavity can be used multiple times for accelerating a single beam particle. The technical realisation of this physical concept comes by the name cyclotron. In the cyclotron particles start from a source in the centre of the cyclotron with basically zero energy. The particles accelerate over the radius in the form of spirals by an AC voltage applied between two neighbouring sectors of the circle as depicted in the left image of Fig. 2.15. While particles are inside one of the sectors, a vertical magnetic field forces them onto circles. Due to the freedom of orbit in the cyclotron and the centripetal force, a faster particle will choose a larger orbit. The gap between

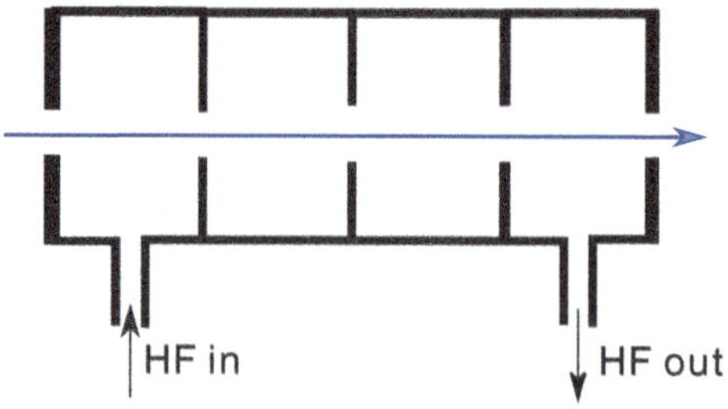

Fig. 2.14 A travelling wave acceleration structure. A wave is coupled in via waveguides. It travels along the individual chambers which are separated by apertures with exits in the last cavity. The size of the cavities and apertures determines the phase velocity of the travelling accelerating wave

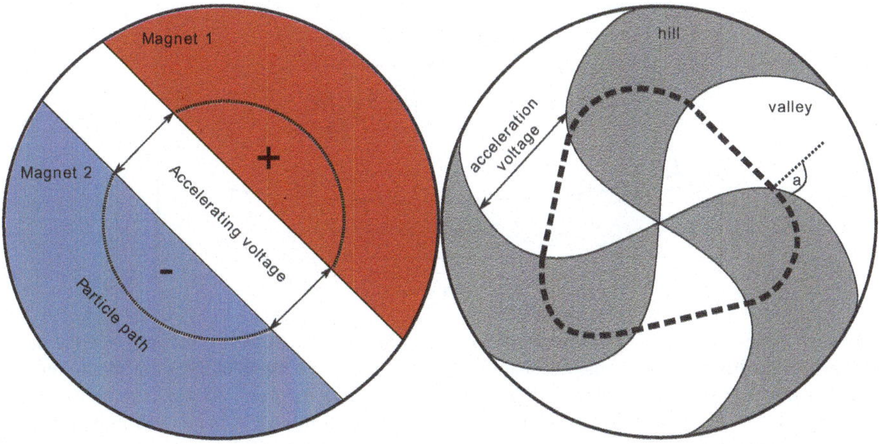

Fig. 2.15 Schematic top-view of a classical two-magnet cyclotron (left) and a modern isochronous cyclotron with 3 magnets (right). The modern magnets have spiral structure to increase edge angles a with radius. For particles approaching the speed of light, the axial field strength has to increase (Fig. 2.16) to compensate for the relativistic mass increase. The defocussing effect increases with radial field gradients, requiring stronger edge focussing at higher radii. In this example, magnets constitute the hill part and the valleys are field free, leading to the dashed particle orbit. This cyclotron operates in $4/3\pi$-mode with the opposite voltages across the valley parts forming the accelerating cavities

two sectors/magnets is the accelerating cavity, where the voltage between the sectors accelerates the particles. The movement time inside the sector corresponds to the phase synchronisation discussed above. Faster particles have to take a longer path, hence particles always pass the gap within the forward acceleration phase. Particles are extracted to the application at the outer radius with their final energy. The name cyclotron derives from the frequency of revolution of charged particles in a homogeneous magnetic field (inside the sectors). This cyclotron frequency $f_{\text{cyclotron}}$ given in (2.9) is independent of the particle energy, as faster particles move in larger circles, compensating their increased velocity by longer path. It only depends on the vertical magnetic field strength B and the particles charge q to mass m ratio. Cyclotrons exploit magnetic fields to guide the beam in a circle, allowing exploitation of the same acceleration cavity over and over again, while integrating drift and acceleration cavity into the same structure.

$$f_{\text{cyclotron}} = \frac{qB}{2\pi m} \tag{2.9}$$

The isochronous cyclotron shown in Fig. 2.15 (right) represents a technical layout variant of the general term cyclotron. Besides other cyclotron types it has established as the state-of-the-art of cyclotron design. The term isochronous takes the constant frequency of the cyclotron idea to the next level by including also corrections for relativistic mass increase. Independent of its energy, isochronous requires a particle

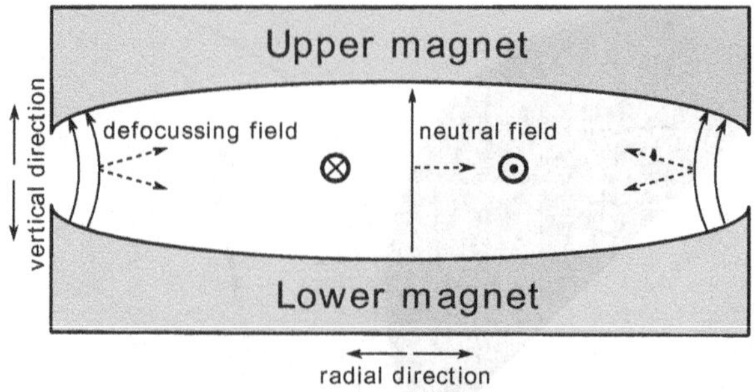

Fig. 2.16 Side-view of an isochronous cyclotron. The beam moves in (x) and out (.) of the paper plane (here positive ion). The magnet shaping densifies the magnetic field lines towards to outside (=higher magnetic field B) compensating for relativistic mass increase. As the Lorentz force acts perpendicular to movement and B-field direction (right-hand rule), the bend field lines lead to a defocussing effect/force as depicted by the arrows with dotted lines for positive ions

to always require the same time for a revelation. This sounds obvious given the independence of cyclotron frequency from energy or velocity of the particle, (2.9), but accelerators bring the particles to extreme speeds, easily reaching significant fractions of the speed of light in a fragile system of for example the phase focussing. Approaching the speed of light, the mass of the particles becomes velocity dependent as noted in (2.5). The trick of isochronism is to compensate for this mass increase by increasing the magnetic field B accordingly in the radial direction via a shaping of the magnetic sectors as depicted in Fig. 2.16. Now, not only the centripetal force defines the faster particles orbits, but also the radius dependent magnetic field strength.

This radial increase in magnetic field leads to a problem of beam divergence. Vertically straight field lines leave the beam shape unaffected, but bent field lines induce force components in the vertical direction, see Fig. 2.16. The Right-hand rule (3 perpendicular fingers) directly visualizes the situation for a single particle. The field index n quantifies this effect (2.10) through calculation of the change of the axial (vertical) magnetic field B_A over the radial direction r multiplied with the radial position R over the absolute local field strength B.

$$n_R = -\frac{\delta B_A}{\delta r}\frac{R}{B} \tag{2.10}$$

Without compensation of this defocussing, the beam will become too large and hit the vessel boundaries removing them from the beam. The highly integrated design of cyclotrons prevents the installation of any additional elements for focussing the beam, we have to find a solution with the parts already present. Let us continue the thinking of the radial field index above. The effect vertically defocusses the beam, but what happens radially? An underprivileged particle with a position slightly outwards from the ideal position in the bunch will experience a stronger magnetic field compared to a particle more in the cyclotron centre, deflecting it more to the inside. On the other hand, a particle displaced slightly to the inside will experience a weaker magnetic field, deflecting it less and bringing it closer to the bunch centre. All in all this describes a radial focussing! Apparently, focussing and defocussing go hand in hand: We have to find an option with axial focussing and radial defocussing. Earnshaw's theorem describes the physical basis for this impossibility to focus in two directions at once. The technology required is the so-called edge focussing. This focussing method was an important aspect for the development of the isochronous cyclotron, but understanding it still remains difficult, even after decades. Edge focussing requires an angle $a \neq 0°$ between the normal of the magnet edge and the beam forward direction at the magnet edge. The bend magnetic field lines connecting the poles of the magnet outside of its gap have a field component perpendicular to the beam direction if a particle enters this outside region displaced from the central axis and at an angle to the edge normal. With the right-hand rule a vertical Lorentz force towards the ideal beam track arises as explained in Fig. 2.17. The combination of radial focussing by field gradients and vertical/axial edge focussing leads to an overall focussing in both axes, as required for stable beam operation.

This kind of magnetic field layout represents an integral part of accelerator design and has a particular importance for AC accelerators, since these reach higher energies compared to DC accelerators. The greater principle of combining different field gradients, each having focussing and defocussing aspects, for generating an overall focussing effect is called strong focussing. Manufacturers exist to deliver 250 MeV isochronous cyclotrons on this basis with a superconducting wiring on a conventional ferromagnetic iron core, reducing the cyclotron diameter to ≈5 m, orders of magnitude smaller than a 250 MeV DC accelerator. The design of these devices follows the spiral structure depicted in Fig. 2.15 (right). The pathways travelled by the particles in circular accelerators are quite long. Still in this example of a 250 MeV

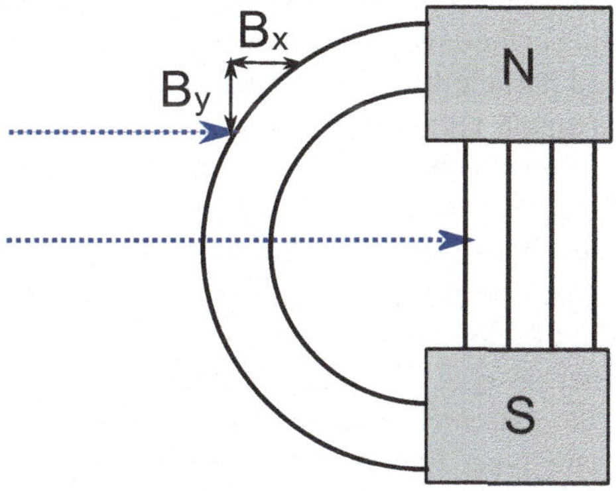

Fig. 2.17 In the depicted situation, two particles enter a sector magnet in x direction (particle coordinate system) at zero edge angle. The ideal beam sees only the B_y field. The displaced beam sees the two magnetic field vector components of the bend outside field B_x (no effect since parallel to charge movement) and B_y (deflects in the same direction as in the magnet centre). Only for a non-zero edge angle (magnet N and S change thickness in the direction of the paper plane) an additional B_z component (bending of field lines into the paper plane) is present, inducing a vertical (axial) force required for edge focussing

beam, accelerated by a 100 kV cavity potential in a 3-fold cyclotron (Fig. 2.15) with 5 m diameter the beam travels about 6.5 km to reach its final energy. Accordingly, adjusting focussing and axial magnetic field in these devices represents a challenging, yet important part of the setup.

As a consequence of this peculiar combination of forces, frequencies, and travelling times, the cyclotron, and AC accelerators in general, are fixed to a certain beam energy and particle type in order to fulfil the resonance and stability conditions. For ions, a limited flexibility exists, if the ratio of mass over charge (q/m) of the ions is maintained (e.g. $^2H^+$ and $^4He^{++}$). The same voltage will accelerate the particle with higher charge proportionally, keeping the cyclotron frequency constant (2.9). Small differences of this ratio to the design value, allow for acceleration with other driving frequencies in the same device, for example $^1H^+$ and $^2H^+$ are possible in many commercial cyclotrons.

Electrons differ in mass drastically from ions, leading to significant differences not only in magnetic deflection, but also in relevance of relativistic effects and Bremsstrahlung. Although electrons could be accelerated in a cyclotron, their low mass would quickly lead to speeds close to c and the requirement of extreme radial field gradients n_R for compensation. Furthermore, power losses by Bremsstrahlung, (2.11), limit the technically reasonable beam energies of electrons in accelerators. In the accelerator context, the Bremsstrahlung emitted upon deflection of beams is called synchrotron radiation, distinguishing it from others origins and spectra of Bremsstrahlung as we will see later.

$$P = \frac{cq^2}{6\pi \varepsilon_0 r^2} * \left(\frac{E}{mc^2}\right)^4 \tag{2.11}$$

All charged particles emit Bremsstrahlung upon acceleration and deceleration. The forward acceleration in the cavities can be neglected compared to the acceleration resulting from a curved particle trajectory of radius r in a circular accelerator. The emitted power P of a single particle depends on the particle energy E, the speed of light c, the particle/electron charge q (=e for electrons), the vacuum permittivity ε_0 and the particles rest mass mc^2. The strong inverse mass dependence in (2.11) highlights the difference between Bremsstrahlung losses of ions and electrons of a factor 10^{13} for protons. The power loss scales equally strong with beam energy, resulting in a hard limit of maximum electron energy for a given device size.

Thinking about other designs of circular accelerators different from cyclotrons makes us appreciate the extremely practical, because highly integrated design of the cyclotron accelerator type. For reaching energies above some hundred MeV, cyclotrons reach their limits even for ions due to the increasing relativistic effects. The amount of material and the size of a cyclotron would become technically unfeasible, other designs are required. For different circular accelerator concepts, we have to start a new way of thinking and go back to the basic cavity concept of Fig. 2.10. Equation (2.11) also highlights the importance of making the system larger, to increase the curvature radii. Let us think about the Large Hadron Collider (LHC) at CERN with its

26.7 km long acceleration ring. Accelerating protons to 13 TeV (and lead ions to 1 148 TeV) certainly required a different way of thinking, besides a large bag of money. At a certain dimension (at the very most the LHC dimension) it becomes unfeasible to construct a single-block type accelerator such as the cyclotron. Separation of the functions integrated into the cyclotron block opens up extra technical design freedom.

We discussed the different ingredients of focussing, acceleration, particle source, compensation of relativistic velocity effects, drifts, and the trajectory deflection. Accelerator language names these separate ingredients *functions*. Each function can be fulfilled by a particular apparatus, such as a dipole magnet induces deflection or a cavity induces acceleration. Combining several such functions to a team/group, represents the unit cell concept. Each unit cell receives a beam, does something with it and releases it to the next unit cell. The design problem reduces to developing the functional elements and tuning them together to form a working unit cell, similar to the object oriented programming. If we tune the unit cell in a way that the incoming beam matches the outgoing beam, all unit cells constituting an accelerator can be identical. This symmetry further reduces the accelerator complexity since only a fraction (say 1/8) of the accelerator needs to be designed, while the remaining parts are just copy and paste. The same way of thinking works not only for circular accelerators, but represents a fundamental concept of thinking for all technical designs. The minimum unit cell we could think of is the combination of deflection, horizontal, and vertical focussing.

This technological approach leads to the Synchrotron accelerator type. This separate function accelerator depicted in Fig. 2.18 requires besides the deflection, focussing, defocussing unit cell additional functional modules. Acceleration cavities are required, but a synchrotron accelerator covers only a part of the total acceleration, not from zero energy at the particle source to the final energy as in a cyclotron. The compensation/limitation of velocity span requires a functional element in the form of sets of different types of accelerators feeding each other with increasing beam energies. In a DC accelerator, the beam extracts directly, but in an AC accelerator with more or less closed path (see Sect. 2.3.1) and resonance condition, the extraction becomes a technical challenge. Cyclotrons with their open spiral shaped beam path can make use of internal targets, i.e. targets at a certain radius, but external targets offer practical advantages and flexibility. In a synchrotron, we cannot extract the beam at a certain radius, because there is only a single radius. The rules of the closed path have to be broken. Changing the charge state of ions using electron stripping (moving towards to lower right corner in Fig. 2.5 by means discussed in Sect. 3.2) in foils or gases is very efficient and allows for the highest beam currents, but requires negative ion beams. If the charge state has to be maintained, so-called kickers (functional element) or septa (thin metal sheet mostly in cyclotrons) periodically change the beam track with a sudden field ramp using fast magnets or electro-static deflection plates to selectively extract particle bunches.

Figure 2.18 shows a unit cell comprised of a deflecting and a focussing module. These functions can be fulfilled also by a single device we already discussed, the "cyclotron magnet" depicted in Fig. 2.16 with its field index $n \neq 0$ focusses radially and defocusses axially. In the isochronous cyclotron, a positive radial field gradient

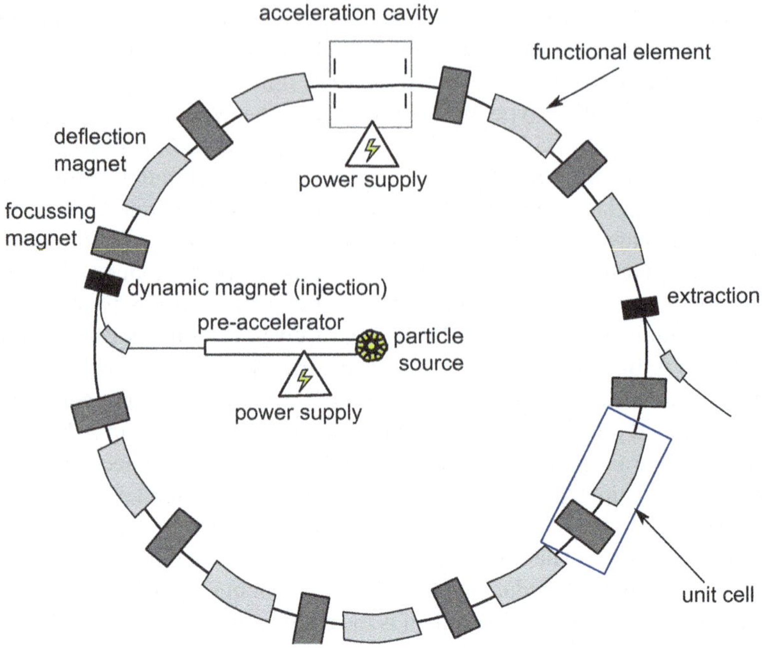

Fig. 2.18 The synchrotron is made from several functional elements and unit cells. The ring is feed by a charged particle source and a pre-accelerator such as a LinAC. For connecting the injected beam to the circulating beam an accurately timed dynamic magnet applies a track switching, like connecting rail tracks. The beam circulates by deflection magnets and quadrupole focus magnets. An acceleration cavity ramps up the beam energy while the magnets strength follows to keep the orbit. Finally the accelerated beam extracts to a target or stays in the ring for particle production

was required for compensation of relativistic effects; in a synchrotron the separation of functions eliminates this restriction and we could inverse the field gradient (radially decreasing field). The magnet would still deflect in the same direction, but an outside oriented field curvature leads to arrows pointing towards the central axis equal to axial focussing (compare Fig. 2.16). Alternating field gradient signs enable a net focussing only by dipole magnets. These alternating gradient focussing dipole magnet arrangements represent a technological option for systems with separated functions. The field gradient n_R determines the focussing strength with $|n_R| >> 1$ (typically $\approx\pm20$) considered as so-called strong focussing. Calculating n_R with real values and (2.10) demonstrates strong focussing with alternating gradient dipoles works only for large accelerators, otherwise technically unfeasible magnetic fields would be required. The technical exploitation of the alternating-gradient technology allows for significantly reduced beam sizes and according reductions in technical investments for large accelerators. Nevertheless, large synchrotrons mostly use alternating quadrupole magnets instead of the combined function dipole magnet. The LHC for example maintains a beam diameter of typically <0.2 mm, much smaller than the ≈45 mm maximum beam tube diameter throughout the whole 26.7 km long ring. This so-called beam optics will be discussed in Sect. 2.3 in more detail.

We can now confine the beam, but what happens after the first round of acceleration? The faster particles demand a longer route, but, in contrast to the cyclotron, the synchrotron offers only a single route. The only other option is starting with a weak magnetic deflection field and increasing the magnetic field strength with every revolution of the beam. What will the next bunch of particle think about this changed situation? They cannot survive under the changed fields, but just like on a roller coaster they have to wait until the first set of beam bunches finishes its ride. Consequently, also our beam output (respectively passenger output) will be discontinuous. The part of filled bunch slots in the AC wave is called the duty cycle, see Fig. 2.11. In contrast to continuous wave (CW) beams such as in DC accelerators, a duty cycle <1 represents an intermittent train of pulses similar to a light beacon. The term duty cycle defines how many of the maximum possible bunches are delivered, e.g. a duty-cycle of 1% delivers one pulse followed by 99 silent cycles or a 100 bunches followed by 9900 silent cycles and so on. The time-averaged beam power and current reduce accordingly, but the peak power and current remain unchanged. Synchrotrons cannot deliver CW beams because they have to ramp up the magnetic field for continued acceleration, representing a major drawback for applications usually requiring beam power, not only maximum beam energy. Their ability to deliver high energies for a relatively low price tag makes them still successful in science, but the energies from which on a synchrotron yields advantages are beyond the ≈250 MeV interesting for applications. Hence besides serving as high-performance Bremsstrahlung based light source (see Sect. 4.3.2) they find little application for particle acceleration outside science.

Low cost per MeV (down to values of about ten thousand € per MeV) represent an important factor in any application, as beam energy defines the types of applications accessible (e.g. certain types of nuclear products). On the other hand, a high average beam power defines the productivity and specific costs of realising this application. Engineering advances and superconductors draw responsible for the success of LinACs with increasing specific acceleration, reaching values up to about 50 MeV/m, as one of the key aspects. The high specific acceleration opens up the MeV scale for electron acceleration for the applications considered in this book and represents a technical option also for ion acceleration. Superior beam quality due to less (error prone) parts, high duty-cycle, and the absence of the limiting Bremsstrahlung losses promoted LinACs to be considered as the state-of-the-art accelerator design in science, except for reaching the highest beam energies. Certain accelerator based particle production facilities, e.g. Free-electron-lasers (Sect. 4.3.3), work with reactions specific to a narrow window around the design energy. Instead of dumping the remaining energy LinACs potentially allow for recovering the kinetic energy, increasing their energy economy. Figure 2.12 explained the importance of being on the right polarity for accelerating the beam, but by refeeding the beam into the cavities and shifting its phase by 180° the beam reinvests its energy into the AC field by travelling on the decelerating phase. Nowadays AC accelerators are applied to introduce nuclear reactions, may it be for investigating fundamental particles, producing isotopes, or medical treatments. Whether a LinAC or a cyclotron fits an application better depends on the required beam power, quality, and outer conditions. For electrons LinACs represent an attractive high power solution for reaching energies above the limits of DC accelerators (≈10 MeV) due to Bremsstrahlung losses in circular machines, therefore a few medical applications use electron LinACs for direct patient irradiation or X-ray production.

All in all the AC accelerators have to fulfil an extensive set of beam stability criteria, making them difficult to design. Over the time only a few designs have prevailed and the above mentioned concepts and aspects are inherent to all designs. Beam optics always receives the major attention for AC accelerator setups and its complex nature will be discussed in Sect. 2.3. This section gave only a brief introduction to the basic concepts of thinking of AC accelerators, due to the complex interplay of physics, geometry and the narrow stability criteria the actual design of an AC accelerator requires a computer solution. Furthermore, the technical feasibility, unavoidable tolerances and the alignment are important points for setting up a state-of-the-art accelerator and calculating its actual performance. Once the accelerator design is finalized the devices are often robust over decades as a tool for providing beams for applications.

2.2.3 Laser and Plasma

The need for miniaturisation of accelerators drives a very recent approach for particle acceleration. The physical idea goes back to one question: How small can the acceleration structure be? There are plenty of self-arranging microscopic structures with high electric potentials in nature. These structures have to obey a certain degree of coherence and external control to accelerate a particle beam in a technically controlled manner. Definitely the waves of a *light amplification through stimulated emission* (Laser) light source fulfil these requirements, but also plasmas feature such structures.

Besides the two established types of accelerators discussed in the last two sections, this defines a few new approaches promising further miniaturisation and increased (length) specific acceleration. These approaches find their physical origin in the extreme electric field gradients related to the small structure sizes occurring in Laser beams and plasmas, offering the prospect of orders of magnitude increases in specific acceleration. The field is still relatively new, but a few successful devices in the scientific context exist, each with its own focus of research. Reference books are not available, hence the reader can only be referred to journal publications and reviews (Macchi 2017; Badziak 2018; Popp 2011; Olsen et al. 2016) to name only a few.

Lasers are themselves beams but with different physics compared to charged particle beams as we will discuss in Chap. 3. Photons cannot be accelerated (speed of light limit) or increased in particle energy (maybe via gravitational waves in the far future) via an accelerator technology. Lasers always have a fixed photon energy defined by a certain laser transition in a source material, e.g. the Nd-Glas or the CO_2 laser. The photon energy defines the oscillation frequency of the electro-magnetic wave associated with light. The photon flux density or intensity, respectively, equals the amplitude (voltage) of these oscillations. Externally pumped/powered Laser cells amplify this initially low beam intensity. The extreme intensity required for Laser accelerators represents a challenge for any optical material used for mirrors and optics. Even dielectric mirrors absorb $\approx 0.1\%$ of the incoming radiation, leaving a relevant heat load on the surfaces considering the extreme Laser power density. A technical setup called Chirped Pulse Amplification uses a combination of spatial and temporal beam widening of the Laser beam to shift this technological limit into the Petawatt (10^{15} W) region with Laser spot diameters <1 mm.

Plasmas represent a quasi-neutral combination of free electrons and ions. They rely on the same principles as the accelerator types discussed above due to their response to electro-magnetic forces. The quasi-neutrality defines along a spatial scale of charge neutrality, the so-called Debye-length λ_d:

$$\lambda_d(n_e, T_e) = \left(\frac{n_e e^2}{\varepsilon_0}\left(\frac{1}{k_B T_e} + \frac{1}{k_B T_i}\right)\right)^{-1/2} \tag{2.12}$$

With the plasma electron density n_e, its electron temperature T_e, and ion temperature T_i, and the physical constants e, ε_0 and k_B. Equation (2.12) represents the e-folding length within which the plasma shields an electric charge to the outside. In contrast to particle beams, plasmas require fulfilling a set of survival criterions such as the Bohm criterion of quasi-neutrality (Piel 2010). These limits in combination with the interactions connected to a plasma being a mixture of differently charged particles introduces resonances/collective behaviours of the plasma when powered by external sources such as electro-magnetic waves, particle beams, or Lasers.

Via this transfer of energy, a Laser can produce a plasma upon impact on a light absorbing material. In fact, a Laser can also induce a plasma in a gas, since light represents a moving electro-magnetic wave the same limits apply as for plasma sources (Sect. 2.4). Physically the absorption of light works via an energy transfer of the electro-magnetic light wave to the electrons present in matter. The light wave accelerates mainly the electrons via its electro-magnetic AC field. In matter, such as our skin, the accelerated electrons quickly stop and we feel only the heat of their energy-loss. We have to exceed some threshold for the electrons and/or ions to become free and remain free, which is the requirement for a particle accelerator.

For this reason, Laser accelerators came up only about 20 years ago, when powerful and short pulsed Lasers became available with the Chirped Pulse Amplification technology (Nobel Prize 2018). The technology of light amplification somewhat limits the energy contained in a single Laser pulse. Reaching a higher power density therefore requires shortening the pulse, since power = energy/time. When fired for example on a thin foil or gas target, the Laser acceleration process works by a combination of up to three different physical mechanisms (Macchi 2017). In any case, the Laser accelerator builds up an all-in-one accelerator system combining particle source and acceleration with the special trait of delivering short-pulse beams connected to the short pulse driver.

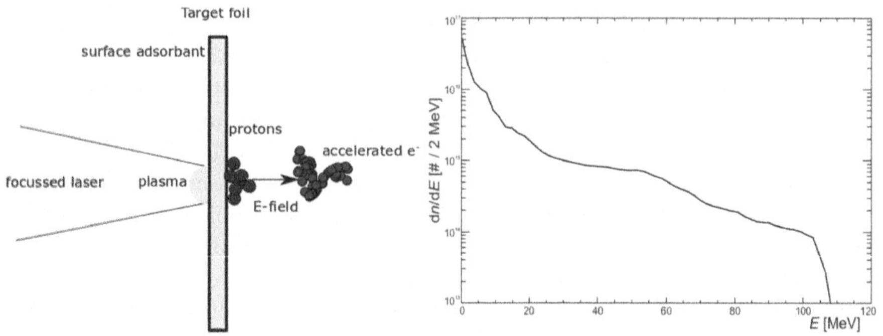

Fig. 2.19 Left: Laser accelerator concept. A laser produces a plasma breakdown on a target foil. Electrons are released, dragging a bunch of protons by their electrical field in the acceleration direction. Right: Simulated proton energy distribution after the process. Courtesy of M. Büscher, Forschungszentrum Jülich GmbH

In the Target Normal Sheath Acceleration (TNSA) mechanism, the laser energy heats up the target material and a plasma breaks down as depicted in Fig. 2.19. The interplay of involved material (target thickness and density) and the Laser energy yields plasma density and temperature. The thermal energy distribution of the plasma produces free charges, but the resulting temperature is far away from the MeV scale. Accelerating a bunch of particles into the MeV range requires the so-called ponderomotive force. This force pushes the charged particles towards the location of lowest AC field strength. This location is naturally behind the plasma as seen from the Laser beam, since the plasma absorbs the light, leading to the "target normal" naming. Since electrons have lower mass than ions, they follow this force first producing a sub-population of fast electrons. These electrons induce an electro-static drag on surface near ions. The magnitude of acceleration depends on the ponderomotive force and with that on the Laser field strength/power density, but with PW class Lasers peak energies up to the application relevant 250 MeV become possible (Esirkepov et al. 2006). Depending on Laser wavelength and the plasma density the plasma changes its optical properties between absorption, transmission, and reflection of the Laser beam. As always for light, the fundamental equation *Reflection* + *Absorption* + *Transmission* = *1* holds true. Transmitted power will not contribute to the acceleration, requiring a fine-tuning of the Laser for an optimal feeding of the fast electrons. For too high plasma densities the plasma reflects the light, for too low density the plasma becomes transparent and the Laser energy cannot be absorbed. The many intermediate steps TNSA requires in between light and particle acceleration limit its energy efficiency. Furthermore, probably due to spatial effects, the resulting charged particle beam features a broad energy distribution with most particles having substantially lower energy than the peak value, see Fig. 2.19 right. These undesired technological limitations call for other mechanisms.

Absorption represents not the only mechanism of energy transfer from light to particles. Upon reflection, the photons induce a pressure onto the reflecting surface by conserving momentum. This mechanism was proposed for space travel with so-called light sails, but works on smaller scales for ion acceleration as well. Plasmas can act as mirror, reflecting light above a certain so-called cut-off wavelength λ_{cut} due to the collective behaviour of the electrons. In this case the light cannot propagate in the plasma, which equals a plasma refractive index of zero (Piel 2010). This happens when the light wavelength is larger than λ_{cut} given by the so-called cut-off electron density $n_{e\text{-cut}}$, see (2.13). The higher the reflection efficiency, the less fast electrons are produced since their acceleration requires an absorption of the light energy in the plasma. In the TNSA parameter range, radiation pressure acts together with TNSA mechanism.

$$n_{e-\text{cut}} = \frac{4\pi^2 \varepsilon_0 m_e}{e^2 \lambda_{\text{cut}}^2} \tag{2.13}$$

In contrast to TNSA, a radiation pressure dominated acceleration first requires a thicker target, which cannot be tunnelled by the light. The Laser then punches a hole into the thick target pushing out a disc of accelerated particles. To find a regime of pure radiation pressure acceleration we require a target thin enough so the accelerated particles at the front (irradiated side) reach the backside of the target before the end of the (short) Laser pulse. The thinner the target, the less mass has to be accelerated with a given Laser pulse energy and consequently the higher the energy per particle, which is what we want. The technological challenge of this light sail mechanism is the balance between a short light wavelength λ_{light} necessary for reflection on the thin targets (reflection requires target thickness $d > \lambda_{\text{light}}$), the target thickness, and a high Laser pulse intensity required for reaching a plasma density above the cut-off density (2.13). The radiation pressure accelerates a sailing quasi-neutral plasma disc from the target. The quasi-neutrality of plasmas makes electrons and ions move together at equal velocity in a bunch of enormous density (compare the density of a solid to typical ion beams by yourself), but it represents an upper limit to the Laser pulse intensity given by (2.14). The right hand side of (2.14) represents the limit of the Coulomb force between electrons (density n_e) and ions keeping the quasi-neutral plasma together (for an ideal plain disc of thickness d). For a light electric field strength E_{light} (equivalent to an intensity I_{Laser}) exceeding this limit, the light electric field allows the electrons to leave the ions. This breaks the quasi-neutrality leaving two highly charged particle bunches which disintegrate due to the inner electro-static repulsion. In conclusion, the light sail mechanism requires a perfect alignment of light wavelength, Laser pulse intensity, pulse length, target, and geometry.

$$E_{\text{light}} = \sqrt{\frac{4\pi I_{\text{Laser}}}{c}} > 2\pi e n_e d \tag{2.14}$$

The third mechanism of Laser accelerators requires a density wave of fast electrons running through the target. The density wave has to be faster than the ion sound speed in the plasma, resulting decoupling of this wave from the plasma ions. This wave runs like a wall of electric-field variation through the plasma accelerating ions like a snowplough accelerates snow. The mechanism is called collision-less shock acceleration since it works via the reflection from an electric-field variation, but its physics are not completely understood, yet.

The beams of Laser accelerators feature very short pulses in the picosecond range with intensities exceeding 10^{13} particles per bunch. On the downside, there is only limited flexibility in these parameters since the Laser pulse has to be short for delivering the power density required for the discussed Laser surface interaction processes. Also the repetition rates, and with this the average beam current, have technical limits in the order of 1000 pulses/s. The quick replacement of targets after each shot, e.g. by rotating foils, gas targets, or liquid droplets represents a yet unsolved technical challenge. The intrinsically high peak flux, potentially broad energy spectra, and low time averaged flux are in conflict with many accelerator applications discussed below due to e.g. detector technology.

Besides these Laser based acceleration methods producing plasmas as a part of the energy transfer process, a plasma can also directly accelerate particles. This acceleration relies on the so-called electron-density wave mechanism, an independent acceleration technology. Think of a cloud of charged particles travelling through a plasma: A wake of charge separation will follow it, generated by the different response speed/inertia (mass difference) of electrons and ions. The faster electrons leave the slower ions behind, forming local charged regions as shown in Fig. 2.20. The separation of charges equals an electric potential similar to the wake of a speedboat inducing a gravitational potential energy by displacing water. This potential structure moves together with the driving cloud, which could be for example an electron beam.

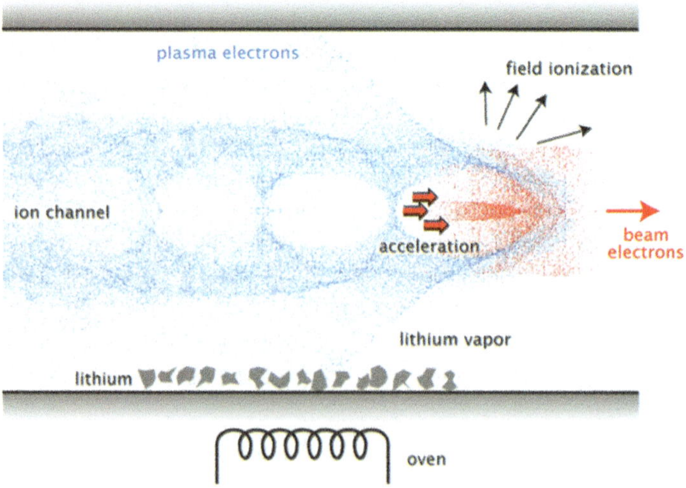

Fig. 2.20 Illustration of a plasma wakefield formed in a lithium vapour plasma. A driving beam (right) induces density waves in the plasma electrons (blue). Particles moving inside these density valleys (arrows) feel the electric field generated by the local electron deficiency. By Rasmus Ischebeck at English Wikipedia, CC BY-SA 3.0, via Wikimedia Commons

A second so-called witness beam (this is the beam we want to accelerate) following the driving beam in the correct phase alignment in the well behind the beam front, as indicated by the red arrows in Fig. 2.20, rides on the wake-field like a water-skier following the speedboat on its stern wave. Residing in this phase equals a continuous acceleration via the constant field gradient as seen in the co-moving frame.

Lasers or particle beams power the accelerating plasma wake-field by travelling through this plasma, see Fig. 2.20. Bunched ion (=Proton), electron, or Laser beams drive the charge separation in the plasma. Proton beams have the advantage of low straggling and higher momentum, better conserving the wake-field properties over the acceleration length. The proton mass on the other hand requires significant energy of at least a few 100 MeV in order to provide a relevant velocity. Electrons and Laser beams offer technological advantages in terms of energy efficiency in this respect, making them more promising for applications outside science. In contrast to the all-in-one system of Laser accelerators, the plasma accelerators form a plasma acceleration cavity in its application similar to AC cavities.

Acceleration gradients > GeV/m become possible, exceeding the current technological limits of AC accelerators by at least an order of magnitude. The AWAKE device (see related publications under https://awake.web.cern.ch/publications) as a pilot project of plasma acceleration technology uses a 200 MeV proton driver beam in a 10 m long Rb plasma heated by a co-axial laser beam for accelerating a 15–20 MeV electron witness beam to a few GeV. By using even higher energy protons (Bracco et al. 2014), this technology opens up an option for producing electron beams in the order of 500 GeV, effectively mitigating the physical (Bremsstrahlung) and technical (acceleration gradient) problems of accelerating electrons to high energies by connecting them with proton acceleration technology. The witness beam currents remain small (nA in the above example) since only a small part of the driver beam energy can be converted to witness beam energy, since the load induced by the witness beam onto the wake-field would continuously reduce its velocity which in turn broadens the witness beam energy spectrum via dephasing. The whole setup with its two supporting accelerators, the Laser beam, and the limited efficiency may be a tool only for reaching higher electron beam energies for fundamental particle research, but it is also only a pilot device.

Besides acting as all in one or as acceleration cavity Lasers can also be used for other functions in AC or DC accelerators. Electron accelerators, for example in FEL accelerators (Sect. 4.3.3), already make use of laser based electron emission in ultra-short pulse electron sources, combining the advantages of both technologies. The compact source region defined by a focussed laser beam leads to high beam quality (low emittance) and the Laser pulses result in bunches shorter than possible with any other method. The release of particles with the Laser surface interaction process connects the charged particle pulse length to the Laser pulse length. Electrical switching and bunch compression in classical accelerators has typical pulse length limits in the range of $\geq$100 ps. Laser source deliver orders of magnitude shorter pulse length. Short pulses are of particular interest for time-resolved investigation of very fast processes with the help of accelerator-based analytics, see Sect. 7.1.

In conclusion, Laser accelerators and plasma cavities could develop to valuable tools of accelerator technology and in a few special scientific case they already are, as discussed above. Extensive developments are still required towards application as established tools, in particular for the Laser accelerators it appears the optimal physical scheme is yet to be found. Whether these tools have the potential to substantially reduce accelerator sizes depends also on the supporting aggregates, e.g. Lasers, and their efficiency. In the next section we will discuss more details on the technological differences between the three accelerator classes discussed. Expecting the same beam "product" from a Laser accelerator as from a classical accelerator is like expecting an electric car to be identical to a combustion car. The Laser accelerator technology is simply a different technology and niches which tolerate its deficiencies (e.g. the broader energy spectra) and value its advantages (e.g. the short pulses) will be found.

2.2.4 Electric and Spatial Efficiency of Accelerators

The efficiency of a process is, in particular in its application in a production environment, the determining factor for its usefulness and practicability. As such, the tool accelerator needs to efficiently deliver charged particle beams. The measure of efficiency strongly depends on the particular application and its boundary conditions. Besides electrical/energy efficiency also dimensional compactness or specific acceleration, respectively, is a decisive factor. Larger accelerators require larger buildings and vacuum systems, inducing secondary problems. Both efficiencies also directly couple to the cost of an accelerator system and its operation and therefore also the specific costs of its service or product. The three main groups of accelerators (DC, AC, Laser) presented in this section, each with a set of sub-groups of devices, are very diverse in their different efficiency aspects.

We take quantitative look into the efficiency. DC accelerators offer the highest electrical efficiency, with typical application values between 70 and 98%. Behind the scenes, DC accelerators also rely on AC frequencies for transformation and electrical/galvanic separation in their initial stages, but the technological restriction in this part are significantly less relevant than for pure AC accelerators. On the other hand these accelerators suffer from the technological difficulty of discharge breakdowns, limiting their specific acceleration. Table 2.3 demonstrated theoretical limits of 80 kV/mm = 80 MV/m, but discharge avoidance means the device grows with beam energy not only in length but also in diameter. Practical difficulties result in only some single MV/m for real devices. Electrical discharge phenomena, most prominently the corona discharge, contribute a base level of electric losses only dependent on the acceleration voltage. For low beam currents at the MV level, e.g. in analytics, the effect dramatically limits the efficiency. For applications below MV the losses become negligible and only the voltage transformation and rectification limits the efficiency.

AC accelerators are a compromise between electrical and spatial efficiency. Modern solid-state generators for the MHz range achieve electrical efficiencies in the order of 25–70%, already substantially better than old vacuum tube amplifiers with values of about 10%. This comparably lower energy efficiency is contrasted by extremely improved spatial efficiency. A modern, superconducting cyclotron can deliver 250 MeV ion beams with a device of about 5m diameter. In this example we achieve an effective specific acceleration of 50 MV/m (although of course the protons travelled a much longer path), but values in the order of 10 MV/m are becoming standard for constant wave devices and up to 100 MV/m become possible with pulsed operation. The technological limitation for AC accelerators lies rather in the electrical resistance and quality factor, compared to DC accelerators.

Power losses inherent to the transport of AC with high frequencies due to the skin effect and other AC physics represent a technological disadvantage of this accelerator type. For a given conductor, the effectively conducting skin layer thickness reduces with the square-root of the AC wavelength. At the same time higher frequencies (=shorter wavelength) reduce the spatial dimension of an AC accelerator due to a proportional reduction of the resonance length (Fig. 2.10). This reduction in resonance length increases the specific acceleration. Higher specific acceleration leads to secondary benefits for the size and cost of beamline, vacuum, and beam focussing. The competition of these scalings in the space between performance and costs could be worse, but only in science, where only specific acceleration (=performance) is relevant, due to its impact on the maximum achievable beam energy, the compromise is easy to resolve (reaching e.g. one GeV with 10 MeV/m = 100 m length). In applications, not only the performance but also the costs and competing technological options have to be considered before deciding for an AC accelerator. Unfortunately, not even superconductors can solve this technological issue completely, as finite conductivity effects occur. A large interest in efficient AC generation at high powers and frequencies for wireless communication will further improve the energy efficiency of AC accelerators, potentially lifting them to the values of DC accelerators, but due to the different loss mechanisms (corona vs. skin effect) energies up to some MeV will probably always be the domain of DC and energies >20 MeV the domain of AC accelerators.

Laser and plasma accelerators represent the other end of the efficiency scale. The technological limits of AC and DC accelerators do not apply to Laser and plasma accelerators, but other limits arise. The devices are extremely small, with field strength of up to some 100 GeV/m. The technology offers completely new applications with a potential for table-top sized accelerators, not considering the extensive aggregates required though. This compactness is on the cost of a low electrical efficiency. Already the initiating laser beams are produced with a maximum of $\approx$30% energy efficiency. On top of that a relevant amount of energy is lost to thermalisation in the beam interaction zone. In Laser accelerators about 10% of the Laser energy contributes to the generation of the required fast electron population. Another large factor is lost to unusable parts in the charged particle energy spectra, which are, so far, by orders of magnitude broader than in AC and DC accelerators. In the end, some 10^{-3}% of electrical efficiency remains for producing a beam similar to what is known from AC or DC accelerators. It has to be admitted, that this technology is still under evaluation and development with high potential in new physical acceleration schemes and increased Laser powers, but the technological aspects of its efficiency losses will probably only be weakened. The physical processes behind Laser and plasma accelerators are not fully understood, therefore their specific technological limits might change in the course of development. With the low efficiency, also only lower average beam powers are reached compared to AC and DC accelerators.

2.3 Ion- and Electron Beam Optics

Charged particle beams, this means ion and electron beams, are ensembles of individually moving and interacting objects. These ensembles can be a few or very many individual particles, each having its own properties, trajectory and energy. A certain similarity or range in particle trajectories and kinetic energies allows us to call them a *beam*, yet there is no clear limitation in these ranges. Similar to a group of people bunched together in a tram (Fig. 2.22), the individuals in this beam ensemble still slightly differ in their properties from the ensemble-averaged values, which we actually call the beam properties. To a certain extent, this individuality is tolerable, yet there is no tram without a solid cabin giving technical limits to it. The cabin is required to restrict the individual movement, keeping the members of the ensemble bunched together. While it would be a catastrophe to lose some people travelling in a tram, certain losses are not critical in particle beams and actually losses are completely unavoidable as a result of the fundamentals of statistical distributions of particle properties. Nevertheless, particle beams need a confinement otherwise, too many particles are lost unused introducing secondary problems. Since charged particles will neutralise and disappear upon contact with solid walls, charged particle beams require a contact free confinement by electro-magnetic fields and a vacuum beam tube diameter significantly larger than the beam diameter. The description and layout of this confinement is the subject of beam optics.

The description of the beam movement either considers the movement of single particles with individual properties or the beam collective with certain distribution functions. In both cases, longitudinal motion describes the components in the beam or acceleration direction, respectively, and transversal motion depicts the perpendicular plane (towards the vacuum vessel walls). A beam is naturally not a point like entity, but for designing the beam path we require a track. Similar to a tram we need a closed track connecting start and destination in an ideal fashion. This track defines the beam centre and, by definition, the ideal particle having ideal starting conditions follows this ideal path. A description of beam optics firstly covers this ideal particle and then considers if and how the non-ideal particles follow the ideal path and how many of them will be lost.

Electric and magnetic fields induced by so-called beam optics such as magnets and deflectors confine the particles in the transversal direction. The longitudinal direction is not always relevant, in particular for DC accelerators, or it is mostly confined by the accelerator itself as discussed with AC accelerators in Sect. 2.2.2. The confining fields keep the overall energy stored in the beam constant. If a charged particle enters a magnetic field its kinetic energy remains constant, but it is only redistributed to different directions. Static electrical fields work in a similar fashion, as we already learned in the beginning of Sect. 2.2.2, Coulombs law tells us a repetitive passage through electro-static fields cannot change the beam energy more than a single passage. For these reasons the forces behind the confining fields/devices are called conservative.

The beam particles oscillate in the transversal directions around these confining fields since conservative forces cannot change or remove the transversal kinetic energy and typically also no other strong friction or damping mechanism exists, respectively. Therefore, sine/cosine-like equations describe the single particle trajectories in an accelerator. The respective equation of motion in the transversal directions x and y is a function of the position in the accelerator s (particle source: $s = 0$) as described by (2.15)

$$x, y(s) = A\sqrt{\beta(s)}\cos(\phi(s) + \phi_0) \tag{2.15}$$

With phase ϕ, phase-shift ϕ_0, an amplitude factor A, and a function β describing the beam optics acting throughout the beam path. The coordinate s is the relative longitudinal coordinate along the ideal track of the beam in the accelerator system starting from any point defined as $s = 0$, for example the charged particle source. For straight tracks this equals a Cartesian coordinate system, but for curved tracks this coordinate transformation reduces the mathematical complexity along with introducing a truly straightforward quantity.

With sufficient understanding and adequate equipment, the accelerator design limits β to avoid values of x and y become larger than the vacuum vessel dimension at any point s. The β function condenses the combined effect of all beam optical elements to a single mathematical function. A physical connection of charged particle beam optics to the geometrical light optics will ease understanding the mathematical concepts but also allow for an improved practical understanding of how to construct an accelerator system. Designing β is the field of beam optical elements. Measuring it requires analytical elements, which will be discussed as the last point of this section. These analytical elements actually determine $x(s)$ and $y(s)$, hence also the α factor can be determined. The above statements on conservative forces leave us basically no freedom for influencing a given α, making it a beam quality factor. In the next section, we will come to know it as the emittance.

In the end, the difference between the ensemble properties and those of the individual particles is just a matter of perspective. The ensemble bears the power, making it more relevant for applications. The single particle motions are still buried behind the ensemble values we are interested in. The superposition of all single particle motions forms the beam envelope, its outer boundary. Consequently, the general motion described by (2.15) applies equally to a single particle and the ensemble. The ensemble represents the integral over all individual properties, namely the individual amplitudes α and phase shifts φ_0, averaging out the individual oscillations. While an AC accelerator allows only for limited ranges of phase shifts due to the resonance conditions, a DC accelerator implies no phase restrictions, leading to a

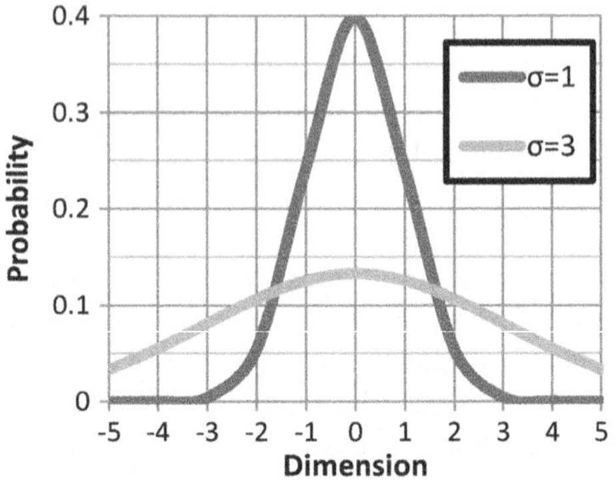

Fig. 2.21 The normal distribution with two different standard deviations of $\sigma = 1$ and $\sigma = 3$. The standard deviation describes how broad the distribution is. For example in a particle beam with larger standard deviation less particles will be found in the centre and more towards the outside. 68.27% of all particles can be found within $-\sigma$ to $+\sigma$, 95.45% within -2σ to $+2\sigma$, 99.73% within -3σ to $+3\sigma$ ...

complete smoothing of the oscillations in the ensemble view. Consequently, the technical freedom we bear for forming the beam to our applications needs relates to constructing the β function and finding a way to start with minimal amplitudes α (see Sect. 2.4).

A beam comprises a large amount of particles and if few of these are lost it will be negligible for the application. Naturally, the large numbers (1 A equals $6.25 * 10^{18}$ particles/s) make beam optics calculations also a statistical problem. Hence, we can never describe all single particles with an ensemble description, but only a certain share of them. The normal distribution describes the statistical distributions of charged particle beams in most cases. Figure 2.21 plots two different instances of this statistical function. The dimension axis could be any properties, for example the transversal dimension of the beam, its energy distribution, or the distribution of starting points of individual particles with respect to the beam centre. It is important to note: The normal distribution is based on exponential functions, therefore we can always find particles outside a certain boundary. When describing beam properties we usually refer to the so-called 1σ ensemble. This refers to all ensemble particles having a property within a range of $\pm 1\sigma$ of the central/ideal value. In other words, only 68.27% of the beam particles are within the given value. If we want to describe a larger fraction (more σ's) of the beam, the range of the value will increase accordingly with a mathematical connection given (ideally) by the normal distribution (see Fig. 2.21).

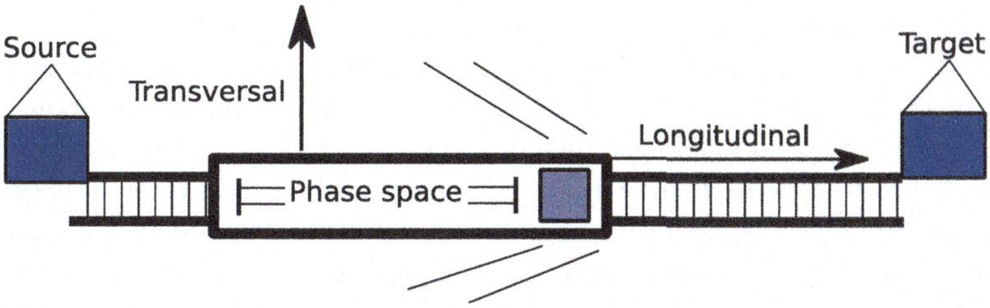

Fig. 2.22 Central beam optics terminology in a train. The train/beam moves along the longitudinal axis on a track defined by the beam optics. Its volume or passenger capacity is fixed and represents the phase space occupied by the beam. The density of seats in this volume represents the inverse beam emittance

2.3.1 Emittance and Betatron Function

The particles travelling together in a beam all occupy their individual position in the so-called *phase space*. The degrees of freedom relevant for the movement of the beam in the electro-magnetic environment of the accelerator define the dimensions of the phase space. The three positional and the three momentum vector components relative to the ideal beam particle span the phase space. The position in the phase space completely describes a single particles movement in a given accelerator with known β-function. The superposition of all individual particles in the phase space yields the beam ensemble phase space corresponding to its size in real and momentum space. The phase space can be understood as a tram moving along the ideal track, see Fig. 2.22. The tram moves mostly in the forward direction, but the passengers inside the tram can still move in all directions and stand or sit wherever they want during the ride, in the given boundaries of the tram cabin size. The combination of this information with the knowledge of the layout of the beam track allows calculating the complete evolution of the particle movement and beam properties throughout the accelerator system. The problem of beam behaviour reduces to knowing two quantities we will discuss in the following.

The phase space cannot be filled to an infinite extent by the particle beam, all particles have to sit in a certain volume of the phase space tram, the confined volume. This is easy to understand with a limiting case example: A particle can sit in the spatial centre of the tram (ideal) at the starting point, but if it has a transversal momentum component it will diverge from the ideal path, increasing its distance to the beam centre. For large transversal momenta it will leave the beam by annihilating with the surrounding vessel walls before reaching the final target. A limited transversal momentum has to be acceptable (less than it takes to leave the tram cabin before reaching the target), otherwise the beam would be unable to contain (the statistically distributed) particles. The same holds true for a particle with zero momentum difference sitting outside the confined phase space. The accelerator system has to be able to cover deviations from the ideal particle for all quantities within its technical limits.

The size of the confined phase space volume (i.e. the tram cabin size), e.g. for the transversal momentum, will be defined by the properties of the beam optical system.

Before the technical aspect, we first take a look at what we are talking about. Phase space can be defined in real space quantities by the position offset (x, y) and divergence (x′, y′) of an individual beam particle in both transversal directions (perpendicular) to the ideal direction of motion (s). In the case of bunched AC beams, the coordinate relative to the bunch centre, and the deviation from the ideal momentum (given by the beam energy and species) have to be considered additionally. For DC beams this coordinate has no meaning since DC beams are by definition longitudinally constant. We could also define the phase space along other quantities, such as transversal momentum instead of divergence, but our choice should be oriented along measurable quantities. Our phase space now becomes a 6-dimensional space consisting of 3 space and 3 momentum-like dimensions forming a vector as given in (2.16).

$$X = \begin{pmatrix} \text{Offset in } x \\ \text{Divergence in } x = x' \\ \text{Offest in } y \\ \text{Divergence in } y = y' \\ \text{Offset in } s \\ \text{Relative momentum offset} \end{pmatrix} \tag{2.16}$$

Each beam particle has an individual vector X, leading to an individual track. Figure 2.23 demonstrates this at the example of a deflecting dipole magnet passed by an ideal and an offset particle with divergence x'. This individual X represents a single point in the 6D phase space. The beam optical and acceleration devices guide all particles of a beam along the accelerator system. This guidance depends on X. The smaller the values of X compared to the beam energy and device size, the more similar the individual tracks will be and consequently the smaller the beam diameter. The individual rows of X can change along the particle track, but since the rows are coupled (divergence leads to displacement and vice-versa) via beam optical

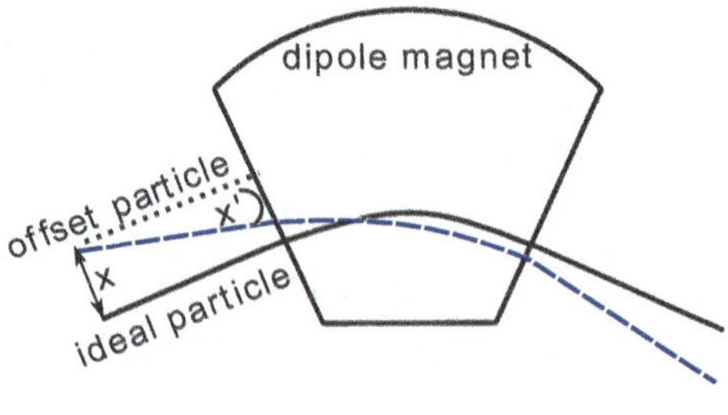

Fig. 2.23 Sketch of the difference of ideal particle to other beam particles with non-zero X. Non-ideal particles start at an arbitrary point with displacement x and angle x' to the ideal particle. In an optical element such as a dipole this leads to different deflection.

elements, the length of the vector X is characteristic for a particle and conserved by beam optical elements, if normalised properly. For example a particle with a high transversal momentum will necessarily also reach higher transversal displacements in its oscillating movement, similar to a pendulum oscillation amplitude is connected to its initial displacement. At the maximum displacement both beam particle and pendulum have the smallest (transversal) momentum due to the conservative nature of the movement controlling fields in both cases.

Mathematically speaking: The beam optical system lets the particles rotate in phase space around the $X = 0$ vector during their movement through the accelerator system. The superposition of all individual tracks equals the beam envelope, the quantity of interest for the application. Up to now we have no idea why and how the individual particles start with different X, or in other words, we have no idea of the distribution function of X. The large number of particles in a beam and the central limit theorem allow for an educated guess, though. The normal distribution usually describes the distribution of X very well, leading to a normal distribution of individual points X in phase space.

Coming back to the analogy of the confined phase space and the tram wagon: People can leave and enter the tram at each stop. A particle beam has only two stops: It starts at the source and ends on the target, where it is used for an application. Consequently, this does not allow any new passengers, or charged particles, to enter the tram after it has left its source. In physics, this is called *Liouvilles Law*, stating that the density in the momentum space can only be increased if the density in real space is reduced and vice-versa. Consequently, beam focussing reduces the beam dimension but increases its angular divergence and vice-versa. Adding new particles to the beam would increase both momentum and real space density and is therefore physically impossible.

Besides the statement that it is impossible to enter or leave the tram during the journey, the tram also has a fixed amount of seats per wagon. A particle beam has a similar quantity called *emittance*. The emittance states the distribution width of momentum and position (vector X) of the individual particles difference from the beam average. This equals the phase space volume covered by the beam. A larger difference equals a lower density of particles in the phase space, a given amount of particles distributes over a larger phase space volume. A low emittance corresponds to a high density of particles in the confined phase space making it generally desirable. High emittances lead to lower beam density, larger devices, and reduced performance. The emittance can also be understood as a beam temperature. Similar to the ideal gas law equation (2.1), a high temperature corresponds to a lower particle density at a given pressure.

Figure 2.24 illustrates a 2D extract of the phase space representing only the first two rows of X (2.16). A certain amount, say 1σ, of all beam particles can be found within the enclosed area. At the given position s in the accelerator system this 1σ of particles feature a maximum angular deviation x'_{max} and a maximum displacement x_{max} from the ideal track. These values define the beam envelope. Focussing the beam would rotate the phase space ellipse, with a minimum size reached with an upright ellipse (a focal point or beam waist).

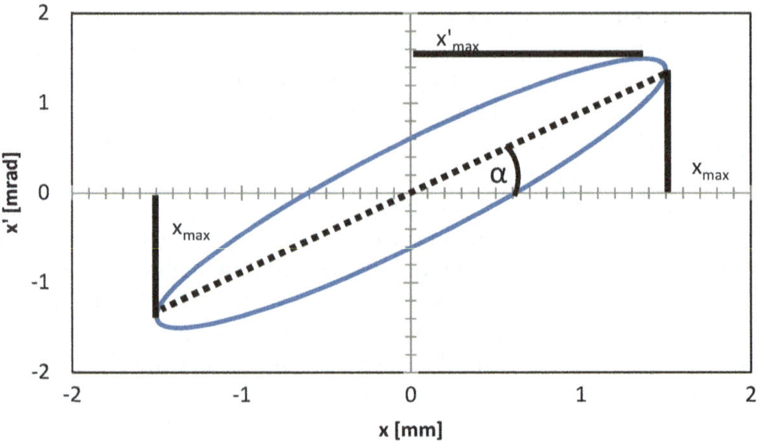

Fig. 2.24 1D phase space volume of a beam with maximum radius $x_{\max} = 1.5$ mm and maximum divergence of $x'_{\max}$=1.5 mrad. The phase space ellipse form depends on the position in the accelerator system, but its enclosed area remains constant

From the beam ellipse we can calculate the phase space volume which corresponds to the emittance ε for a certain share of $n\ \sigma$'s ($n = 1$ equals 68.27% for each axis, see Fig. 2.21) of the particles via the following equation:

$$\varepsilon_x^{n\sigma} = n^2\sqrt{x_{\max}^2 x_{\max}'^2 - x_{\max}^4 \tan^2(\alpha)} \tag{2.17}$$

The emittance is a measure for the deviation of the particles phase vectors from the beam average, hence increasing the beam average makes the individual deviation less dramatic. It is a bit like a train moving at relativistic speeds in one of these popular science books about Einstein's relativity theory, just that moving at relativistic speeds is very possible for a charged particle beam. From the laboratory point of view, the time dilatation slows the beam particles relative motion. In the co-moving frame of the beam nothing changes in between the particles, hence this effect is called Adiabatic cooling (no exchange with the surroundings). If our tram in Fig. 2.22 moves faster, it reaches its target faster and we have less time to reach the cabin boundary with a given transversal velocity, a larger transversal velocity becomes acceptable. The normalized emittance cancels this relativistic effect by considering the particle velocity v:

$$\varepsilon_{\text{Norm}} = \frac{v/c}{\sqrt{1-(v/c)^2}}\epsilon = \frac{v/c}{\sqrt{1-(v/c)^2}}\sqrt{r_x r_y} \tag{2.18}$$

In addition to this passive method of reducing emittance, also active methods were developed to reduce beam emittance. Usually the application defines the beam energy, hence the possibilities of adiabatic damping bears no technical freedom. Three main classes of methods were invented, but due to their complexity they are seldom found outside science. In medicine, industry, and most scientific applications, the emittance problem is literally addressed at its source. The concepts of active emittance cooling are specific to the beam species and accelerator type. Good elaborations of the details can be found in beam optics literature, e.g. (Hinterberger 2008). For the cooling of ion beams, the ions are directed through a bath of cold electrons. When jumping into the bath, the ions have to have a very small relative velocity difference to the bath, otherwise it is water skiing. Therefore, the bath is actually a beam of fast, yet low beam energy (due to small electron mass) electrons. If the temperature of this electron beam is lower than the ion beam temperature, thermal equilibration cools the ion beam. Obviously, this bathing cooling method is not applicable to electron beams, since these are already the lightest charged particles. Electrons on the other hand easily reach highly relativistic velocities where they effectively emit bremsstrahlung (=synchrotron radiation) upon acceleration (2.11). Bremsstrahlung occurs naturally in bending magnets, but it can also be deliberately applied as we will see in Sect. 4.3. The emission power of this Bremsstrahlung strongly depends on the kinetic energy and hence the faster part of the thermal spectrum releases more energy than the slower part, inducing a cooling effect, the radiation damping. Finally yet importantly, the phase space nature of emittance enables reducing emittance not only from the velocity side, but also from the spatial deviations from the ideal beam path. The so-called stochastic cooling exploits this by actively kicking particles which are off track back into their ideal path.

The emittance limits many accelerator applications to a certain extent in minimum device size and maximum beam power density. The conservation of emittance makes it an important yet hard to optimize quantity. Still we can change its technically relevant result, the beam envelope size. This important connection is described by the betatron function $\beta(s)$. The betatron function does not describe whether the particles have to move left or right, it rather connects the abstract idea of the 6D phase space with beam dimensions in real space. Figure 2.25 displays an exemplary betatron function in a series of beam optical elements and the s-coordinate resolved transversal (β_x and β_y) betatron functions resulting from these optical elements. The beam envelope then derives from the given betatron function, the emittance, and (2.19). Beam optical elements define the betatron function, allowing tuning between size and angular divergence of the beam (smaller size = larger divergence due to conservation of emittance). Hence, a significant aspect of accelerator development relates to the construction of low emittance charged particle sources (see Sect. 2.4) and the conservation of emittance.

$$x_{\max}(s) = \sqrt{\varepsilon * \beta(s)} \tag{2.19}$$

Finally, we have all concepts at hand to describe and influence the technically relevant quantity, the beam dimension. As demonstrated in Fig. 2.26, a beam tube can only fit a certain beam size as described by (2.19). Considering the beam particle offset in the transversal direction to follow a normal distribution (Fig. 2.21), a share of the beam will always be lost to the vessel walls. The beam tube acts as an aperture

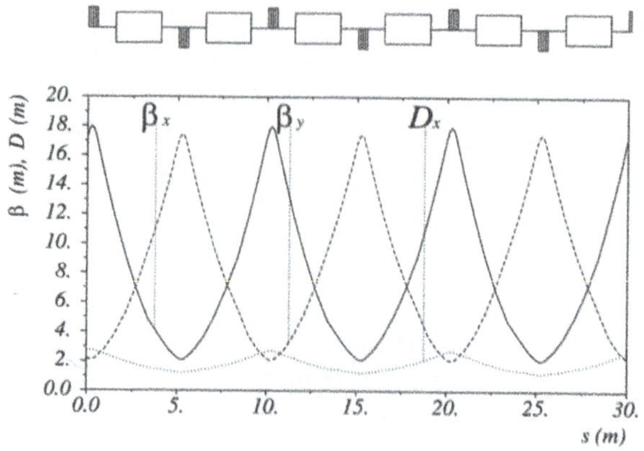

Fig. 2.25 Horizontal and vertical betatron function β and dispersion D in a synchrotron accelerator unit cell. The filled upper rectangles represent horizontally focussing and the lower rectangles horizontally defocussing quadrupole magnets. The open rectangles are dipole magnets. The betatron function clearly shows the connection of focussing in one and defocussing in the other direction already discussed for cyclotrons, but also typical for quadrupole magnets. Reproduced from Hinterberger (2008), page 280 with permission by Springer

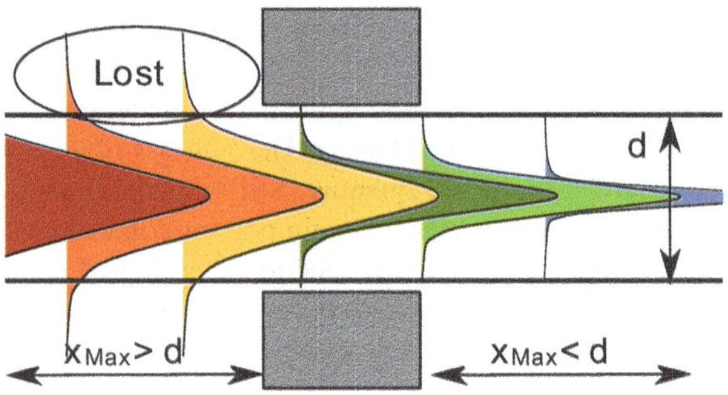

Fig. 2.26 Transversal beam dimension in a beam tube of diameter d with a focussing element in the centre (grey rectangles). On the left hand side a significant part of the beam lies outside the chamber limits. This part is lost, reducing the beam emittance to the acceptable limit, but also reducing its current. Focussing reduces the beam diameter, but it increases the divergence requiring another focussing element after 2 times its focal length

cutting down the beam to fit inside. This means a loss of beam current, but a relatively larger tube reduces the losses. The accelerator system acceptance states the maximum emittance the beam system can transport with a given betatron function (solving (2.19) for ε). Usually there are only a few critical points where the betatron function is largest or the beam tube diameter is smallest. A deliberately placed aperture limits the x_{max}, fixing the location of this cut of the phase space to a technically adapted device. For example, a beam tube diameter $d = x_{max}$ will cut everything outside the 1σ range $(100\% - 68.27\% = 31.73\%)$ of the beam distribution, if x_{max} was specified with a 1σ emittance. Higher emittances or smaller beam tubes will further increase losses to the system boundaries/vacuum vessel walls. Since ε only defines a certain share of the beam particle ensemble, losses are practically unavoidable. This has quite important implications for radiation safety, see Sect. 2.7, and long beam-lines.

The losses also provide us means to measure the emittance. A variable beam tube diameter, for example with movable aperture sizes or by moving a wire through the beam, allows measuring the beam emittance by recording x_{max} and the beam current lost to the analytical device. Several points are required to be able to solve the equations due to the many parameters required for outlining the phase space. Measurements at several points on the other hand require knowledge of the transfer functions (the effect of beam optics) connecting the different points. More details will be discussed in the following sections.

In this section, we discussed only the simplified so-called linear approximation of beam optics. For example, we assumed a dipole magnet to feature only dipole effects, neglecting possible multipole components (quadrupole, sextupole, octupole…). Furthermore, space-charge effects originating from beam ensemble self-interaction due to intra-beam scattering were neglected. Finally yet importantly practical aspects have an important impact on beam optics. Alignment errors and stray fields induce additional deflecting components and track offsets leading to non-ideal behaviour of the beam optical elements. A beam optical calculation allows for an assessment of the possible magnitude of these effects. Larger scientific accelerators foresee correcting optical elements in the beam path. In accelerator applications the accelerator device length are often limited and higher order beam optical aspects only reach a certain relevance for very small or large beam diameters or high beam densities/currents.

The above discussion of emittance and betatron functions shall grant the reader the grace to accept the things that cannot be changed, the basis to change the things that can be changed, and the wisdom to distinguish the one from the other. The beam optical treatment in this book intentionally remains superficial. Only years of studying and practical experience enable to contribute to the state-of-the-art of beam optics or any other field, but a basic understanding already allows appreciating an expert's work and understanding layout concepts of accelerator systems. The mathematical complexity explodes for beam optics of real accelerator systems, requiring anyways computer codes for evaluation. Numerous codes exist, ranging from older scientific variants such as TRANSPORT over newer version. Also commercial products exist, for example the specific code Simion. Besides this, also a few general electro-magnetic simulation tools such as OPERA or COMSOL allow for charged particle electro-magnetic simulations. These finite-element based codes are rather suitable for assessing the true properties of a single optical element rather than a whole accelerator.

2.3.2 Beam Optical Elements

Having defined the mathematical properties describing a beam, the next task is to establish a control of the track and the shape of the beam. Basically, all accelerator applications require the beam to be directed onto a target, changed in diameter, or transported over long distances without relevant beam current losses. Consequently, we require devices for deflecting and de-/focussing or in general manipulation of charged particle beams, the so-called beam optical elements.

Until now no details on the beam particle species were mentioned. The particles can be electrons, protons, heavier, or multiply charged ions or even some less common particles from the standard model. Differences between charged particle species restrict to quantitative differences in input data required for this formalism. Regarding control of the beam, the main differences between all these different charged particle species lies in the differences in particle mass and mass to charge ratio (e.g. -1838 for electrons to protons). This factor comes into play when the beam direction or dimension is to be changed by a force, e.g. the Lorentz force. Heavier particles require stronger forces and/or longer application of the forces compared to lighter particles. Physically more complex aspects such as changes in ion charge or bremsstrahlung are strongly connected with particle species, but are not covered in this purely beam optical treatment.

The name beam optics originates from a similarity of mathematical formalism with geometrical/ray optics of light. In geometrical optics, rays travel in straight lines through homogeneous media and deflect at optical elements/lenses. The term focal length completely describes a lens, making its behaviour concrete. Because of the aforementioned geometrical aspects, all optical elements can be treated independently. From these conditions, a matrix-formalism for the description of optical elements derives. The effect of an optical system is completely described by knowledge of the initial beam properties in the form of a vector or matrix, the matrices of the optical elements comprising it, and a matrix multiplication of the beam vector with all optical element matrices.

This formalism was adapted for charged particle beams. Similar to light optics, this formalism only approximates the reality, neglecting not only self-interaction effects such as space-charge or intra-beam scattering but also higher order aspects of the optical elements originating from finite length, technical imperfections, and so on. A correct treatment therefore requires computer codes. Nevertheless, geometrical beam optics is a good approximation and allows understanding several important features and physics of charged particle beam optics. In the last section we defined the particle property vector X with six quantities according to (2.16). Extending this concept to the beam ensemble requires a larger mathematical object. There are two equivalent ways for obtaining this object. Either we extend the 6D single particle vector X to a 6×6 matrix M_{Beam} describing the beam as an ensemble of X. This matrix describes the 6D phase space ellipsoid of the beam with the determinant of the matrix representing the phase space volume. Every beam optical element changes this matrix, but conserves its determinant, which equals the emittance. Alternatively, we have to find a function describing how the beam envelope changes throughout the accelerator system, similar to the considerations done with Fig. 2.24. We can find 3 functions, the so-called Twiss parameters but for the details of this Ansatz the reader is referred to specific literature. As discussed in the last section the 6D phase space volume represents an important quantity representing the population of the individual beam particles. Multiplying the single particle vector X or the beam matrix M_{Beam} at position $s = 0$ with optical element matrices R yields the vector or matrix after passing the element at position s, or in other words the translated quantity:

$$X(s) = R(s) * X(0) \tag{2.20}$$

$$M_{\text{Beam}}(s) = R(s) M_{\text{Beam}}(0) R^T(s) \tag{2.21}$$

with R^T the translated matrix to R. Many beam optical effects are easier to understand in the single particle picture with its relative coordinates and, due to the mathematical equivalence of treating the whole beam or its independent constituents, we will stick to the single particle view. The simplest example of an optical element is the drift-tube of length L with the transfer matrix R_{Drift}. Equation (2.22) tells us the drift-tube effect is given by the off-diagonal elements, which are all proportional to its length L. These elements state the spatial growth/shrinkage of the beam via its divergence and momentum width. The momentum width Δp describes how much the particle momentum differs from the ideal particle momentum corresponding to the accelerator design energy. These size changes are proportional to the tube length L, as geometrical optics dictates by the straight path rule.

$$R_{\text{Drift}} = \begin{pmatrix} 1 & L & 0 & 0 & 0 & 0 \\ 0 & 1 & 0 & 0 & 0 & 0 \\ 0 & 0 & 1 & L & 0 & 0 \\ 0 & 0 & 0 & 1 & 0 & 0 \\ 0 & 0 & 0 & 0 & 1 & L\left(1 - \left(v/c\right)^2\right) \\ 0 & 0 & 0 & 0 & 0 & 1 \end{pmatrix} \tag{2.22}$$

Applying (2.20) and (2.22) to an arbitrary particle yields the particle properties after a drift of length L:

$$R_{\text{Drift}} * X = \begin{pmatrix} x + Lx' \\ x' \\ y + Ly' \\ y' \\ z + \Delta p L\left(1 - \left(v/c\right)^2\right) \\ \Delta p / p_0 \end{pmatrix} \tag{2.23}$$

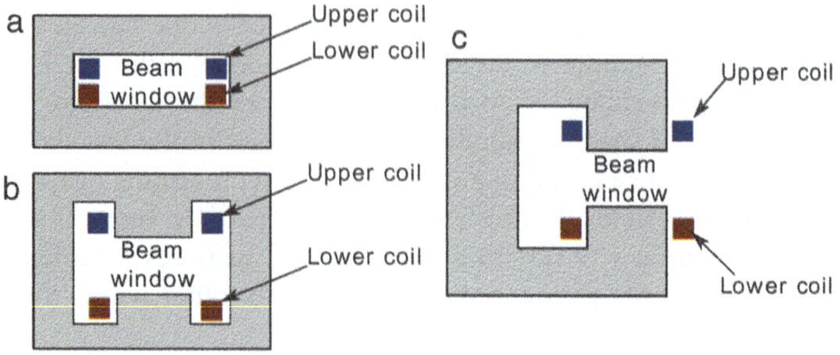

Fig. 2.27 Three construction variants of a dipole magnet. The window-frame magnet (**a**) offers a compact construction for large beam windows and high magnetic field homogeneity. The H-magnet (**b**) eases the coil winding compared to the window-frame but requires more magnet iron (grey). The horseshoe magnet (**c**) eases installation with one side being open, but the open side increases the field inhomogeneity resulting in less ideal behaviour

The multiplication of (2.23) describes a growth of beam diameter in x and y direction by the drift length L and the divergence x' and y'. The longitudinal (z direction) length grows with the momentum spread, if the beam is not too relativistic. In the limit case of $v = c$ the contribution to the longitudinal spread of the beam cancels out, because different momentum only changes the effective mass in the relativistic limit, not the velocity itself.

The first active beam optical element is the steerer, deflection, bending, or dipole element, respectively. Figure 2.27 depicts several technical options for a magnetic dipole element. In its basic functionality these elements change the direction of the beam by a certain angle corresponding to a curvature of radius r_0 inside the element. Technically, an electro-static deflector (a construction equivalent to a parallel-plate capacitor) or a dipole magnet can induce the deflection; hence accelerator physicists often call it just dipole. The deflection depends on all elements of the particle vector X and the construction of the deflection element (Fig. 2.23). Charged particle beams naturally feature certain energy and spatial width', the phase space volume discussed in Sect. 2.3.1. The deflection element generates different deflections for different X. In other words, the dipole deflects particles, for example, from the left and the right end of the beam to different end-points. Already this simple element demonstrates the importance of looking beyond the "zero-th order" of a single particle. Depending on whether the deflection element works by electro-static or magnetic fields it acts differently on charged particles with different energy and momentum, see the Lorentz force in (2.24).

$$\boldsymbol{F}_L(E, B) = \boldsymbol{F}_E + \boldsymbol{F}_B = q(\boldsymbol{E} + v \times \boldsymbol{B}) \tag{2.24}$$

A homogeneous dipole magnet (no radial field gradient n or edge focussing) for deflecting in the x-axis by an angle α is represented by (2.25). Here $r_0 = \rho_0$ is the ideal track radius in the element centre. The magnet acts as a drift of length L in axial/y-direction, hence column 3, 4, and 5 of the matrix look exactly like the ones from the drift-tube (2.22). The $\cos(\alpha)$ dependence on offset and divergence (first two diagonal elements) tells us, this optical element induces different deflection for the ideal particle and particles with non-ideal coordinates (non-zero X). From (2.24) we can directly see a momentum offset $\Delta p \neq 0$ (=velocity $\Delta v \neq 0$) will lead to different magnetic forces F_B and hence different deflection. For offset or divergent particles, longer or shorter path' through the magnet will arise, leading to different effective deflection angles α for these particles. The larger the deflection angle α, the stronger the effect.

$$R_{\text{Dipole}} = \begin{pmatrix} \cos(\alpha) & \rho_0 \sin(\alpha) & 0 & 0 & 0 & \rho_0(1-\cos(\alpha)) \\ -\sin(\alpha)/\rho_0 & \cos(\alpha) & 0 & 0 & 0 & \sin(\alpha) \\ 0 & 0 & 1 & L & 0 & 0 \\ 0 & 0 & 0 & 1 & 0 & 0 \\ -\sin(\alpha) & -\rho_0(1-\cos(\alpha)) & 0 & 0 & 1 & \rho_0\left(\sin(\alpha) - \left({}^{v}\!/_{c}\right)^2\right) \\ 0 & 0 & 0 & 0 & 0 & 1 \end{pmatrix} \tag{2.25}$$

The technical freedom in the design of dipole magnets regarding edge angle and field gradient plays an important role for accelerators, in particular for cyclotrons (Sect. 2.2.2). For the functioning of the cyclotron, the dipole elements literally require a cutting edge. The edge focussing was a requirement for the working of the isochronous cyclotron with its radial field gradient. Chamfering the entrance and exit edge in a way that particles with transversal offset see a different path length inside the deflection element leads to a focussing effect. In addition, also the distance of the magnet pole shoes can be varied in the direction perpendicular to the beam. This induces a field gradient inside the deflection element, which in turn changes the path dependent deflection. The focussing characteristic is pre-defined by the construction and therefore relatively inflexible, but the dipole magnet is the only single magnetic beam optical element able to focus a beam in both transversal directions.

Quadrupole magnets represent the single function element for beam focussing. A quadrupole features four magnetic poles with the opposing poles having the same field-polarity. Their ideal matrix R_{Quad} involves the strength factor of the focussing magnetic field given by k (2.26) with the quadrupole tip field strength B_0, the radius of the open diameter r_{ap} (=distance centre to quadrupole tip), and the magnetic rigidity $B\rho$ which is equivalent to the particle momentum p divided by its charge q.

$$k = \frac{B_0}{r_{ap}(B\rho)} = \frac{B}{r_{ap}} * \frac{q}{p} \tag{2.26}$$

With (2.26) the transfer matrix of a quadrupole element of length L and strength k reads

$$R_{\mathrm{Quad}} = \begin{pmatrix} \cos\left(\sqrt{k}L\right) & \frac{\sin\left(\sqrt{k}L\right)}{\sqrt{k}} & 0 & 0 & 0 & 0 \\ -\sqrt{k}\sin\left(\sqrt{k}L\right) & \cos\left(\sqrt{k}L\right) & 0 & 0 & 0 & 0 \\ 0 & 0 & \cosh\left(\sqrt{k}L\right) & \frac{\sinh\left(\sqrt{k}L\right)}{\sqrt{k}} & 0 & 0 \\ 0 & 0 & \sqrt{k}\sinh\left(\sqrt{k}L\right) & \cosh\left(\sqrt{k}L\right) & 0 & 0 \\ 0 & 0 & 0 & 0 & 1 & L\left(1-\left(v/c\right)^2\right) \\ 0 & 0 & 0 & 0 & 0 & 1 \end{pmatrix} \tag{2.27}$$

A single quadrupole magnet focusses only in one direction. In (2.27) the x-axis features a negative contribution for the x' component (2nd row) and a positive for the y' component (4th row), namely it reduces the divergence in x and increase it in y. In order to focus the beam in both transversal directions x and y we require two quadrupole magnets rotated by 90° against each other (=inverted current flow direction). The four non-zero x and y components switch for the 90° rotated element. Each quadrupole will focus in one direction and defocus in the other direction. The sum result is a focussing in both transversal directions, since the defocussing effect is slightly smaller than the focussing effect (sin vs. sinh), see (2.27). This focussing system is called a quadrupole doublet or also the FODO (focussing, drift, defocussing, drift) structure as shown in Fig. 2.28.

Thinking in the form of optics we can derive the focal length f in the focussed axis of a quadrupole magnet from (2.26) and (2.27), giving us a practical estimate for layouts of focussing systems. Equation (2.28) tells us the focal length f shrinks with stronger, longer, and smaller open diameter quadrupole magnets and grows for heavier particles.

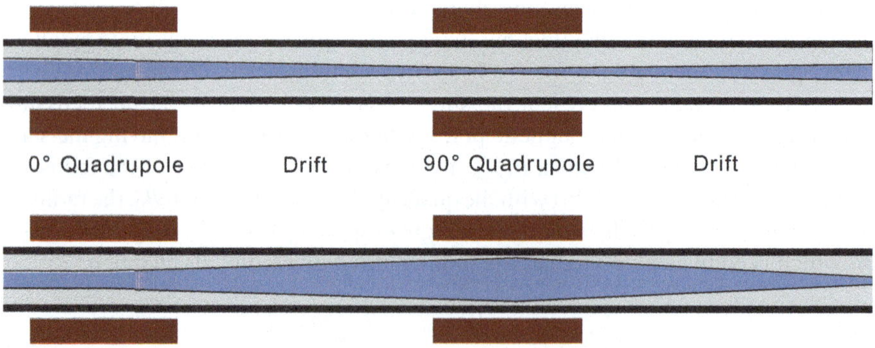

Fig. 2.28 The quadrupole doublet or FODO structure. The upper line shows the horizontal, the lower the vertical plane. Particle beams naturally feature a non-zero divergence, hence require beam optics for confinement. Focussing, defocussing, and drift parts form the charged particle equivalent of an optical lens. The FODO structure as a part of a unit cell allows for a confined or loss-less transfer of the beam over long distances, e.g. several turns around a synchrotron storage ring

$$f = \frac{1}{\sqrt{k}\sin\left(\sqrt{k}L\right)} \approx \frac{1}{kL} = \frac{r_{ap}}{B_0 L}\frac{p}{q} \tag{2.28}$$

The minimum beam spot radius achievable with a quadrupole and a given beam type depends on the beam emittance ε and the quadrupole focal length. Analytical applications require small spot sizes for scanning a sample surface with high spatial resolution and production application reach higher beam power density with smaller beams. As emittance (remember its units: m (length) times rad (angle)) is conserved, the focussing basically squeezes the phase ellipse in Fig. 2.24 as much as possible into the upright position. This decreases its size, but increases its divergence x' due to conservation of the surrounded area, leading to a natural connection between small spot size and large divergence which equals a short focal length. To derive the minimum spot size we create a FODO/doublet structure by multiplying the transfer matrices

$$R_{\text{doublet}} = R_{\text{Quad-Hor}} * R_{\text{Drift}} * R_{\text{Quad-Ver}} * R_{\text{Drift}} \tag{2.29}$$

Equation (2.29) assumes we start with a certain beam size $x_{\max}$ and divergence $x'_{\max}$ (according to Fig. 2.24) at the FODO entrance, both identical quadrupole magnets (both focal length f) focus in different transversal direction and have a distance d between their centres ($d \geq$ size of quadrupole magnet) and a distance s of the last magnets centre to the focussing point. We assume a diagonal beam matrix M_{Beam} equivalent to a beam waist at the FODO entrance for simplicity and consider only the x-direction with the first matrix element $x^2_{\max}$ and the second diagonal element $x'^2_{\max}$

$$x_{\max}(s) = \sqrt{x^2_{\max}\left(1 - \frac{d}{f}\right)^2 + x'^2_{\max}\left(d + s\left(1 - \frac{d}{f}\right)\right)^2} \tag{2.30}$$

We could continue in finding the minimum for s followed by solving for d or f, but this becomes only mathematically exhausting. The main conclusions can already be drawn from (2.30): The minimum spot radius $x_{max}(s)$ is defined by the beam emittance (see Eq. 2.17) and the magnet strength. Stronger magnets will reduce f, but we have to keep the distance d close to f in order for minimizing $(1 - d/f)$. Technical reality adds up, since if we decrease f by building longer magnets (L in Eq. 2.28) we will increase the lower limit of d accordingly, but d has to be minimized too. Small spot sizes therefore require high magnetic field strength and small beam apertures (which is equivalent to smaller beam currents). If we would enter the FODO setup with a minimal divergence ($x'_{\max} \approx 0$) the aperture changes the emittance by defining $x_{\max}$, but the beam current reduces via the ratio of the areas of initial beam to aperture weighted by the beam shape (e.g. normal distribution). The technological limit of spatial resolution (=small spot sizes) comes down to technological limits of the focussing strength and the beam quality. Additionally, practical devices never feature only pure quadrupole fields, but due to manufacturing and alignment, aberrations increase the real minimum spot size above the one of an ideal lens discussed here.

All lenses steer: A general rule for understanding practical aspects of beam optics. A beam entering a quadrupole lens off-axis will see a dipole field aspect due to its proximity to one of the pole-pairs: The more off-axis, the stronger the dipole aspect. This dipole aspect steers the beam in addition to the intended focussing. In extreme cases, the beam shape becomes distorted, in critical cases resulting in a sickle shape. A slight misalignment remains practically unavoidable; hence the steering effect belongs to a focussing lens like cheese to pizza. The next section will discuss how to become aware of this and how to find information for correcting the problem.

So far we discussed only magnetic optical elements. In applications, these are mostly found with ion beams of at least some MeV, since magnets offer higher deflection strength required for heavier and faster particles. Electron applications usually feature lower energies, which in combination with the lower particle mass allows for applying electro-static elements. The workings are similar to magnetic elements, but electro-static elements deflection strength decreases with particle energy (as electro-static fields change energy but electric breakdown limits their voltage potential) compared to the deflection strength decrease with particle momentum/rigidity p/q (momentum/charge) experienced with magnetic elements (2.24 and 2.26). We will see an example highlighting the difference in the next section. Their relevant technical advantage is their low power consumption, since electro-static fields are generated by static charges (=zero current), while magnetic fields are generated by moving charges (=current). The electro-static equivalent to the dipole magnet is the parallel-plate capacitor like deflection plate. Electro-static focussing requires the so-called Einzellens, see Fig. 2.29. The Einzellens focusses in both transversal directions via a deceleration-acceleration structure. The two outer cylinders are grounded, while the focussing voltage is applied to the central cylinder. The bend electric field lines in combination with the changing particle energy and the energy dependent deflection strength leads to a net focussing/defocussing of equivalent strength in both transversal directions.

Arranging the many different beam optical elements in a useful manner represents a science on its own, but a set of standard configurations have established for the common tasks of focussing and beam manipulation. Figure 2.30 depicts an example of a standard configuration of an electron microscope. Modern electron and ion

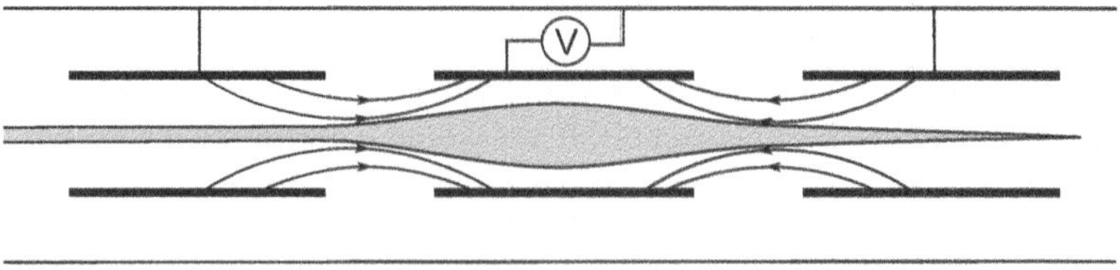

Fig. 2.29 Sketch of an Einzellens with grounded outer and biased inner electrode rings. The central electric field (lines) first decelerates the beam, broadening it. When leaving the lens, the beam is accelerated and focussed by the same potential

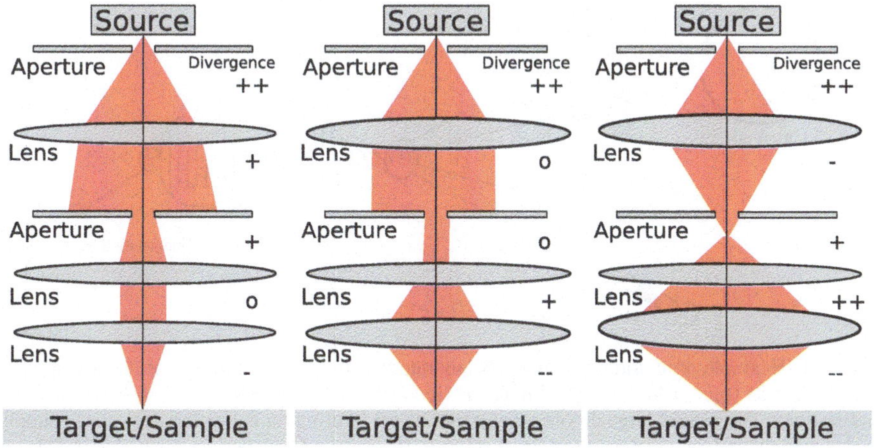

Fig. 2.30 Example of a common beam optics system with two apertures and three lens systems. This system is common for analytical devices, e.g. electron microscopes. Varying the lens properties/settings allows for a high degree of flexibility in the beam properties on the target/sample indicated by the 3 configurations. Towards the right, the focal length of the final lens reduces, decreasing the minimum beam spot size on the sample (2.30) at the expense of requiring a higher focussing strength. The divergence symbols indicate the required electric and/or magnetic field strength in the lens. When focussing with quadrupole magnets, each lens has to be understood as a doublet

microscopes use these systems with electrostatic (3 Einzellenses in this example) and magnetic components up to about 1 MeV. At higher energies, magnetic elements are employed, but the technical concept remains the same, just the devices are exchanged. The variability of the system allows for several different operation modes. The left mode starts by forming a parallel beam from which a cut-out comes into the lower part. Here a defocussing followed by a focussing allows for a minimum spot size, since the stronger the focus, the smaller the minimum spot size. The second and third settings produce an intermediate beam-waist, allowing for a variable and potentially higher beam current after the second aperture. The lower part either defocusses and focusses for a maximum resolution or double-focusses for a more parallel beam on the target in order to decrease angular beam spread equivalent to an increased depth of field. Depending on the analytical requirements, the electron microscope can be set for maximum spatial resolution, depth of field, or sample current/signal intensity by varying only the focal length' to one of the presented settings.

Laboratories applying higher energy charged particles often use several end-stations for different purposes sharing a common accelerator and particle source. Each chamber is optimized for a specific question or task. A combination in one lab allows for reduced down-times and a better exploitation of the expensive accelerator, reducing the overall costs. The ion beam analytical laboratory depicted in Fig. 2.31 uses a DC accelerator for material analysis. A central dipole magnet switches the beam to the individual analytical stations on the left hand side by varying

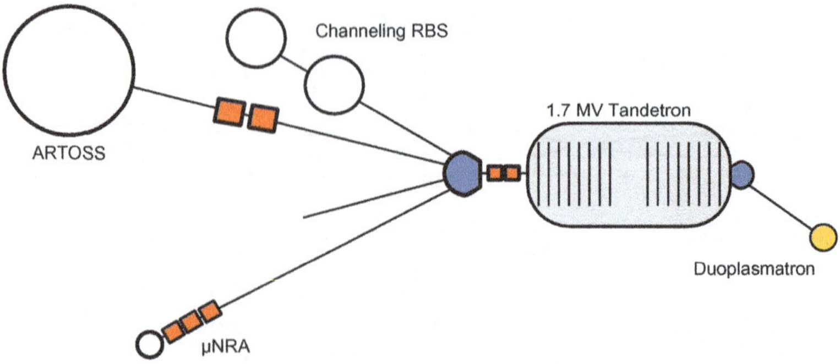

Fig. 2.31 Example of an ion beam analysis laboratory with a Cockcroft-Walton DC accelerator, dipole magnets (blue) and quadrupole magnets (red squares) connected to three different analysis end-stations. The central large dipole magnet can be set to different deflection angles α to supply the beam to the individual experiments/end-stations. Reproduced with permission by IAEA (Mayer 2019)

the magnetic field strength and direction. Each end-station has its own beam optical system optimised for small spot sizes, heavy ions, or high beam current/transmission.

2.3.3 *Diagnostic Elements*

Accelerators are complex and often rather large devices with tight tolerances. The installation and alignment of beam optics consequently becomes time consuming and delicate. The first tests after start-up aim at fine tuning the alignment of beam optical components to enable beam transport from source to target in the form of a non-zero current on the target. In cases of intense focussing, long beam path, or high currents, beam optics have to be aligned to the ideal beam path within a few µm over distances of several metres. Forces induced by the magnetic fields, creep of materials over time, or temperature changes can furthermore alter alignments dynamically during operation. An example: Laboratory temperatures vary by a 1–3 K over the day due to sunlight and working activities, without temperature stabilisation. The thermal expansion of steel of 17 µm/Km then results in a height variation of a steel beamline support of 1.5 m height of up to 51 µm, a lot for microscopic applications with sub-µm beam spots. Therefore, the beam position and shape have to be frequently monitored and adjusted and sufficient amounts of correcting elements have to be foreseen to be able to handle these effects, if beam optics, end-station, and accelerator rest on different supports.

Diagnostic/analytic elements provide the eyes and ears to the accelerator operator by measuring the actual beam position and its 6D phase space properties in the form of the beam vector X (2.16) or at least parts of it. The beam energy is usually known from the accelerating voltages, but a more precise determination might be required. Regarding physics, this translates to finding physical mechanisms to make these quantities accessible for a measurement. Technically, we have to find feasible solutions and align them to an accuracy such that the ideal physical mechanism we thought of is reproduced to a sufficient accuracy. The measurement, in most cases, only indirectly determines the sought after quantity, therefore a physical model is required to recalculate the measured quantity. Following this, the accuracy in determining the quantity is limited by the accuracy of the model (technical challenge of alignment ...) and the measurement accuracy as dictated by the laws of error propagation (discussed in Sect. 7.1.1 e.g. in (7.3).

An example: We measure the current density ρ_B of a beam using a Faraday cup. Do not worry its physical concept will be discussed quickly hereafter. We need some kind of ampere-meter for measuring the current I, say for the range of μA to mA, since we expect $I_D \approx 1$ mA/mm^2 and we want a 1 mm spatial resolution. Here standard ampere-meters offer an accuracy of 0.1% or 10^{-3}, respectively, plus some measurement range dependent fixed uncertainty, say for a range of 0–1 mA we have $\Delta I = 0.1\% \pm 0.1$ μA. Measuring a current of $I = 1$ μA therefore leads to an uncertainty of 10%, due to the fixed uncertainty contribution! Accordingly we design the Faraday cup to feature a sensitive area (opening) large enough to collect a current I of 0.1–1 mA (magnitude of lowest relative uncertainty). Technically, this sensitive area has a manufacturing diameter tolerance of say $d = 1 \pm 0.01$ mm (H7 tolerance for drill-holes). In total this (simplified) example allows us to measure the beam current density with an uncertainty ΔI_D given by

$$\Delta \rho_B = \sqrt{\left(\frac{4\Delta I}{\pi d^2}\right)^2 + \left(\frac{-8I\Delta d}{\pi d^3}\right)^2} \tag{2.31}$$

If the actual $\rho_B = 1$ mA/mm^2, the device would measure a ρ_B of 1 ± 0.02 mA/mm^2 (=0.98–1.02 mA/mm^2) according to the error propagation used in (2.31). The manufacturing tolerance of the Faraday cup entrance dominates the uncertainty in this case. Reducing the uncertainty would require better tolerances or a larger sensitive area. We could increase the opening diameter at the expense of spatial resolution to reduce the relative manufacturing uncertainty, as a cost efficient method to reduce the relative uncertainty. As soon as the manufacturing tolerance becomes small against the current measurement accuracy, we will reach a technical limit. Increasing the integration time of the ampere-meter would allow for better noise filtering reducing its uncertainty, at the expense of its time resolution. Buying more precise (and expensive) ampere-meters and manufacturing tools would also improve the device accuracy. In conclusion, the technical design of this Faraday cup, and actually all diagnostic devices, comes down to finding an optimal point of minimum uncertainty with respect to all uncertainty contributions, the aims of the measurement, and the available resources.

The Faraday cup, named after the concept of the Faraday cage, is a tool for accurate beam current measurement. Its design allows suppressing the emission of secondary electrons released upon impact of a particle beam on solid matter (Sect. 4.1). For positively charged ions these electrons provide an opposite current flowing in the opposite direction, hence increasing the apparent beam current. For electron and negative ion beams, the effect reduces the apparent beam current. Secondary electrons are released by collisional energy transfer from beam particles to bound electrons at the impact surface. The cup suppresses this emission by catching the electrons in a cage around the impact location, as depicted in Fig. 2.32. The ratio of length to open diameter represents an important layout parameter. The geometrical suppression efficiency derives from the ratio of the half-space solid angle over the opening solid angle weighted with the emission shape. This emission shape follows approximately a cosine of the angle to the surface normal (emission mostly towards surface normal). An inclined target surface therefore further improves the geometrical efficiency by pointing the surface normal to a side-wall. By applying a positive bias voltage of typically some 100 V on the cup against the vacuum chamber ground or a surrounding cage, the suppression efficiency further increases. The secondary electron energy spectrum is broad, but its maximum lies at a few ten to hundred eV. Only the combination of geometric and electrostatic suppression catches close to 100% of the secondary electrons, enabling a true beam current measurement.

Besides the beam current, the Faraday cup yields also positional and dimensional information of the beam via its own position and open diameter. More precise information is the domain of beam profile and beam position monitors (BPM). Their technological goals are determining position, dimension, and divergence (the 6D phase space) of the beam. In large circular accelerators deconfining resonances are avoided by online tuning of the beam according to BPM data. In applications, BPMs visualize the beam, enabling adjustment of the optical elements for maximum beam transmission in particular for daily start-up or after technical changes. In any case we aim at providing a high level of reproducibility of the beam conditions at a relevant position (where the beam is used) while the conditions change at other non-relevant positions (e.g. by thermal drifts, maintenance …) in the accelerator system. BPMs feature a particular broad range of technological options.

The simplest realisation of a BPM is a white paper which blackens upon beam impact, see Fig. 2.33. The reversible, yet more costly, alternatives are scintillator

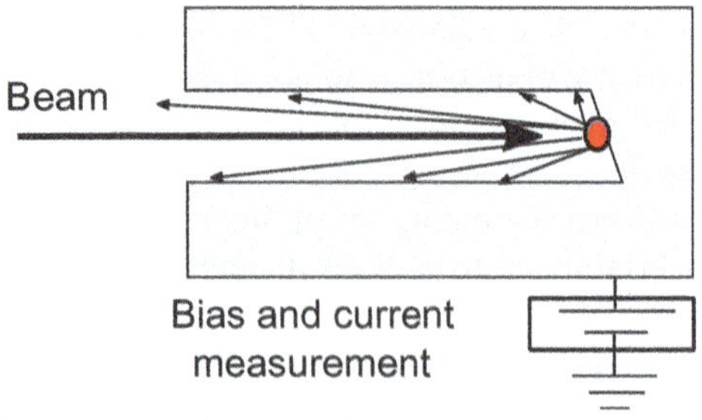

Fig. 2.32 A Faraday cup, measuring the beam current by geometrical and electrostatic secondary electron suppression. The small arrows represent secondary electrons emitted in the open half-space

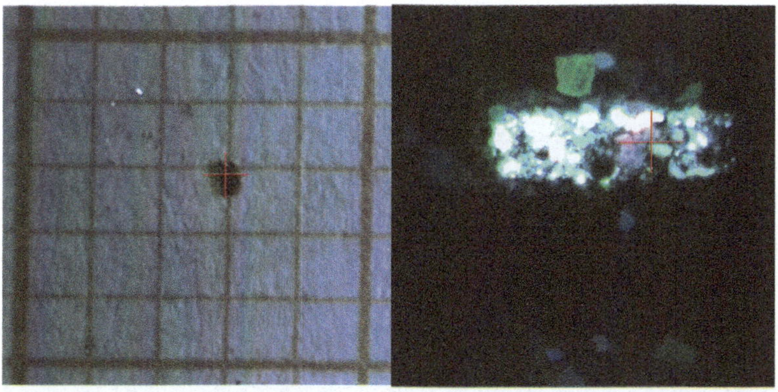

Fig. 2.33 Left: 100 μm diameter 3 MeV ^{3}He beam spot visualised by heat induced changes in a millimetre paper. Right: Beam spot ($\approx 1 \times 0.2$ mm^2) visualized by scintillating ZnS particles (<100 μm) on an adhesive tape. The individual ZnS particles blur the impression of the rectangular beam spot

targets with camera observation. The problem of the paper is its coarse material structure consisting of about 100 μm thick fibres blurring the obtained image for small beam spots. The same problem arises with polycrystalline materials, since internal reflection lights up the individual grains as a whole, even if only part of it is hit by the beam. Larger single crystal scintillators avoid these resolution limit factors, reaching <10 μm resolution. The same is true for amorphous materials, but they usually emit less light per incident beam power than single crystals, making them more suitable for high density beams. Scintillating materials can be found among numerous material classes, ranging from classical detector materials such as NaI:Tl^+ over many crystalline and amorphous glasses (ZnS, $LiAlO_2$ ...), even float glass (yes the material used for windows) scintillates under beam impact. One particularly creative option for beam visualisation is the application of dry ice (solid CO_2) for high-resolution beam profiling. Dry ice sublimes directly (no liquid phase below 5.185 bar) and has low thermal conductivity, hence a beam impacting on it will drill a hole with a local depth according to its current density distribution (assuming a constant beam energy distribution). The evaporation rate is proportional to the beam power and analysis of the crater yields the beam distribution and position.

Figure 2.34 shows two examples for the BPM based alignment of a magnetic quadrupole triplet used to focus a 3 MeV proton beam. A scintillator provides a live image of the beam profile. Based on this profile and the alignment advices developed from the presented knowledge on quadrupole magnets, the magnet positions are corrected. A steering effect upon varying individual magnet field strength' indicates individual magnet positional alignment errors. Rhombohedral shape distortions indicate rotational misalignments between the individual magnets of a focussing lens setup.

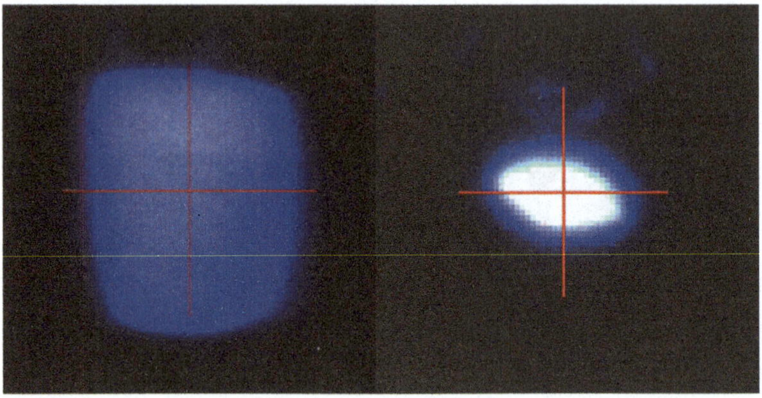

Fig. 2.34 Picture 3 MeV two proton micro-beams impacting on $LiAlO_2$ for quadrupole magnet alignment. Left: after 1 mm aperture. Right: After 1 mm aperture and 10× demagnification. The rhomboid beam shape indicates a rotational misalignment of the quadrupole triplet

Instead of converting the charged particles to photons the beam current can be measured directly. Simplifying the Faraday cup idea to a bare wire makes it technically more flexible, but due to secondary electrons a wire only yields an arbitrary current somehow proportional to the true beam current. Moving the wire through the beam and measuring the time/position resolved current yields the beam profile. A more intrusive yet more robust variant are profiling apertures; moving an aperture into the beam while measuring the apparent current on the aperture face yields a profile integrated from the outside to the current position. These current collecting methods all feature the same drawback of averaging over the beam profile, may it be over a line, circle, or square. This works fine with a known Gaussian beam profiles as shown in Fig. 2.21, but what do we see if the beam is rhomboid or even sickle shaped? Depending on the current collecting area and its angle to the sickle we see Gaussian profiles of identical or strongly different width. The 1D/line BPMs cannot provide enough information points of the transversal plane for a non-ambiguous identification of the beam shape. This may lead to false conclusions and advices for beam optics for non-Gaussian beam shapes.

The absorbed beam power density limits these beam stopping BPM options. Non-stopping options exploit the electro-magnetic field of the beam. Electro-magnetically the beam is identical to a conducting wire, since a current (DC or AC) flows along the beam direction. We experienced electro-magnetism as a central physical concept for accelerating and controlling the beam, but in the idea of action = reaction, the beam necessarily emits electro-magnetic fields itself. The picture of *beam* = *wire*, yields the magnetic field B around the beam as

$$B(r) = \frac{\mu_0 I}{2\pi r} \tag{2.32}$$

With the beam current I, the vacuum magnetic permeability μ_0 and the radial distance to the beam centre r. The effect is weak: reaching for example the field strength of the earth magnetic field of about 50 μT requires a beam current of 8 A measured at 33 mm distance. The transformer principle and the Hall effect allow measuring such magnetic fields. A similar effect is present for the beam electro-static field. The electro-static force shifts charges on adjacent surfaces, generating voltages. Placing several of these contact-free BPM sensors around the beam allows for a beam localisation via the known distance laws (e.g. $1/r$) and a triangulation. These devices allow for determining the beam position to an accuracy of 10^{-3} to 10^{-5} of the beam tube diameter, with practical values down to the μm range (Forck et al. 2009). High intensity beams (≥kW beam power) require contact-free electro-magnetic methods for beam analysis due to limits of heat loading and problems with radioactivity induced by the beam impact on the analytical elements. On the other hand, they provide enough current for the rather insensitive contact-free methods.

Analytical elements on the one hand provide information about the beam for beam control, but they also allow selecting parts of the analysed/separated beam for further applications. Devices selecting certain parts of a beam or separating a beam according to a beam property are called analysers. Putting an ampere-meter to a through hole already distinguishes a beam diagnostic from an analysing element. Most beam optical elements, as discussed in the last section, have certain analytical features. A dipole magnet will analyse a particle beam according to its momentum component perpendicular to the dipole field via the gyroradius r_g (a.k.a. cyclotron radius).

$$r_g = \frac{m * v_\perp}{q * B} \tag{2.33}$$

Particles starting at a fixed point will draw different radii, allowing a selection using an aperture (or detector) at a selected radial point. An electrostatic deflector has the same effect but it selects particles according to their energy

$$r_E = \frac{m * v_\perp^2}{q * E} \tag{2.34}$$

Combining these two ideas, we obtain the Wien filter, Fig. 2.35. The Wien filter consists of an electric field and a perpendicular magnetic field. In this configuration, magnetic and electrostatic force act in the same direction. As (2.33) and (2.34) state, the central axis is defined by a point where the Lorentz force (2.24) is zero

$$\begin{aligned} F_L &= 0 \\ \Rightarrow v_0 &= E/B \end{aligned} \tag{2.35}$$

Accordingly, only particles with a fixed velocity v_0 (and hence kinetic energy) pass the Wien filter undeflected. In reality a certain velocity acceptance Δv exists, depending on the exit aperture size and the field strength'. The Wien filter acts as

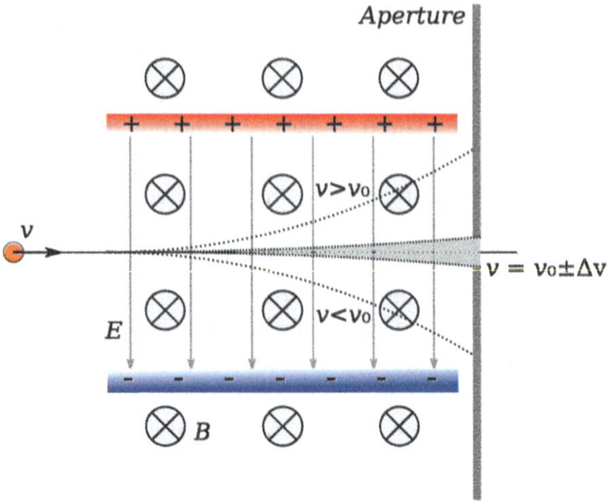

Fig. 2.35 Schematic of a Wien filter. A vertical electro-static field combined with a static magnetic field directed into the paper plane results in a velocity filtering. The filter acceptance (grey band) Δv around the design frequency v_0 depends on the aperture opening, field strength, and the filter length. Original work by Miessen, CC0, via Wikimedia Commons

an aperture for longitudinal momentum. Wien filters are used in high resolution electron microscopes for decreasing the beam energy spread/emittance by orders of magnitude, allowing for improved imaging quality.

As with all measurements, limitations and implications are connected to beam analytical elements. Limitations of high intensity beams due to power loads are very similar to beam targets which will be discussed in Sect. 2.6. A beam features, from a mathematical perspective of an equation system, 7 degrees of freedom (the X vector plus the beam current) and hence the full characterisation requires at least 7 independent measurements sensitive to these points. Accurate results typically require at least 3 times more data points. Fewer points require potentially wrong assumptions as discussed with the sickle shaped beam profile. Last but not least: The beam profile is not fully defined by the beam optical device settings! Different charged particle source settings, in particular the current density, result in different emittances and self-interaction effects or bad vacuum conditions can reduce the beam energy. Figure 2.36 illustrates the dramatic impact of current density on the beam focussing. In the left half of the figure, the beam remains unchanged, regardless of beam current density. On the right part after the accelerator the higher current density results in a defocussing effect due to the high charge density, shifting the second focal point from the inside of the accelerator several metres to the right. An analytical measurement at low currents will not cover this effect, requiring both high power and low power BPMs for discovery.

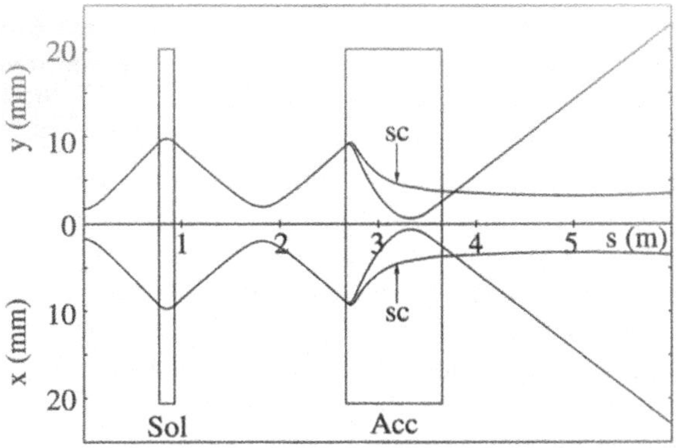

Fig. 2.36 Visualisation of space charge effects of a DC accelerator with 860 keV final energy and 15 mA beam current. "sc" marks the space charge affected beam diameter at high current, showing a clear focussed behaviour. The situation without space charge effects or at low currents, respectively, leads to a divergent beam due to the missing de-focussing effect of the space charge. Reproduced from Hinterberger (2008, p. 376) with permission by Springer

2.4 Electron and Ion Sources

With the knowledge of conservation of emittance, the importance of the first origin of the particle beams, the ion or electron source, becomes obvious. A possible conclusion from beam optics says: An accelerator system can be understood as an imaging optic projecting the source onto the target. Furthermore, the beam current finally arriving in our accelerator application naturally must be $\leq$ the current generated by the source with numerous losses on apertures and walls limiting the transmission of the optical system. The source therefore defines the limits of beam power, current density, and beam size. For extended reading, the reader is referred to (Zhang 1999) regarding ion sources. Regarding electron sources information can be found mostly in the context of their application in electron microscopes and scientific accelerators (Orloff 2008; Williams and Carter 2009; Jensen et al. 2010). Besides the terms ion and electron source, several different terms are common in different contexts such as gun, injector, or emitter. Except for minor differences these terms depict the same apparatus and hence here only the term source will be used.

The types of sources developed for electrons and ions differ quite substantially in their technical realisation and capabilities. Electrons being present in any material and having only one species, the electron, can be extracted by several means from solid materials. The main differences to ions lie in the variety of ion charges and species, and the ion containing resources. This implies totally different physical source concepts for ions and electrons. Hydrogen in its purest form exists as H2 gas (the resource) from which hydrogen ions (protons) can be generated. Other hydrogen containing gases such as methane (CH_4) additionally contain other constituent (here carbon). The elements will not disappear upon removal of the hydrogen from the compound, but lead to detrimental effects such as layer deposition or corrosive processes, making the purest resource generally the best choice. Extraction of ions from solid resources, e.g. polyethylene ($–CH_2^-$) for hydrogen, requires an intermediate sputtering process for producing free protons, due to the strong binding in the solid. Sputtering consumes the solid, resulting in a limited lifetime of the resource. The use of hydrogen ions, and most other ions, therefore also implies a vacuum system to cope with the gas introduced by the resource and a consumption of resources requiring a refilling mechanism such as a gas injector. In contrast, in electron sources the resource (electrons) is constantly "resupplied" from the power supply.

Before we discuss the different source technologies and realisations, we have to align our expectations. Technologically, no limit of the beam current/intensity exists for any source type, the source can always be made larger or multiple emitters can be used. Instead, a technological comparison requires a measure of its output density in the idea of beam emittance (Sect. 2.3.1). Imagine having a small aperture through which we see light. The factor we are missing is the difference between putting a light bulb or a laser behind this aperture. In both cases, we can achieve the same brightness directly behind the aperture, but in 1 km distance to the aperture, the situation will be very different as the laser beam is less divergent. The photons emitted from the laser originate from a smaller spot and emit into a limited directional cone, while the light bulb emits from an extended source wire into all directions (4π solid angle).

Charged particle sources emit charged particle beams (of low energy though), therefore the emittance concept and its conservation, discussed in Sect. 2.3.1, already provide a relevant quantity when evaluated at the source exit. The ion or electron emission properties of the source define this quantity just like in the example of the light source. Knowing only the output current incompletely describes the produced beam, a second quantity such as the emittance is required. In Sect. 2.3 we learned how to use beam optics for changing the beam density and divergence using focussing beam optical elements, therefore also a current density is insufficient. The beam emittance summarises the angular and spatial extent of a certain fraction of the beam (see Eq. 2.17) without stating how many particles this phase volume contains absolutely. Therefore, only beam emittance and source exit current (density) together define the source quality.

For describing the technological quality of a charged particle source we start with the beam current density ρ_B

$$\rho_B = I/A \tag{2.36}$$

with beam current I distributed over the beam area A. Considering the beam optics in between source and target we have to add the beam divergence for a meaningful source performance definition. The term *brightness* combines the beam current density with its divergence. Quite a few definitions exist in the different physics disciplines, but the physical basis makes no difference between light, electron, and ion beams.

$$\text{Brightness} = \frac{I}{A\Omega} = \frac{\rho_B}{\Omega} \tag{2.37}$$

With the divergence angle (in steradian) Ω at a given beam waist in the source (e.g. the exit aperture) or the alternative definition by its transversal emittance $\varepsilon_{x,y}$. The term brilliance adds an energy window, for example 10^{-3} of the beam energy, within which the brightness is evaluated. The brilliance mostly finds application with photon beams, but this longitudinal energy spread also has an importance for accelerator applications, in particular for focussing beams. For ion beams below a few MeV the energy width contribution of the source of typically about 10 eV can dominate the total beam energy width.

The physics leading to the brightness of a certain source type connects with the particle production and energy transfer mechanism releasing the particles from their resource. Increasing brightness requires either emitting more particles from a smaller area (emission density), or releasing them with a smaller energy and angular spread, or both. The thermal energy spectrum according to the Maxwell-Boltzmann distribution, (2.38), represents a physical minimum for the kinetic energy width of emitted particles. Thermal energy E_{Th} evenly distributes along all spatial directions with an increasing distribution width with temperature T. Increasing source brightness therefore translates to increasing density and decreasing temperature of the ensemble of charged particles.

$$p_{\text{Thermal}}(E_{\text{Th}}) = \sqrt{\frac{4E_{\text{Th}}}{\pi}}(k_B T)^{-3/2} e^{-\frac{E_{\text{Th}}}{k_B T}} \tag{2.38}$$

Before the release of free electrons or ions from a source, the particles feature the thermal energy distribution of the resource they are bound to. The binding energy of electrons to the solid is given by the work function with values in the range of 2–5 eV. The electron source needs to provide this energy in order to emit free electrons. Thermionic emission takes this binding energy from the high energy tail of the thermal Maxwell-Boltzmann distribution, (2.38). The part of this exponential tail above the work function contains a sufficient population for emitting several A/cm^2 only at elevated temperatures. The conflict of mechanical stability and the required temperatures leaves only a few material options for such a source type. Tungsten with its high melting point of 3695 K and high work function and a set of hexaborides, in particular CeB_6 and LaB_6, with their low work function and intermediate melting point (LaB_6: 2483 K), feature feasible combinations of properties. Small single crystals as shown in Fig. 2.37 provide long lifetime with high current density and low emittance due to the compact source region, while larger polycrystalline plates and wires are applied for higher absolute currents at lower densities. Continuous sources apply heating via electrical currents, but the same physics allows also for extremely short electron pulses using pulsed laser beam heating.

$$\rho_{\text{Th}} = AT^2 e^{\frac{-(E_W - E_F)}{k_B T}} \quad \text{with } E_F = \sqrt{\frac{e^3 V}{4\pi \varepsilon_0}} \tag{2.39}$$

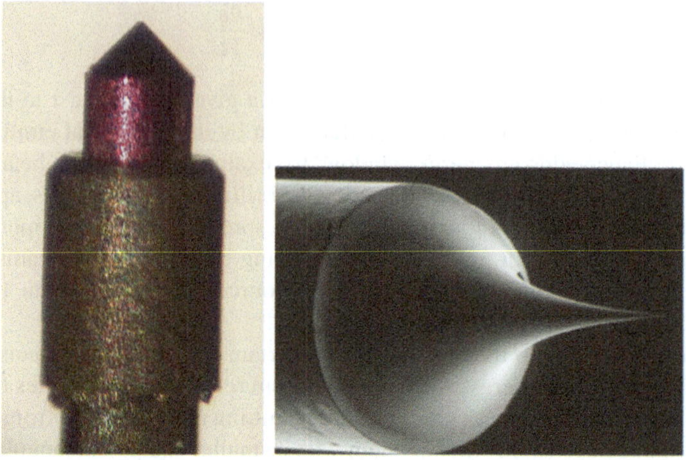

Fig. 2.37 Left: LaB_6 single crystal thermionic emitter cut to a cone with a flat top of 15 μm diameter on a graphite heater. Courtesy of Kimball physics Inc., USA. The crystal is cut in a way to have the <100> direction (lowest work function) face in the beam emission direction. Right: Electron field emission tip made from a 2 mm diameter tungsten wire. The extremely sharp, point-like tip leads to high electric field strength and electron emission with minimal angular spread of the extracted beam

Physically the thermionic emission current density ρ_{Th} depends exponentially on the temperature, as depicted in (2.39) with the temperature T, a material specific constant A, the work function E_W, the electric field strength V, the electron charge/elementary charge e, and the vacuum permittivity constant ε_0. Besides the temperature and the material choice, no additional degree of freedom exists for increasing the emission current density of thermal emitters. Adding an electrical field to the hot emitter enables a further increase in the emission density via a reduction of the effective work function by the amount E_F, resulting in the so-called Schottky emission. Sharp tips are required to obtain relevant electric field strength V. Tungsten represents the ideal material due to its temperature and mechanical strength to resists this combined load. Increasing field strength V further increases the emission current density until another effect becomes relevant at very high field gradients. The high field gradient enables the so-called cold field emission (CFE). CFE is more than a quantitative increase of the Schottky emission, but it relies on the tunnelling effect, a completely different physical mechanism described by the Fowler-Nordheim equation. The electrons bound via the work function always have a certain quantum mechanical chance of escaping the solid, in spite of their binding. The applied voltage acts as a metaphorical obstetric care enabling the electrons to tunnel the potential barrier. Rather than the birth situation we all went through it is more comparable to giving birth to a truck through a birth canal as long as the distance between earth and moon. The extreme field strength in the order of GV/m requires atomically sharp tips for CFE as shown in the right of Fig. 2.37. This combination of small source region and high current density result in the lowest emittance electron sources. The total current provided by the electron source is given by the sum of thermal and field emission, but since CFE works independent of thermionic emission both emitter types are rather disjunct. With CFE, electronic voltage switching potentially allows for pulsed electron beams in the nanosecond range. In the Schottky regime significantly shorter pulses can be triggered via ultra-short pulse lasers in the order of picoseconds. The extraction voltage is kept slightly below the threshold for cold field emission. The laser impact then heats the surface, leading to thermionic and photon assisted emission in a small time-frame of the laser pulse duration. Besides the emission current, also the energy distribution function of the emitted particles plays a major role for the electron source properties with a clear advantage for *cold* field emission, leading to their superior brightness and brilliance.

The three types of electron sources all have their specific application cases. CFE offers the best beam performance required for the highest resolution electron microscopes (see Sect. 7.1.3), but they suffer from high cost and limited stability due to adsorption layers (Sect. 2.1) degrading their performance rather quickly (depending on vacuum pressure). Thermal emitters in contrast offer about 5 times smaller drift rates in their output current and provide certain economical scaling advantages for larger sources, making them the technology of choice for non-analytical applications where performance/emission current per cost counts. Schottky emitters represent a performance compromise with the stability of pure thermal emitters, yet a higher beam-quality. They offer the best compromise of performance and cost, making them the most successful electron source for a broad range of analytical applications.

The emission and physics of ion sources significantly differ from electron sources. A thermal release of ions from solids is not possible due to their stronger binding compared to electrons (e.g. 8.7 eV vs. $\approx$4.5 eV for W) leading to emission of mostly neutral atoms. Plasma physics forms the foundation of free ion production in most ion sources. Plasmas generate positive and for some elements also negative ions. Gases, vapours, and surface sputtering can act as resources for the ions. The book (Piel 2010) provides an easy yet efficient introduction into plasma physics. To a large extent modern plasma physics research relates to the development of nuclear fusion reactors which generated a solid, but not complete, understanding of plasmas. In particular, large databases and codes on ionisation cross-sections and processes of various species were produced, e.g. (Reiter 2019).

Special interest lies in the production of hydrogen ion beams due to on the one hand basic physics of hydrogen allowing producing in particular intense hydrogen ion beams compared to other elements, and on the other hand for the excellent nuclear reaction capabilities of hydrogen isotopes, namely protons and deuterons. Most ion sources are therefore designed and optimised as hydrogen ion source, although the basic considerations readily transfer to other resources.

The ion flux density Γ_i (in particles per area) as a measure of the source current density of a plasma relates to the plasma ion density n_i, electron temperature T_e, ion temperature T_i, ion charge Z_i, and ion mass m_i via (2.40). The flux density mostly depends on the plasma ion density and only weakly on the plasma temperature and ion species/mass. The plasma density and temperature set according to the equilibrium of input power to power-losses of the plasma. The plasma power losses have a strong scaling with the plasma temperature and a linear scaling with plasma density. Consequently, the emitted ion source current density increases with input power.

$$\Gamma_i(n, T_e) = n_i e * \sqrt{k_B(Z_i T_e + T_i)/m_i} \tag{2.40}$$

Energy efficiency plays a major role for realising high density plasmas due to technical limitations of cooling and the high temperatures of plasmas in the order of 100,000 K dissipating the heat. The ion flux available for extraction to the accelerator will equally flow in all directions. In addition the plasma needs to maintain its quasi-neutrality, leading to an equivalent flux of electrons. The sum of both, the so-called transport losses, is controlled by the electron and ion motion which can be affected by electro-magnetic fields. Similar to the plasma confinement in nuclear fusion reactors, magnetic fields effectively reduce the transport losses as depicted in Fig. 2.38 by restricting the motion of plasma particles, resulting in increased plasma density and flux per input power.

Due to their low mass and therefore high velocity at a given plasma temperature, electrons contribute the major part of transport losses but are at the same time easiest to confine with magnetic fields (Lorentz force, (2.24)). Besides transport losses, also photon emission from recombination of electrons and ions and from non-ionizing excitation reactions contribute to the power losses. The photon emission strongly scales with the plasma temperature. Both, low photon radiation and transport losses require low plasma temperatures. A low plasma temperature, which is still high enough for sufficient ionisation cross-sections, together with a high plasma density delivering a bright ion beam therefore characterize the peculiar point of optimisation of the plasma based sources. The fine tuning of the plasma temperature requires the right gas/vapour pressure and a clever magnetic field design confining electrons and ions while allowing for the effective extraction of ions from the ion source. Currently, the so-called Halbach array (Fig. 2.39) represents a clever solution for

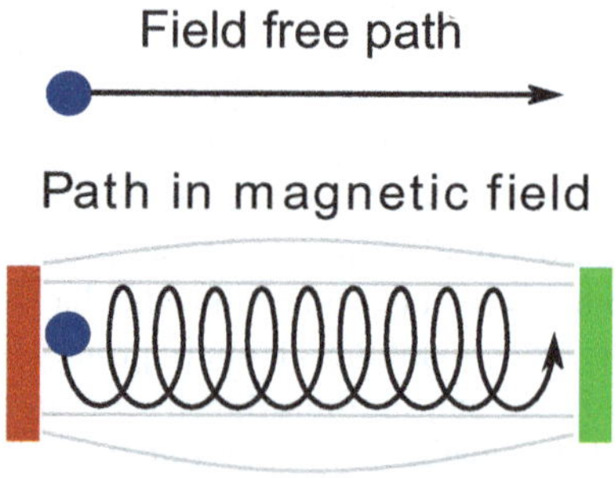

Fig. 2.38 Introducing magnetic fields into a plasma reduces transport losses by forcing charged particles into longer gyration pathways. The velocity components perpendicular to the magnetic field line determine the gyration radius according to (2.33). High ion extraction voltages accelerate the confined ions, practically breaking the magnetic confinement

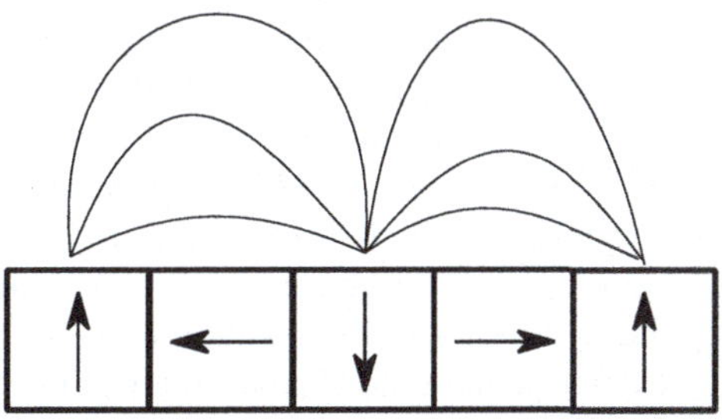

Fig. 2.39 Magnetic Halbach array with magnetic field orientation (arrows) and magnetic field lines. The orientations of the magnets rotate by 90° with each magnet, resulting in a field concentration on top of the array with only little field strength below. A plasma sitting at the top of the figure will experience the field as a transport barrier, increasing its density

this confining magnetic structure by providing a vessel wall near transport barrier due to the magnetic field effect depicted in Fig. 2.38.

The production of ions from injected neutral resources atoms and molecules (e.g. H_2) depends predominantly on the electron impact ionisation process and its cross-section. This cross-section is a strong function of the electron kinetic energy given by the plasma temperature or discharge voltage, respectively, see Fig. 2.40. The electrons travel through the gas/vapour from cathode to anode via the discharge voltage. Alternative to DC power, the plasma electrons can be driven by a high-frequency AC electric field resulting in oscillating movements of the electrons through the plasma. Typically, higher densities are obtained with a dedicated electron source of the thermionic type discussed above. The limited emitter lifetime represents a drawback, but a controlled emission current and acceleration voltage can be set with an emitting cathode.

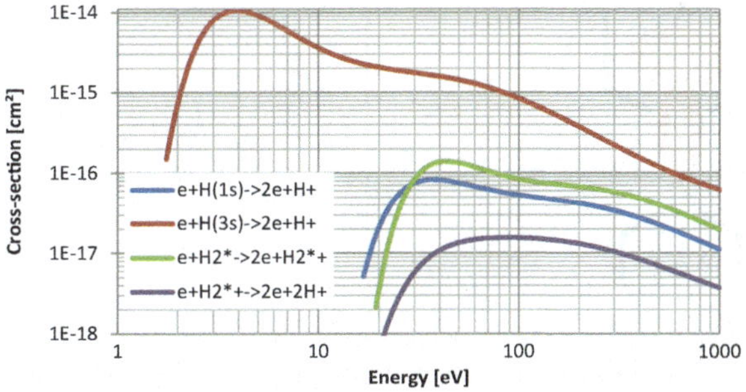

Fig. 2.40 Ionisation cross-section of hydrogen versus electron impact energy. The graph shows the cross-section for the ground state (1s) and excited states (*) of the atom, the molecule and the molecular ion. The ion source receives gas (H_2 molecules) and has to convert these stepwise to molecular (H_2^+) and atomic (H^+) ions. A maximum in splitting cross-section of the H_2 molecule lies at about 50–100 eV, corresponding to typical discharge voltages in the ion sources. Data from Hydkin (Reiter 2019)

In hydrogen ion sources, as the most important example, first the H_2 molecule needs to be separated by the electron impact. A set of different species are generated from this first step. Atoms (H), atomic ions (H^+, H^-), molecular ions (H_2^+, but also H_3^+) and excited variants of each species populate the plasma along with the free electrons after this step. Figure 2.40 demonstrates the importance of intermediate excited states in reducing the energetic barrier for further ionisation indicated by the two orders of magnitude difference in ionisation cross-section between the 1s ground state and the 3s excited state of the H atom. Interestingly, in most ion source setups the initial ionisation of the H_2 molecule limits the whole ionisation process, requiring discharge voltages of >50 V (→ >50 eV electrons) for effective ionisation. Noble gases such as helium have simpler ionisation schemes since no molecular configuration is involved, allowing for direct ionisation.

A possibility for reducing the conflict of excitation and ionisation is the spatial separation of the different interaction steps by staging the plasma temperature for example in an extended high temperature low density plasma zone with efficient dissociation and a second lower temperature but higher density plasma zone for atomic ionisation and ion extraction. The Duoplasmatron picks up this idea of separate function plasma zones (Fig. 2.41). Its modern derivative, the multi-cusp source (Fig. 2.41 right), represents the state-of-the-art with improved confinement by applying the Halbach array. Permanent magnets outside the whole plasma chamber introduce a close mesh of arcs of magnet field lines, leading to a drastically improved confinement of electrons as depicted in Fig. 2.38. Additional magnetic filter fields

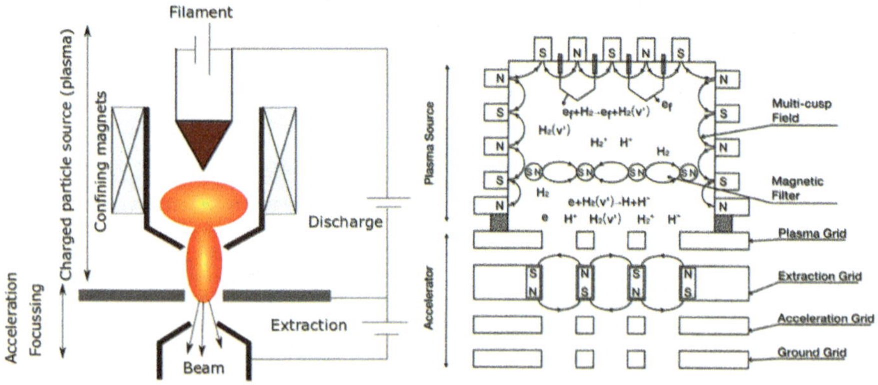

Fig. 2.41 Basic sketch (left) of the Duoplasmatron ion source. The plasma expands from the filament chamber towards the anode, forming plasmas in two regions. The plasma focusses in the lower region, leading to increased ion density. An extraction voltage draws the ions from the plasma and accelerates them. Basically all charged particle sources whether electron or ion source consist of a production feature (emitter or plasma), an electrostatic beam focus, and an acceleration stage similar to the Duoplasmatron source. Right: the multi-cusp source adds a Halbach array (cusps) around the chamber, reducing transport losses and staging reaction zones (Kuroda 1997), Reprinted with permission by Elsevier

improve the plasma staging for further increased efficiency or specialisation to negative ion extraction.

The Duoplasmatron physics were intensively studied with some interesting general conclusions for the maximum extracted ion current density ρ_B (Lejeune 1974). The input power in the form of the discharge voltage and current linearly increase ρ_B up to a hydrogen pressure dependent limit. At this point, the ionisation degree reaches a threshold and additional power will be invested rather in the plasma temperature than its density, resulting in a detrimental scaling with input power. The neutral gas pressure of the plasma discharge represents the main degree of freedom for tuning between plasma temperature and density in a given design with values usually in the fine vacuum range (Sect. 2.1, Table 2.1). Increasing pressure/gas input rate will reduce the ionisation degree and allow for further ρ_B increase up to the power handling limit of the source. The magnetic field strength increases ρ_B with a square-root proportionality due to the reduction of transport losses with increasing magnetic field strength.

With an optimised source design for heavier elements, the source typically generates lower ρ_B due to their higher mass and correspondingly smaller flux (2.40). These heavier elements can also be extracted as multiply charged ions, but as the higher ionisation steps typically require greater electron impact energies, the conservation of energy demands even lower ρ_B. Last but not least, the dimensioning of the plasma via the source geometry affects ρ_B. A narrower bottom plasma, see Fig. 2.41 left, produces higher ρ_B due to increased compression of the power from the top plasma, but at a certain ratio (typically 1:10) a maximum is reached since shrinking the bottom plasma also increases the transport losses due to the reduced wall-distances.

The injected resource gas flows through the source exit into the accelerator system together with the extracted ions. The gas pressure in the beam line should be as low as possible with typical values in the UHV range. A recovery of the lost neutral gas is difficult due to the mixing with other pumped species, leading to a gas consumption of the ion source. The recovery represents an interesting aspect when rare isotopes (e.g. He^3 or ^{18}O) are used, considering usually <10% of the gas atoms actually convert to ions. The flux of neutral particles per time n_N through the extraction aperture is given by the neutral gas temperature T, their particle mass m (together defining the average velocity), the exit aperture area A, and the pressure difference over the aperture of p_{source} according to (2.41). The flux of ions is given by the beam current divided by their charge. Accordingly, the brightness determines the resource efficiency of the ion source.

$$n_N = \sqrt{\frac{8k_B T}{\pi m}} A * \frac{p_{\text{source}}}{k_B T} \tag{2.41}$$

In Sect. 2.2 we learned about the advantage of negative ions for the DC tandem accelerator, but also for extraction of beams from AC accelerators via charge exchange. The existence of stable/bound negative ions is a prerequisite for plasma sources to contain them in a usable density. Hydrogen atoms (and a few others such as fluorine and oxygen) can form stable negative ions by filling of their valence shell with a second electron. Negative hydrogen ions are efficiently produced under certain plasma temperatures with densities in the order of 1–10% of the positive ion density. The advantages of negative ions often compensate for their reduced extraction brilliance. For other elements, such as helium, the formation of negative ions relies on non-equilibrium processes in charge exchange canals. These canals contain a vapour of an alkali metal, a group of elements with a single weakly bound outer electron. The alkali metal donates this electron to the incoming ions, forming negative ions. The process efficiency is only about 1 negative ion per 1000 incoming positive ions and has a maximum in the region of a few ten keV ion energy. The heavier the alkali metal the higher the efficiency, but also the more difficult the practical handling of the metal.

Besides the discussed plasma based sources, other source types restrict to special applications. These can be grouped in two types: Surface ion sources (sputter, Liquid tip (heavy) ion source) and field emission sources. Sputter ion source deliver only small Γ_i, but they offer the valuable addition by adding any solid as ion emitting resource. In these types, a secondary (plasma) source generates fast particles for sputtering ions from the resource which are then extracted into the accelerator. Here an electro-magnetic separation filters out the sputtering ions. This source type is of particular interest when non-gaseous ions are required and for analysing the resource by accelerator mass spectrometry (Sect. 7.1.6). The field ion source works similar to the cold field emission electron source with an atomically sharp tip where ions are extracted from. This type can be an ion source, forming for example a high brilliance He ion beam, or it can be an integrated analysis tool for the atoms contained in the tip (Sect. 7.1).

Table 2.4 compares the performance properties of different realisations of the discussed charged particle source technologies. The source properties span orders of magnitude, but in general electron source have slightly better properties than ion sources. The excellent properties of electron sources are the basis of the success of many high-resolution applications such as electron microscopy and free-electron lasers. Analytical applications rely on the best source properties, since, due to the imaging nature of accelerators, the source defines their resolution. In production

Table 2.4 Comparison of different particle source types and their typical performance properties

Source type	High frequency	Duoplasma-tron	Multi-cusp	Cs-Sputter	Thermal LaB_6	Cold field emission
Particle type	Ions	Ions	Ions	Ions	Electrons	Electrons
Typical 1σ emittance (π mm mrad)	1.5	6	0.5	4	Not found	7.5
Current density (H^+, e^-) (A/cm^2)	0.02	1	10	0.002	10	100

The numbers represent rough values of realisations on the market, but in the end depend on the technical details and medical applications improved efficiency and compactness derive from better sources, but the overall relevance of source properties is less demanding.

2.5 Charged and Neutral Particle Detectors

In this section, we are slowly sliding into the practical aspect of what we can actually do with the particle beams produced by the accelerators discussed in the last sections. The main aspect of accelerator applications is inducing reactions with something hit by the beam. These reactions are not silent, at least metaphorically. The "noise" produced are the numerous different types of secondary particles emitted from these reactions. Detecting the particles produced by the beam can serve several tasks by listening to their message. Most of the reactions, as we will see in Sect. 3.3, follow a very specific process leading to a specific message in the emitted particle species and particle energy. Detectors catch this message forming the first step in analytical applications. For more specific literature on particle detectors, practical aspects, and fundamental physics of the detection processes see e.g. (Lutz 2008; Knoll 2010; Tsoulfanidis and Landsberger 2015; Abbrescia et al. 2018). Besides these rather similar specialised detector books also the literature on applications usually contains detector basics specific for the given application.

Nowadays, practically any detection relies on transferring the particle or quantity to detect into an electrical signal, namely a current or a voltage, which electronics and computers can evaluate and store. This pathway offers the advantage of direct data storage (for later evaluation) with high speed and reproducibility, but in the translation process from particle to electrical signal, transforming elements are required and information can be lost. This transforming element is the detector and its connected electronics. The loss of information relates to the terms resolution and identification. Two main groups of particle transformations currently constitute the detector standards. Charge separation detectors form the first and scintillation detectors the second group, see Fig. 2.42 for some technical examples. Furthermore, in special applications also calorimetric detectors can be useful. These detector types convert the particle energy into heat by stopping or absorption of the particles. Since the transport of heat is by orders of magnitude slower than the transport of light and electrons, calorimetric detectors are slow and therefore of minor relevance for the applications discussed here.

A transfer of the particle energy to shell electrons of a gaseous or solid medium by collisions separates these negative charges from the positive charge of the nucleus. In gases this results in ionization the atoms or molecules in the detector. In solids, namely semiconductors, this generates an electron-hole pair. The probability of these transfer reactions depends in most cases somehow on the atomic/proton number Z of the elements forming the detector, since a higher Z equals more electrons as collision partners. In other words, a detector made from a heavier material, e.g. germanium, will absorb the radiation more effectively, requiring less detector volume than a detector made from a lighter element, e.g. silicon. The free charges present after

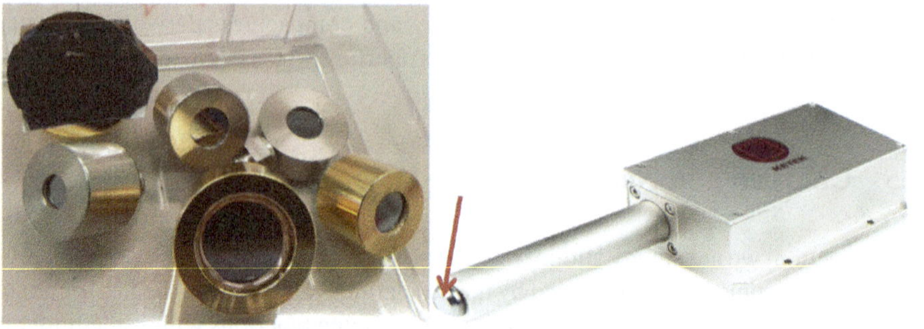

Fig. 2.42 Left: Different sizes of commercial depleted silicon detectors for charged particle detection. The central window represents the sensitive silicon crystal of here 25, 50, or 150 mm^2. All but one detector feature a backside connector. The top-left is covered by a foil making it more dE/dx sensitive but reducing its resolution. Own work. Right: a silicon-drift-detector (arrow) for X-rays integrated with electronics into a compact package. Courtesy of KETEK GmbH

the ionisation are then separated by a voltage to collect them before recombination, which would equal an information loss. Each ionisation requires an energy in the eV range. With keV to a few 10 MeV particle energies this results in 10^3 to a few 10^6 electrons per detection event in typical analysis applications. The ionisation process makes the released charge proportional to the deposited particle energy. To be able to measure this separated charge quantity in the form of a current, the detector material has to be electrically isolating, e.g. a gas, an isolator, or a depleted semiconductor.

The by far most common material for detectors is silicon. Silicon features a band-gap of 1.1–1.2 eV. A silicon detector is built similar to a diode with reverse-biasing. In a semiconductor, this configuration results in a depletion of the charge carriers in the diode resulting in a high resistance. This high reverse-biasing resistance equals a very little current flow through the diode, the so-called dark current. Impacting charged particles generate clouds of separated charges via ionisation in the reverse-biased diode. These positive and negative charges will separate via the applied voltage. The particle interaction transfers electrons from the valence to the conduction band where they are only quasi-free, but as the name "conduction band" indicates, they can contribute to a current. This current flows through the diode with a total charge equal to the number of separated charges. The quantity of the silicon band-gap allows for a room temperature operation without excessive dark currents due to the reverse-biasing voltage. Due to the band-gap visible light (photons of 1.6–3 eV) can already induce charge separation just like high-energy particles, the detector needs to operate in zero-light conditions. The second common material is germanium. Its lower band gap of about 0.7 eV requires liquid nitrogen (LN_2) temperatures ($\approx$77 K) to suppress the dark current, but its higher atomic number leads to improved detection efficiency in particular for the long-ranged photons. For the lower range of charged particles silicon-based detectors suffice. Due to the practical drawbacks of LN_2 operation germanium is exclusively used in γ-detection in the form of so-called high purity germanium detectors (HPGe). The statistical nature of the beam-interaction, conversion electronics, and residual conductivity somewhat limit the energy resolution of semiconductor detectors to about 10^{-3} of the particle energy. On the other hand, the detectors are relatively cheap, robust, and compact.

Converting the energy of the detected particle to a photon instead of a free electron yields the working principle of scintillation detectors. The scintillation effect relies on de-excitation of electrons after beam-induced excitation and the corresponding line emission. The particle energy information transforms into a proportional light intensity at a wavelength specific to the scintillator material. This light has to leave the scintillator towards a further analysis step, remember in the end we have to have an electrical signal. Consequently, the scintillator has to be transparent to its emitted radiation. Scintillation is indeed a very common phenomenon. Nowadays numerous oxide ceramics, plastics, and liquids with this combination of properties are known. Depending on the application different scintillators with optimized resolution, detection efficiency or costs are available. A simple and common example of a position sensitive scintillation detector is the outdated CRT (cathode-ray tube) display. In this device, a fixed energy electron beam hits a scintillator at the front window, inducing the visible light which constitutes the display functionality. The large scintillator allows for spatial resolution. In principle, it even allows for an energy resolution, if the electron beam energy would be variable with fixed beam optics settings. In fact, this happens upon switching the CRT display on and off, which induces a change in the illuminated area of the display by charging/discharging the electron high voltage. In the case of the display, the light gets detected in our eyes, but as mentioned earlier electronic systems always require the signal to be a voltage or current. We can directly remember the fact that silicon detectors are sensitive to light and hence they can be connected to the scintillators for the second conversion step. For the amplification of very weak photon signals so-called photomultipliers are applied. Due to the two-stage detection, scintillators typical feature worse energy resolution than semiconductor detectors, but they offer cost and detection advantages for larger detector thicknesses required for higher range particles (photons). Weakly interacting particles such as high energy photons or electrons typically pass over 100 mm of silicon in order to reach some 10% of absorption efficiency. Semiconductor detectors require a single crystalline structure with a certain dopant concentration, making it increasingly difficult to produce a single detector with large detection volume. In contrast, a scintillator allows for cost efficient upscaling due to its reduced requirements in terms of material properties.

Physical, technical, and statistical effects limit the energy resolution of detectors. The conversion of particle energy to a signal intensity also misses the species and mass information of the detected particle, the obtained information remains incomplete in most cases. Take a moment and think about what we learned in Sect. 2.3 to develop a solution to preserve this information. A hint: In the discussion of particle beam optics the terms of particle energy and momentum were of central importance. Maybe you can find a workaround for the limited resolution by yourself, if not just think about using beam optical elements as a pre-selector for the detector. Beam optics

(Sect. 2.3.2) can separate charged particles by energy (by electrostatic elements) and momentum (by magnetic elements) with much better values than 10^{-3} by bringing the particles on curved pathways with a detector at the end. The resolution improves with size and field strength of the optical element. An electro-magnetic analyser combines two technologies to mitigate each other's weaknesses.

In all detectors, particles not related to the actually observed process can and will be detected and also random events/noise will be interpreted as detected particles. Both effects form the so-called background. In almost any case, the cosmic background with its highly penetrating particles forms an ultimate limit, but also other backgrounds such as the background from naturally present and man-made radioactive isotopes and their decay can be present. A prominent example is ^{137}Cs from the Chernobyl accident often visible in γ-detectors. Natural radioactive traces in concrete or metals can provide a significant background for low level measurements. The main strategies involve increasing the signal strength to improve the signal-to-background (and signal-to-noise) ratio, carefully selecting the materials around the detectors (e.g. by using old or high-purity materials) and special low self-activity shielding containers which feature an opening towards the intended particle source while blocking background sources.

The signal-to-noise or background parameter is a fundamental aspect of detection physics and applications among a few others. In order to detect a particle and quantify the flux of similar particles it has to be absorbed in the detector with a known probability (efficiency). In order to quantify the particles energy, a known amount of this energy needs to be deposited in the detector and the response function needs to be known. The composition, thickness, and dimension of a particle detector form the central quantities for detection efficiency and particle energy deposition. Photons and massive particles like ions fundamentally differ in these two parameters as we will see in Chap. 3. For practical calculations free tools exist (Sect. 3.5). Figure 2.43 depicts the energy loss/deposition efficiency of protons and electrons and the detection efficiency of photons in different silicon detector thicknesses. The strong variations in energy deposition make the 100% efficiency point desirable for avoiding an efficiency calibration.

In many reactions, particles originated from a strongly localised point, for example a sample. If this source dimension is small compared to the source to detector distance (far sample limit), the source can be treated as a point source. The emission of a fixed flux of particles from a point source follows the concept of the solid-angle defined in (2.42). The detector solid angle defines the lateral detection efficiency, adding up to the energy detection efficiency discussed above. Here Ω is the solid angle, Θ the opening angle of the cone from source to detector edges, A the detector active surface area towards the source, and r the source-detector distance. The particle flux through a fixed area will vary with distance of this area from the source, but the flux through a solid angle remains constant. Hence, the solid angle, not the detector size, properly describes the lateral detection probability of a detector for a point source. In other cases with non-negligible source dimension, such as a large reactor or a near volume-sample illuminated by a probing beam, the solid angle description becomes less and less adequate. Finally, in the large sample limit, the particles to be detected are best

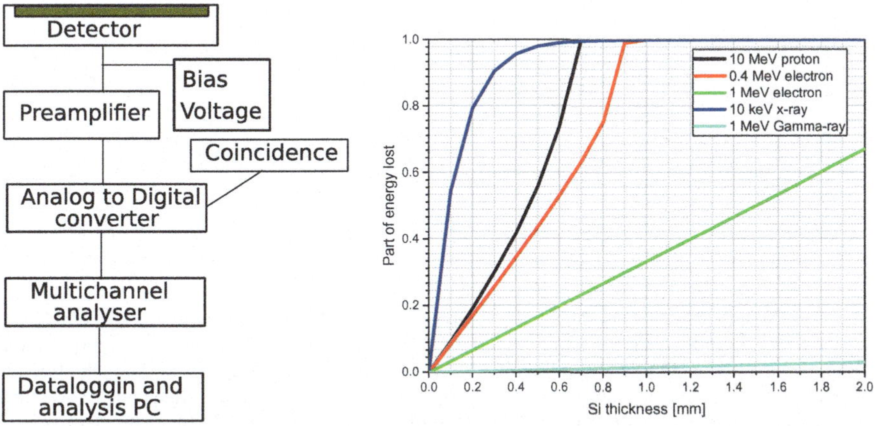

Fig. 2.43 Left: typical chain of devices for energy resolving silicon detectors with optional coincidence filtering with a second detector. Right: energy deposition fraction of protons, electrons, and photons versus the thickness of a silicon detector. Different corrections have to be applied to each particle species due to the energy reducing absorption of protons and electrons, and the quantity absorption of photons. 1 MeV photons and electrons will not be completely absorbed in detectors <2 mm thickness, but the other particles have a 100% detection efficiency around 1 mm detector thickness.

described by a flux density, e.g. particles/m^{2}s. The detector size hence becomes the determining factor for the detection rate.

$$\Omega = 4\pi \sin^2\left(\frac{\theta}{4}\right) \approx \frac{A}{r^2}, \quad \text{with } 0 < \Omega \leq 4\pi \tag{2.42}$$

Detector layout essentially depends on these considerations. Detector counting rates and detection efficiency are technically limited. Signal amplification, electrical capacities, and digital processing speeds limit the maximum count rates. Modern electronics achieve typical count-rate limits of 10^6 single particle events per second. The energy of the particle is derived from the amount of charges released in the detector via a so-called pulse-height-analysis. Since each event should represent a single particle, the detector should not receive a second particle during the charge collection and analysis time. This collection time is called dead-time and ultimately limits the count rate. Current electronics allow setting charge collection times in the order of 0.1–30 μs. For 1 μs collection time, a maximum of 10^6 events are detectable per second. Practically, the incoming particles do not wait in line like decent English men, but they arrive chaotically and the dead-time has to be limited to values <50% due to the loss of events and the detrimental effects of high dead-time on the detection quality. Figure 2.44 demonstrates this continuous saturation of the detection system. The left plot shows the levelling of processed output count rate in relation the received particle count rate. Ideally, both should equal forming a straight diagonal line, but the dead-time continuously reduces the output count rate

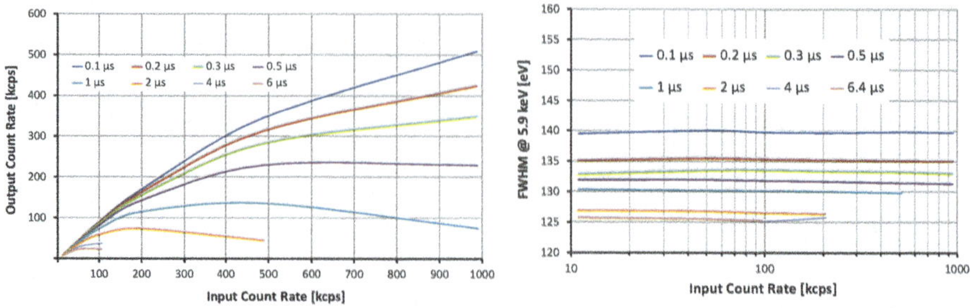

Fig. 2.44 Plots of detector parameters for a Ketek silicon drift detector for X-ray detection with transistor reset amplifier. Left: Plot of count rate versus signal integration time. The ratio of input to output count rate is the dead-time ratio. At excessive particle input and dead time the effective count rate can even come to a negative slope. Right: Impact of count rate and signal integration time on the energy resolution. Courtesy of KETEK GmbH, Hofer Str. 3, 81737 München, Germany

with increasing input events. Shorter integration times reduce the effect, but as the right plot shows this also reduces the energy resolution.

The detector is not literally dead during the collection time. If a second particle will arrive during the collection time, its energy will be added or piled-up to the energy of the first particle. This pile-up effect strongly complicates the data interpretation, giving a good reason to limit the dead-time ratio to even smaller values. A longer collection time typically yields better energy resolution, if the detector dark current integral remains small compared to the signal, but on the other hand dead time and pile-up increase. Therefore, any detector setup requires a detailed analysis and optimisation of expected count rates, detector size, required resolution, and system complexity.

Calculating the uncertainties of these counting statistics marks the first step of a data analysis. Data analysis extends beyond stating a result, it also has to be clear how credible this result is. We have to consider that many of the discussed detector techniques are able to detect single particles with a high probability, but they provide incomplete information about the particle with limited resolution, angular coverage, and they add up noise and background to the actual signal. For the detection of a single peak a certain expectation value for the event or counting rate exists, but from statistical considerations this expectation value will only be reached with infinite counting time. From a real experiment only probabilities for a certain expectation value can be derived. The uncertainty or the error, respectively, derives from the Poisson distribution to the square-root of the measured quantity of events N, see (2.43). Frankly, this uncertainty only marks a 68.3% (1σ) corridor of finding the true expectation value. In other words, if we count 100 events of a certain reaction, we have a 1σ uncertainty of 10 events, therefore the only statement we can make is that our expectation value lies between 90 and 110 count with a 68.3% probability. By increasing the counting statistics to 10,000 the uncertainty increases to 100, but the relative uncertainty drops from 10% (10/100) to 1% (100/10,000) and in conclusion the result becomes more precise.

$$\Delta N = \sqrt{N} \overset{yields}{\rightarrow} \frac{\Delta N}{N} = \frac{1}{\sqrt{N}} \tag{2.43}$$

So far we considered only single detectors at a fixed position. Due to the higher number of read-out channels the technical setup complexity, size, and cost of position sensitive detectors exceed those of point detectors. On the other hand, the additional effort for positional sensitivity provides more information on the reaction kinematics, offers better resolution, and allows for localisation of reactions via triangulation methods. Besides increasing the number of detectors, also the movement of detectors enable for a (virtual) increase of position sensitivity. The best examples are medical applications where a multi-detector setup moves along a patient or the patient moves along the detectors to enable a full body scan. The disadvantages of these scanning detectors lie in the increase in analysis time and the potential pitfall since the analysed situation can change over the individual detector runs.

As usual, fundamental research provides the extreme examples of detectors. Several large detector systems are attached to the LHC accelerator for investigating the nuclear processes and particle physics at extremely high energies. Due to the large number and variety of particles released from the TeV reactions, a full angular coverage and identification of the particles is required. One of these 4π (covering a solid angle of 4π) detectors was named ATLAS (**A** **T**oroidal **LHC** **A**pparatu**S**) and contributed significantly to the identification of the Higgs boson. ATLAS separates charged particles by momentum and charge using a 2 T magnetic field present inside the detector. The inner detector shell consists of about 140 million silicon pixel detectors. These detectors are surrounded by eight layers of larger silicon detectors. Around this sits a layer of polyethylene blocks for producing secondary X-ray photons which are detected in embedded xenon-based gas-detectors. The outermost layer consists of several hundred thousand calorimeters based on liquid argon and metal absorbers. Overall, the detector system forms a cylinder of 22 m diameter and 45 m length. The complex ATLAS detector structure allows a geometrical tracking, energy, charge, and momentum detection of the particles emitted from the reaction zone for a precise identification and quantification of all reactions, e.g. the formation and decay of the Higgs boson.

In the example of ATLAS hundreds of individual detectors were combined to a single system. These detector systems are common in elementary particle physics, but also, in a reduced fashion, in accelerator applications. Systems usually comprise only of a few detectors with different detection characteristics for separating different secondary particles (photons, ion, electrons ...) or positional detection in a similar fashion as ATLAS. Detector systems for positron emission tomography (6.1.2) exploit the annihilation of the positron and the subsequent emission of two correlated 511 keV photons. Through knowledge of the reaction kinematics and the incidence delay of the photon detection between two opposing detectors, the origin of the annihilation reaction can be localized. The corresponding system is called a coincidence and tracking system. Exploiting coincidences through knowledge of the involved reaction kinematics allows for tracking, but also for identification of particles and background suppression since the probability of a random coincidence from the background reduces quadratically with the coincidence window width. The identification is possible by certain angular relations between different products of a nuclear reaction or by their energy-loss per unit length in so-called dE/dx detectors (energy-loss E per detector thickness x). These detectors are sensitive to the particle stopping power, see Sect. 3.2. By stacking of several thin detectors, similar to the onion structure of ATLAS, the particles lose only part of their energy in each detector. Knowledge of the detector thickness, its material, and the energy deposited in it yields extra information on the particle passing it following the idea of Fig. 2.43 (right). Furthermore, detector systems also improve the signal-to-noise ratio by suppressing uncorrelated background reactions and providing more information on the detected particles for event selection.

2.6 Targets

In the beginning of this section we discussed the term technology and the difference between technological and technical challenges. In this section we are at a good point to further develop our understanding of technical solutions. A central term for a first assessment of how to technically improve a given technology for a given application is by investigation its *limitations*. Consider driving a car with the task to be as fast as possible. Let us further detail the surrounding situation, our application scenario. We are on an autobahn in Germany, no speed limits. Our speed will probably be limited by the power of the car's engine. A slight change of the scenario to a nightly trip will change this limitation to the illumination range of the cars headlights and our personal response time (assuming a will to survive). The darkness limits our vision range, but as we need to control and steer the car with our limited reflexes, the limitation of the attainable speed changed. A further slight change to a winter situation with snow on the road again changes the limitation of our maximum speed to the grip of the car's tyres defining the stopping distance.

In all three cases we apply the same technology to different situations. From the point of view of a developer looking too improve a certain parameter (in that case the velocity) all three situations lead to different technical development approaches. In the first case we would aim at increasing the engine output, in the second at improving the light density of the head-lights and in the third case better tyres would be the way to go. The other options offer only negligible gain in each situation, since they are not addressing the actual limiting factor. The reason to discuss this aspect here is, with the power of modern particle sources and accelerators, the limitations of accelerator applications start to shift from the accelerators towards the targets and detectors.

The term target defines the part where the particle beam hits its final destination (the terminal station in the image of the tram). In particle physics, usually two opposing beams collide with each other in the target chamber. This is explicitly not the situation discussed here for accelerator applications. Still not all application accelerators feature a target, as we will see for accelerator based light sources in Sect. 4.3

Fig. 2.45 Left: A 4-axis nano-manipulator for ion beam analysis capable of handling 1.5 W of load on a $10 \times 10\ \text{mm}^2$ sample (red box). Right: a cyclotron target with graphite mask, copper separator, and W sample to be attached on the KF16 water-cooled header by a special copper nut. The target can handle about 1 kW on a 12 mm diameter target disc. Courtesy of Rahul Rayaprolu

or later in Sect. 8.3.2. The concept and limitations of targets apply to several other technical arrangements in accelerator systems as well. Beam diagnostics (Sect. 2.3.3) such as apertures and Faraday cups have similar technical challenges as targets since they represent at least an occasional beam terminal. Apertures and vacuum vessels are basically targets which are not frequently replaced or follow an intended use, but technically very similar concepts apply to them. In addition, the dimensions of targets vary substantially, ranging from mm sized manipulators for high-resolution microscopes (Fig. 2.45 left), over 100 mm sized production targets (Fig. 2.45 right) to constructions of several meters in diameter for spallation (Sect. 8.1).

In this section, fixed targets will be discussed for both isotope production and analytical purposes. The term "fixed" denotes a target (with particles of mass m_2) resting in the laboratory inertial system. As the beam particles (mass m_1) are moving with kinetic energy E_0, the centre-of-mass (CMS) of targets and projectiles (the individual particles of the beam) also moves in the laboratory frame, containing the kinetic energy E_{CMS} according to (2.44) in the non-relativistic case (Nastasi et al. 2014). This centre-of-mass energy states the main quantity for beam-matter interactions since it represents the energy available for reactions. E_{CMS} is always smaller than E_0 and it is smaller in the case of a fixed target compared to the colliding beam target. Energy has to be conserved, hence the remaining energy of a fixed target situation will be transferred to the product particles, potentially leading to large secondary particle energies with implications on radiation protection (see Sect. 2.7) and information properties for the detection of secondary particles as discussed in Chap. 7.

$$E_{CMS} = \frac{E_0 m_2}{m_1 + m_2} \tag{2.44}$$

In applications the target should provide a low maintenance frequency/long lifetime, low amounts of unwanted activation (e.g. of structural parts), and tolerable operating temperatures. The operating temperature arranges according to the balance of input power (practically the beam power P_{Beam}) to power removed by cooling. Three mechanisms of power removal exist: Thermal radiation, conduction, and convection. Figure 2.46 demonstrates a target construction for solid targets, exploiting convection in a water coolant pipe, conduction in the metal structure fixing the targets beam impact area to the coolant tube, and radiation from the target surface into the vacuum system.

All matter emits photons with an emission power proportional to the fourth power of the absolute temperature (Kelvin) due to fundamental physical properties of matter. The Stefan–Boltzmann law describes this process. Solving the Stefan–Boltzmann law for the temperature with the surface area A of the target object, its material dependent emissivity ε $(0 < \varepsilon < 1)$ and the Stefan–Boltzmann constant σ_B yields:

$$T_{\text{Max}}(P_{\text{Beam}}) = \sqrt[4]{\frac{P_{\text{Beam}}}{A\varepsilon\sigma_B}} \tag{2.45}$$

Practically, radiation-cooling represents the weakest contribution. Therefore, radiation and (2.45) provide the upper temperature limit T_{Max} a target can reach, if no other heat removal process is effective. For ideal heat conduction from a target with area A irradiated by a particle beam of power P_{Beam} to an effective heat sink, such as a water cooling system, over the material thickness d of a solid (central part in Fig. 2.46) with thermal conductivity κ to an effective cooling liquid flow, a temperature difference of ΔT between target surface and cooling liquid will arise. Equation (2.46) allows for calculation of the maximum heat load or the surface temperature of a target, respectively. In practice, complications of limited contacts, for example between a sample and its holder, or limits of heat removal by the coolant lead to a temperature in between the values given by (2.45) and (2.46). The quality of target construction determines which of these limits is closer to the real setup.

$$\Delta T \approx \frac{d\,P_{\text{Beam}}}{A\kappa} \tag{2.46}$$

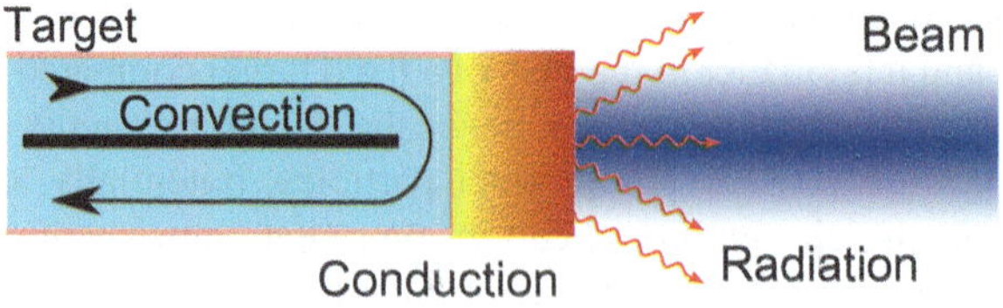

Fig. 2.46 Up to three mechanisms cool a target irradiated by a particle beam (incident from the right)

Targets for isotope production or particle beam induced modification receive high thermal loads for reaching maximum beam current and therefore productivity. Loads of $P_{\mathrm{Beam}} > 100\ \mathrm{MW/m^2}$ are easily possible in modern accelerators, but don't be misled by small numbers: A 1 MeV * 1 μA = 1 W focussed beam of 1 mm^2 already delivers 1 MW/m^2 onto the beam area.

The convection represents the most effective cooling mechanism due to the constant removal of the heated material. Assuming a perfect exchange between the hot part and the cooling liquid the temperature increase calculates according to (2.47) from the specific heat capacity c_P [J/(kg K)] of the coolant, the beam power P_{Beam}, and the mass flow of the coolant F_C for example in units of [kg/s]. Typical coolants are water and air and in special cases of high temperature and oxidation prone materials also helium. In some cases, the coolant can also be the target, for example in the production of ^{18}F from ^{18}O enriched water for PET diagnostics (Sect. 6.1.2).

$$\Delta T_{\mathrm{conv}} = \frac{P_{\mathrm{Beam}}}{F_C * c_P} \tag{2.47}$$

The thermal properties are a crucial property of accelerator targets and hence state-of-the-art engineering techniques are applied for their development. The construction by computer-assisted design (CAD) allows a direct integration with thermo-mechanics by finite-element method (FEM) simulations. The modelling of coolant flows or fluid dynamics in general (gases and liquids) requires so-called computational fluid dynamics (CFD). Given the right parameters, CFD tools can simulate numerous quantities such as heat transfer between fluid and solids, phase transitions (evaporation), turbulent flow barriers, and even chemical reactions. Modelling allows a more precise estimation of sample and structure temperatures and heat flow beyond the simplified analytical models discussed above. The combination of thermal properties with the mechanics and geometry enables identifying possible weaknesses in the design. A feedback to the target construction and sample handling will lead to increased power handling capabilities and hence productivity of the targets. The central weakness, especially for solid targets, is the thermal contact area in between two separate parts. Pressing a sample onto a holder will in general not lead to a full surface contact, but in the worst case only to small contacts at the edges or in the centre, due to slight surface warps. Thermal gradients induced by the particle beam impact lead to bending of material, potentially further separating sample and cooling structure like a growing pancake on a pan.

Figure 2.47 shows a FEM calculation of 1.5 W beam induced heating representative for focussed ion beams of a few MeV or electron beams in the keV range applied in material analysis (Chap. 7). In the approximation of (2.46) the 1.5 W of this figure would yield a temperature increase of 2.5 °C (=62.5 °C) above the heatsink temperature, when distributed homogeneously over the 100 mm^2 sample surface. For a sample area as small as the beam spot, 3596 °C would be reached with (2.46). The simulation yields 164 °C for the given situation, highlighting the importance of a precise beam area/size definition for target design. In any case, the target will bend

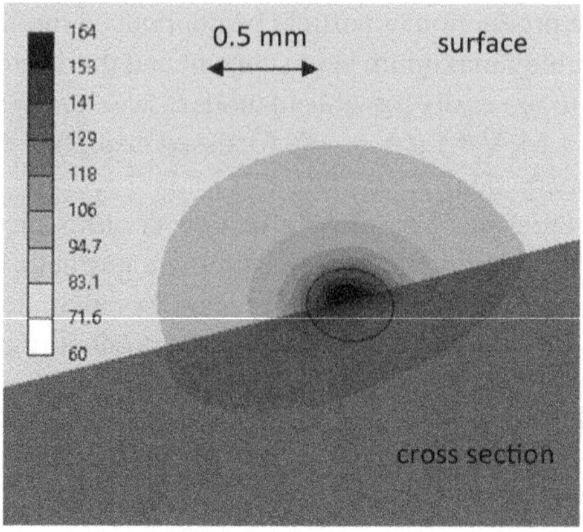

Fig. 2.47 Sliced plane view of a FEM simulation of a 10 × 10 m^2 steel sample (κ = 30 W/m K) irradiated with a 0.3 mm beam spot with 1.5 W of beam power (see Fig. 2.45 left). The top part shows the exposed surface and the darker part on the bottom a cross-sectional area into the depth of the sample. 60 °C water cools the 5 mm thick target from the bottom (not shown), but the spot centre reaches 164 °C

upwards due to the temperature gradient and the resulting inhomogeneous thermal expansion, affecting the backside contact.

For powerful accelerators and small spots, the conduction of heat, even with copper, approaches its technological limits, allowing only for a few millimetres of material towards the coolant/heatsink. Convection, the heat transport via movement of hot fluids, allows for more efficient heat transfer and flexible thermal contacts. Gas and liquid targets feature this specific advantage over solid targets at high beam loads with the drawback of requiring an additional vacuum barrier. In this case the target itself represents the coolant, reducing the heat barrier to the physical minimum and excluding the less effective cooling by conduction and radiation. Optimal exploitation of accelerator and target equipment requires both to match in capabilities.

While the situation is easy in the example of Fig. 2.45 (right) where a massive copper heat exchanger separates cooling liquid and vacuum, liquid and gaseous targets require a vacuum barrier, which is transparent for the particle beam. These so-called beam windows consist of thin plastic or metal foils. The foil needs to have a thickness d with

$$d \geq p * \frac{R}{2\sigma_S} \tag{2.48}$$

where p is the coolant gas/liquid pressure, R is the open radius of the foil and σ_S is the foils (temperature dependent) tensile strength. On the one hand, higher gas/liquid pressures are advantageous for power removal via forced flow and reduce the beam range in gases. On the other hand, the foil needs to be as thin as possible to minimize losses of the expensive beam energy in the foil via the so-called stopping effect discussed in Sect. 3.2. This, in combination with (2.46) and (2.48), implies a limitation to window materials and their σ_S, respectively. The other technical freedom lies in the open radius R. Intuitively we would choose R according to the beam radius. Focussing down the beam allows for a smaller R, allowing for proportionally smaller d, but the foil temperature increases inversely proportional to the beam area due to the increasing power density with focussing (2.46) and hence scales quadratically with R. The optimal balance depends on the temperature evolution of σ_S and the target fluid, but in any case σ_S will reduce with increasing temperature. A trick for extending this technological limit lies in adding a supporting mesh with thin webs, which are long in the direction of the beam. This way the foil is supported leading to a reduced R, but 10–20% of the beam are obstructed by the webs.

A smart target design and detailed modelling of its properties represents a solid starting point. Practical imperfections, such as the contacts, and the ravages of time necessarily lead to a difference between theory and practice, which becomes the more relevant the more the target is pushed to its limits, which is what we should always want. The above models aid in extending these limits starting with simple considerations of material thickness and coolant flow rates up to complex 3D models. In the end an expensive MeV accelerator production device should not be limited by the water flow rate in the target.

Target diagnostics provide valuable information on the live target status. The most common diagnostic is the beam current measurement (which also yields the beam power) via an ampere-meter connected to the target. The target has to be electrically isolated to the device ground and other electrical systems for this measurement. Measuring the true current is more difficult than it seems: Secondary particle emission as discussed with the Faraday cup in Sect. 2.3.3 of mostly electrons induce extra currents which add up to the true beam current. For secondary electron suppression either special target constructions or electrostatic suppression via a biasing voltage of a few 100 V are required, see wiring example in Fig. 2.48.

Online temperature measurements enable process control and protection against failure. Contact sensors such as the resistance temperature detectors of the Pt100 type or thermocouples such as the Type N represent cost efficient sensors with accuracy in the 1% (of the Kelvin value) range. The strongly localised heat deposition of focussed charged particle beams in combination with their limited range challenges the credibility of their readings. Figure 2.47 demonstrated this difficulty with only a few K temperature increase outside the beam spot (undetectable considering sensor accuracy) but 164 K increase in the beam spot impact point. Where could we attach a temperature sensor without obstructing the beam, yet measuring the peak temperature? Infrared emission observation circumvents these issues, but infrared emission efficiency, the so-called emissivity ε, depends on the surface morphology/roughness which changes under beam irradiation (see Sect. 5.3.3 or Sect. 7.1) and also with temperature itself. More accurate devices therefore analyse the emission at several wavelength, for example 2 or 4, but even here derivation of the temperature from the emission intensity requires the assumption of a wavelength independent emissivity. 2D cameras on the other hand yield hundreds of data points over the sample area, but measure only in a single wavelength band. Without clever calibration, e.g.

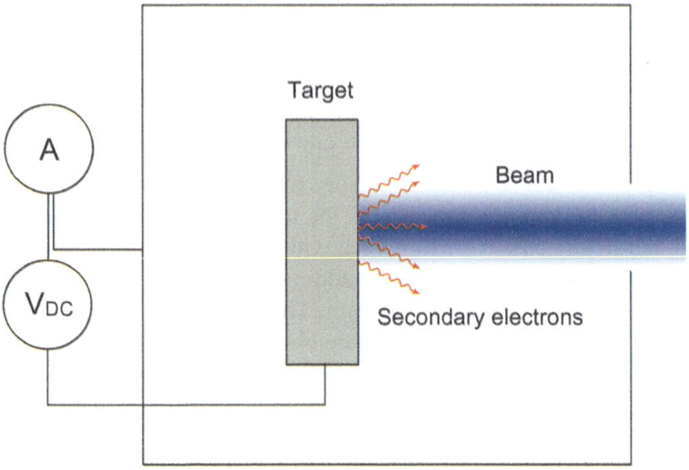

Fig. 2.48 Wiring for biased target current measurement with triaxial cables for noise suppression. This simple setup allows for secondary electron suppression by converting the vacuum chamber to a faraday cup

(Möller et al. 2017), or emissivity data uncertainties of infrared analysis reach a few 10 K. The vacuum pressure analysis also provides information on target temperature via outgassing effects, but careful, the outgassing reservoirs deplete (Sec. 2.1.2), therefore only pressure increases can indicate temperature excursions.

Figure 2.49 shows an example of morphological changes induced by beam impact on a beam dump. The implantation of 2.2 MeV protons into tungsten lead to the

Fig. 2.49 Blistering induced by proton implantation into W. At low temperatures and high beam fluxes some materials cannot desorb the hydrogen fast enough and subsurface pressurized cavities (bubbles) appear. The bubbles represent a thermal barrier and potentially burst, removing material from the target. Reproduced from Segev et al. (2017) with permission by Elsevier

formation of mm sized blisters with a few 10 μm thick caps split-off the bulk (Segev et al. 2017). The accumulation of hydrogen implanted by a proton beam forms pressurized voids, if the temperature is too low for efficient outgassing. Bursting blister caps form potentially radioactive dust (^{184}Re formed by (p, n)) and limit the component lifetime. The induced radiation might even change the chemical composition of liquids and gases by ionisation, generating acids, ozone, or other highly reactive chemicals. This chemically challenging of target materials by corrosion changes their material properties potentially leading to so-called stress-corrosion cracking. Numerous other mechanisms induce fatigue of targets, limiting their lifetime.

For analysis targets, partially different limitations and goals apply than for production targets. Sensitive samples require temperature control, while detectors usually benefit from higher beam power. Analysis targets additionally require high accuracy alignment and the operation of sensible detectors close to the beam impact point. While for production targets rigid constructions with mm tolerances are acceptable, the alignment of analytical targets require 3 or more stages of adjustment possibilities with increasing accuracy down to the μm-scale, representing a cascade of strong thermal barriers. An analytical beam requires at maximum some 10 nA currents with loads usually in the mW range. Strong beam focussing down to nm spot sizes and sometimes low thermal conductivity samples anyways lead to relevant power densities and situations described by (2.45). Assuming a quadratic beam spot of 1 μm side length with 1 mW load and an emissivity $\varepsilon = 1$ we obtain a maximum temperature of 17,232 K. A situation possible, e.g. when analysing nano-powder particles, although the radiating area A is slightly larger and the absorbed beam energy slightly smaller in this case to be honest. In this example, the analysis definitely alters the analysed sample, a situation, which has to be avoided for meaningful measurement results. In addition, irradiation damage and particle implantation (ions) alter samples (Sect. 7.3). Therefore, achieving strong detector signals with minimum beam current becomes a central task of analysis targets. The targets need to become compact in order for the detectors to achieve a maximum catching efficiency/solid angle for all the emitted particles containing information about the target. A technological limit arises, since with given machining tolerances, the uncertainty of the analysis geometry increases the smaller the target becomes.

2.7 Radiation Protection

Accelerators are dangerous devices. This strong statement is easily understandable for everybody, after calculating for the first time the dose rates for example emitted from a modern radio-pharmacy target or from an analytical device working with nA currents and deuterium ion beams. The following sections will point this out, but they will also point out how these dangers can be very effectively mitigated. In the end every accelerator user, whether it's a scientist, a technician, or a student, will profit from a basic knowledge of the dangers and avoidance strategies in radiation protection by awareness and readiness in unforeseen situations and incidents. Furthermore, knowledge will decrease natural fear from the invisible threat of radiation and help improve the working motivation of all employees in these areas. This book can only introduce the basic ideas to encourage the reader to undergo a more sophisticated education. The reader should not overrate the knowledge presented here, but rather understand it as an introductory lesson for a real radiation protection course.

Radiation protection, or better protection of living things from ionising radiation induced harm, is a science for itself and a strongly regulated legal domain alike. Basically, every country in the world has more or less similar regulations which are mostly recommended by international organisations such as the international atomic energy agency (IAEA) or the international commission on radiological protection (ICRP). The ICRP releases updated datasets and models based on the progress of science and understanding every 10–20 years. National regulations usually adapt these recommendations as the state-of-the-art. Consequently, the knowledge of radiation protection has a certain half-life and older literature has to be taken with care. This professional driven legislation lead, to the belief of the author, to generally efficient and adequate regulations implementing a reasonable balance of productivity and safety. For further reading dedicated books teaching the physical basics and providing tables and diagrams for the required data exist, here a few examples (International Commission on Radiological Protection 2007; Obodovskiy 2019; Faw and Shultis 1999).

The basic working principle in radiation protection can be condensed to four words, the four A's in German language, also depicted in Fig. 2.50. Firstly, reduce the activity/intensity of the radiation source to the reasonable minimum, which is required for your specific application (Aktivität). Then take technical efforts, by placing for example shielding around the radiation source to reduce its intensity where it's not required (Abschirmung). Furthermore, keep distance to the source, as radiation intensity reduces with the square of the distance to the source (Abstand).

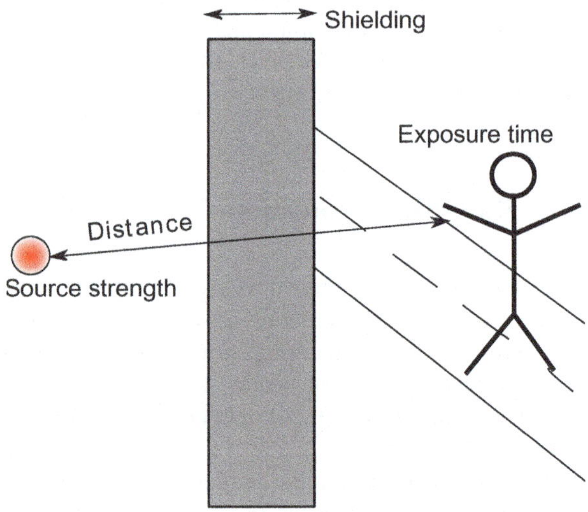

Fig. 2.50 The four A's in a figure

Lastly, reduce human exposure time to the absolute minimum required time, as dose accumulates over time (Aufenhaltsdauer). In English language the term ALARA (As Low As Reasonably Achievable) could be considered as equivalent, but it is not as much on the point as the four A's. Figure 2.50 depicts the four A's as the most condensed technical handbook for radiation safety in device layout and everyday work in radiation exposed areas.

Exposure time and source strength proportionally increase the dose, dose scales with the square of the distance r. Equation (2.49) relates the dose received at two distances r_1 and r_2 from the source. This distance scaling considers the origin of the radiation to be a point like source. The point like property requires the size of the radiating object being small compared to the distance r to the object. Naturally, every radiation source has an extent, e.g. given by a particle beam diameter or the size of a radioactive sample. As we come closer to the source, the distance law invalidates, hence touching the source will not result in an infinite dose to the finger. In spite of this it makes a large difference whether a radioactive sample is handled with fingers or 100 mm long tweezers.

$$\mathrm{Dose}(r_2) = \left(\frac{r_2}{r_1}\right)^2 * \mathrm{Dose}(r_1) \tag{2.49}$$

As the hazardous radiation remains invisible to the human perception and the negative effect usually only set in long after exposure, e.g. radiation induced cancer, it becomes very important to define a quantity/dose for the damage. The quantity Sievert (Sv) is now internationally accepted as radiation dose. It represents an integrated value of the dose rates experienced by the specific individual during his/her work. Its sum will accompany any radiation exposed worker and help to understand if any real risk of health issues exists.

Analysis devices and personal dosimeters allow for determining this quantity, which is otherwise not accessible from the irradiated person by any means. Figure 2.51 demonstrates the practical situation for a radiation protected scientist. Dosimeters represent the sixth sense for visualizing radiation. Different materials exist for integration during radiation exposure and later evaluation. This type of detectors cannot provide direct feedback of the received dose or dose rate, but requires a laboratory for later evaluation similar to an old photographic film. Electronic devices with detectors based on proportional counters or scintillators provide this direct information. Electronic dosimeters have the drawback of a dead-time, a technological limit leading to reduced apparent dose rates in the situation of very high real dose rates (see Fig. 2.44 left) e.g. in pulsed AC beam situations. For all dosimeters the detection of neutrons represents a special challenge. So-called Albedo dosimeters implement special materials with strong neutron interaction cross-section. Since neutron dose rate problems are very specific to accelerators aiming at nuclear reactions, their use is not always standard in established institutions with mostly fission reactor related experience where radiation dose rates rather originate from the radioactive decay of produced isotopes.

Fig. 2.51 Radiation exposed workers such as the author require several devices for dose monitoring. From left to right: Electronic dosimeter for direct readout of γ and β dose and dose rate, albedo-neutron-dosimeter, photon film dosimeter

In accelerator applications, we experience two types of radiation sources. On the one hand, the accelerator itself produces radiation by particle impact onto matter, we call this the *beam on* radiation. For electrons, this is mostly bremsstrahlung and X-rays, unless energies in excess of 10 MeV are applied. Ion impact emits mostly neutrons for energies above at least 1 MeV with negligible doses for lower energy ions. These radiation types vanish if the accelerator is switched off. On the other hand, mostly ion impact and neutrons produce radioactive isotopes by nuclear reactions, the *nuclear inventory*. In many applications, e.g. nuclear medicine, the production of these radioactive materials represents the main goal, but parasitic production of isotopes, e.g. in beam optical elements or vessel walls, is unavoidable if a certain beam energy is required. This radioactive inventory remains active if the accelerator is switched off. Inventory accumulates over the operational hours, but also continuously disappears via the radioactive decay (Sect. 5.1). Nuclear reaction products mostly emit photons (γ-radiation) and to a lesser extent electrons and positrons (which finally emit photons too). The choice of effective radiation protection measures and the importance for productivity depend on the radiation origin and type, hence this has to be kept in mind in the following sections.

The advanced reader may wonder why he or she never heard about the distinction between beam on and inventory in (fission) reactor safety courses. In fact, things are very different between fission and accelerators (and also fusion if it ever comes up). First of all, accelerators are part of laboratories and not sophisticated, self-powered boilers! Nobody will enter a boiler, but a laboratory is a place of constant work and development. Second, the isotopes and radiation types are completely different. Remember Fig. 1.1? Neutron/fission produced isotopes mostly lie above the line of stability, while accelerators produce isotopes mostly below it. An example: Irradiating a steel sample in a fission reactor will produce the very dangerous Co-60 isotope from the natural Co-59 impurity in the steel via (n, γ) reactions. Irradiating the same material with a few MeV protons will also produce radioactive Cobalt, but this time Co-57 and Co-58 due to (p, n) reactions with the natural Fe-57 and Fe-58. The isotopes feature completely different spectra, half-lives, and specific dose rates.

Due to the large amount of involved particle species, spatial aspects, and spans of the relevant quantities over about 20 orders of magnitude, computer models are key tools for accurate radiation protection. All computer models are only as good as their input and their user. This involves nuclear cross-sections and decay data, but also information on the present materials, their impurities, and their geometrical shapes. In the last decade a lot of nuclear and radiation protection related knowledge became more and more accessible by public and private projects and websites in the internet. A few examples: the National Institute of Standards and Technology (NIST), the International Atomic Energy Agency Nuclear Data Services (IAEA-NDS), the Nucleonica GmbH (hosting the famous Karlsruhe Nuclide Chart), or the python PyNE package. Usually these data are included in the computer model code packages. For all nuclear reaction cross-sections theoretical extrapolations and interpolations by sophisticated physical models are available which are often within a factor 3 of the real values (check JANIS database (OECD Nuclear Energy Agency (NEA) 2017) and the TENDL library (Koning et al. 2015). Due to the vast amount of nuclear reactions and their angular and energy dependence maybe only <1% of the reaction data were actually measured, the rest originates from semi-empirical models.

None of the existing computer models available for radiation protection calculations fully cover the whole spectrum of required analysis. The most complete packages are FLUKA (CERN 2020; Böhlen et al. 2014; Ferrari, Sala, Fasso, and Ranft), GEANT4 (GEANT Collaboration 2020), and MCNP (Los Alamos National Laboratory 2019). While these codes work on full 3D models and cover all particle types and nuclear reactions they are weaker in beam optical calculations, full nuclear housekeeping, and a full treatment of particle stopping. These two models work on the basis of Monte-Carlo calculations, hence they follow the track of many individual particles. This solution type is very flexible, e.g. with regards to geometry, but is also computationally slow and suffers from statistical uncertainties. Analytical codes are faster but somewhat restricted to 0D parts of the real world. The fundamental physics behind these codes will be discussed in Chap. 3. The FISPACT code (UK Atomic Energy Authority 2018) is such an example, covering nuclear inventory housekeeping of hundreds of isotopes over arbitrary time intervals in seconds of calculation time, but the code does not include geometry or the complete beam energy-loss mechanisms discussed in Chap. 3. In general, only the coupling of codes and a basic understanding of radiation protection physics and programming will allow a qualified planning of radiation protection.

On the one hand, the availability of models and data significant ease the work for radiation protection officers and other interested people. On the other hand, by far not all required information is available or easily accessible from digital sources and also the knowledge for a correct processing and interpretation of the data has to be available. Training for proper use and judgement of the data remains the domain of university education and specialised (and expensive) radiation protection classes. Many countries require such certified trainings for legally binding appointments of radiation protection officers, the responsible for protecting the public and the employees of any company, hospital, scientific institute from the dangers of radiation.

2.7.1 *Hazards for Man (and Machine)*

With great (beam) power comes great responsibility (Peter Parker principle)

In order to be able to understand and assess the hazards of radiation we first need to gain some experience in judging the quantities of dose and dose rate. The dose rate ranges possible in the accelerator context span over about 10 orders of magnitude. Understanding of a certain radiation situation requires understanding these orders of magnitude. Figure 2.52 explains them by comparing different doses received in a set of commonly known situations to each other, ranging from totally irrelevant to deadly values. In the very first block, we see that even the human body emits a small radiation dose rate due to its natural content of radioactive isotopes (mostly ^{40}K). As we go down in the first group (blue), the amount of blocks increases dramatically for flights or X-ray examinations. Interestingly, the entire dose depicted in the first group becomes negligible when looking at the second group. The lower left of set the second group depicts the unavoidable dose everybody receives from natural background. Everyone receives this natural level of 2–4 mSv per year since the first homo-sapiens were born. The radiation originates from cosmic radiation, radon gas emanating from the ground, radioactive isotopes in stones and concrete, and the ones present in our everyday consumables such as tobacco or bananas (see first group). Due to this fact the 4 mSv represent only an average value, which is exceed by up to a factor 10 in regions of high levels of natural radiation. Medical treatment doses are usually excluded from the radiation dose accounting of the patient, since they are assumed to deliver more advantages than disadvantages to the person's health. The regulations in many countries in the world allow only very little extra radiation (1 mSv in Germany) above this level by the technical devices discussed in this book. Only people registered for working in radiation environments are allowed to be exposed to about one order of magnitude more (20 mSv in Germany). In the current understanding these levels induce health risks invisible within the health risks of our

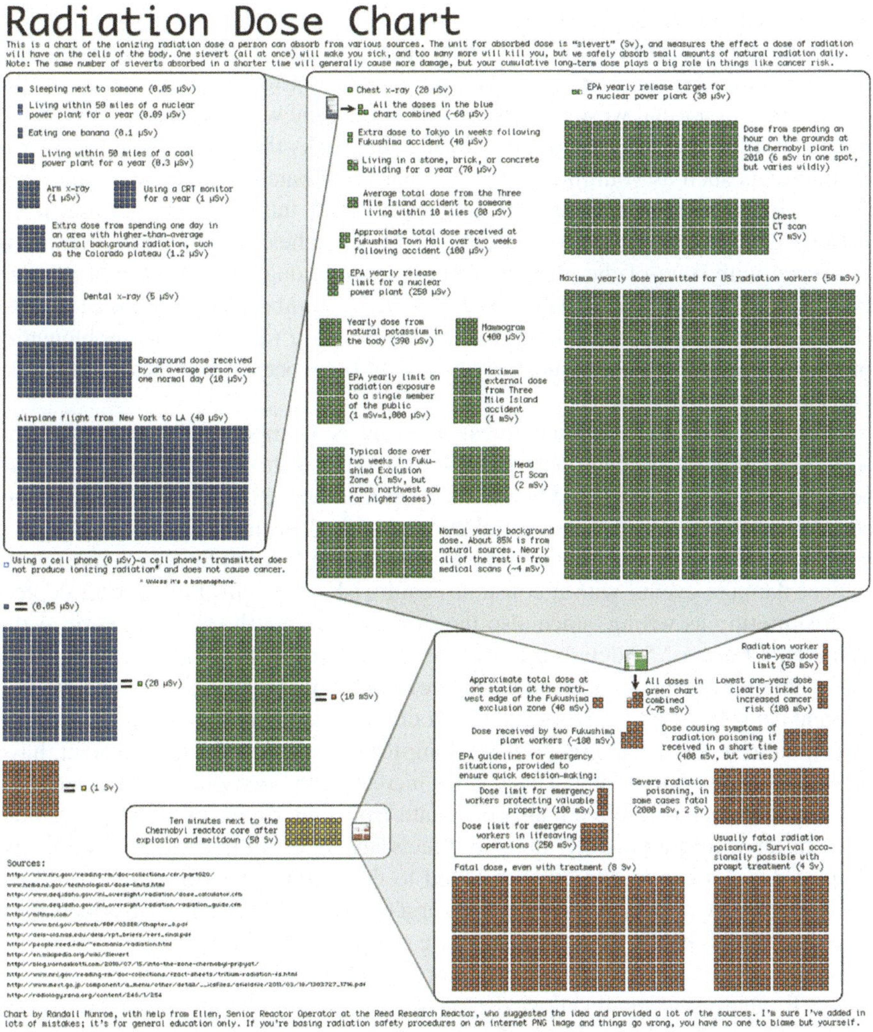

Fig. 2.52 Illustrative comparison of radiation dose levels received in numerous situations. Radiation is everywhere (top-left), but until it reaches critical levels for human health (bottom right) a lot has to happen. Reproduced from: xkcd.com/radiation, public domain

everyday life. Hence, only in the third group radiation doses become dangerous. This group depicts several health limits and doses received in certain nuclear accidents, but luckily the examples become very scarce here.

Irradiation sickness starts with changes in the blood, since the blood producing cells (and other cells with high metabolic rate) suffer the most from radiation. For this reason, radiation exposed workers participate in regular blood tests, not only for identifying radiation poisoning but also for knowing the persons regular blood status for differential diagnosis in case of exposure. Upon equivalent short-term doses of 0.4 Sv, the damage received by the body starts to become instantly relevant. The amount of cells damaged reaches a critical density. 0.4 to 1 Sv also represents the lifetime dose in normal western conditions, but the time within this dose is received makes the difference. High dose rates bear higher risks than low dose rates. Direct health effects such as vomiting, diarrhoea, burn-like patches on the skin, and pain increase in probability and severity with dose. First fatalities occur above 1 Sv. Towards 7–10 Sv instantaneous dose, the lethality reaches close to 100% after a few days, even with state-of-the-art treatments. The individual physical conditions and irradiation conditions lead to a large scatter in these numbers. In particular, the most relevant range of some mSv to Sv with moderate dose rates remains problematic, since only long-term effects such as cancer can be expected, but these potentially originate from other reasons.

What does Sievert (Sv) actually mean and what is its importance for assessing the hazards of radiation? Radiation dose accounting starts with a very simple approach to radiation hazards by seeing the human body as a calorimetric detector. This depicts the amount of energy of ionising radiation absorbed in the body in units of J/kg (=1 gray). Taking this quantity as a measure of health risk implicitly assumes independent damage events with a constant health risk per event. Figure 2.52 depicts this assumption as wrong, since also the time and certain thresholds appear in the third group. Figure 2.53 details the cascade of events in a living biological organism responsible for the thresholds and damage evolution. Generally, we end up with two different categories of health effects, the stochastic and the deterministic. Stochastic effects, namely the development of the long-term results cancer and cataract, have a certain chance of occurring. This chance increases linearly with the received dose rate, but the stochastic effects also originate from other reasons and most people never suffer from them. Deterministic effects allow for a direct connection to the received dose. It can be directly attributed to the irradiation. This could be a skin irritation, a necrosis, radiation poisoning, or a few others. Deterministic effects have a dose threshold; below this value, the dose only has a stochastic effect. Above this threshold, the effect severity increases exponentially following an error function (S-curve). Typically, the legally allowed exposure limits of radiation workers are chosen to be well below the deterministic threshold.

For a more precise judgement of the health risks related to radiation doses, the radiation protection science decouples the Sievert quantity from the physical J/kg by applying several corrections. This makes Sievert the protection quantity for human exposure, but it also becomes immeasurable in contrast to the dose quantity J/kg or radiation flux density measured by detectors and dosimeters. The first step of this conversion to a biological protection quantity includes considering the radiation type, see Table 2.5. These radiation quality factors represent the relative amounts of damage induced by a single particle of the corresponding type. The value loosely connects to its energy transfer per length (see Chap. 3 and Sect. 6.2). Photons define the reference value with a quality factor of one. This factor remains independent of the photon energy since a higher energy photon can potentially transfer more energy to the body but its interaction probability also reduces with energy. This

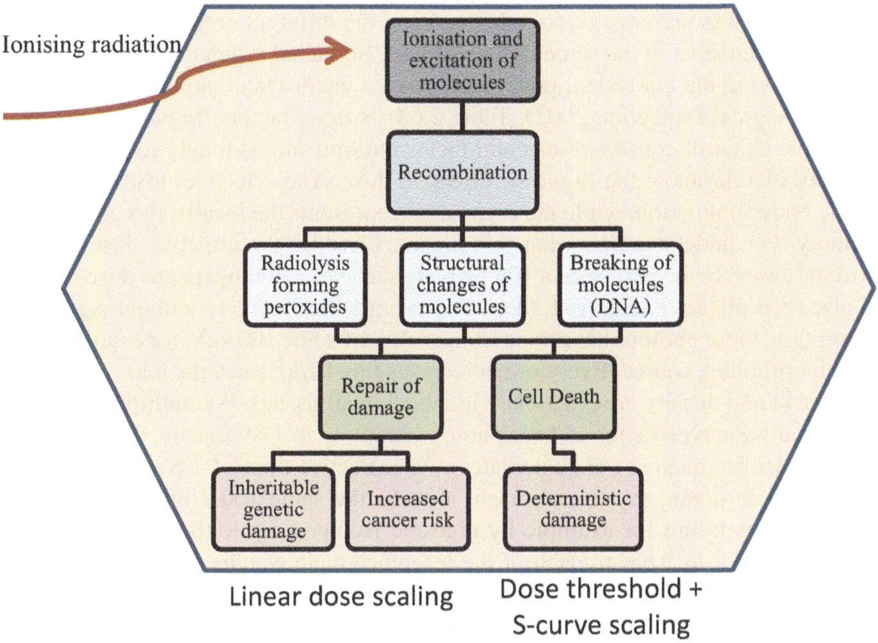

Fig. 2.53 The chain of events and outcomes of ionising radiation on living biological organisms and cells

Table 2.5 Table of biological dose quality factors for all particle types as recommended in ICRP (2007)

Particle	Photons	Electrons/ positrons	Thermal/>50 MeV neutrons	1 MeV neutrons	Protons	Other ions
Quality factor	1	1	2.5	22	2	20

Due to their long range in matter electrons and photons have a factor of one, while heavier particles induce up to 22 times the dose impact per particle entering the body

cancellation is only an approximation, but it simplifies the working of radiation protection. Similarly, electrons (and positrons) are treated. Neutrons on the other hand carry additional damage potential by nuclear reactions and resonances, which are in particular harmful around 1 MeV, leading to the highest quality factor for neutrons in this energy range. Lower energy neutrons bear only little damage energy, while higher energy neutrons less probably undergo interactions with matter, hence these particles represent lower quality factors. Ions, as heavy charged particles, feature also a high quality factor of 20, but their significantly lower range compared to the other particle types limits their impact for outside body source usually to the layer of anyways dead skin cells, except for deliberate situations in radiation therapy (Sect. 6.2).

As if that was not enough complexity, even the different organs of the body are differently sensitive to the same radiation dose. So-called organ doses consider this by multiplying the equivalent dose with a tissue factor (International Commission on Radiological Protection 2007). Table 2.6 lists these factors. Figure 2.54 depicts the whole chain of considerations and factor multiplications finally resulting in the quantity of relevance, the organ or effective dose. The effective dose depicts the whole body summation, while the organ dose represents the localized organ specific quantity. For understanding what this means, consider the effective dose in three extreme cases: In the first case, a 100 kg body receives a homogeneous dose of 100 J of absorbed photon energy (=1 J/kg). For penetrating the body without significant absorption, these photons have to have an energy of a few 100 keV, for example from a Bremsstrahlung source. Every organ receives this 1 J/kg since the mass absorption coefficient and density are practically identical in all tissues. We multiply the 1 J/kg with a radiation type factor of 1 for photons resulting in 1 Sv. Lastly, the sum of the tissue weighting factors (=1) calculates to an effective dose of 1 Sv. In the second case, a single organ, say the stomach, absorbs the same 100 J of photons due to a localised exposure for example by ingested isotopes, while the rest of the body receives zero dose. The stomach of the reference male weighs 140 g (ICRP 2002) resulting in a dose of 714 J/kg. The photon factor of 1 result in 714 Sv. The effective dose calculates from the organ factor of 0.12 times 714 Sv (Table 2.6) and the zero doses of the other organs resulting in 85.7 Sv effective dose. Calculating the same example for the skin with its mass of 3.3 kg (ICRP 2002) results in 0.3 Sv for this relatively insensitive organ. These examples demonstrate the high relevance of the type of irradiation, its exposure path, and the affected parts of the body. Keeping the

Table 2.6 Recommended individual organ/tissue sensitivity factors

Organs	Tissue factor
Bone marrow, lung, breast, colon, stomach, all others	0.12
Gonad	0.08
Bladder, liver, thyroid, oesophagus	0.04
Bones surface, brain, skin, salivary gland	0.01

The sum of the weights of these 15 organs results in a total weight of 1. Data from ICRP (2007)

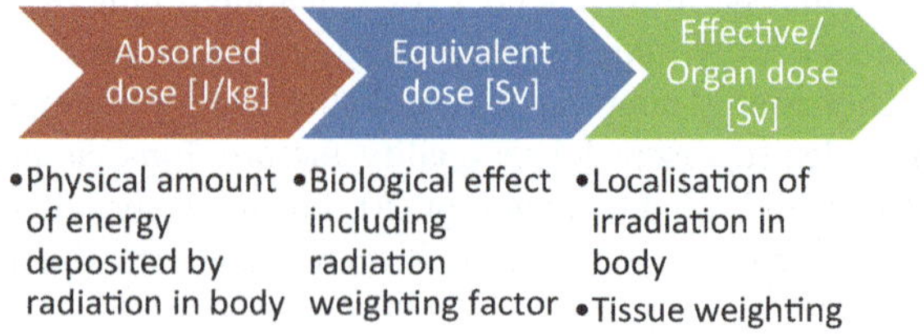

Fig. 2.54 Calculation pathway from the physically absorbed energy to the biological/health risks. The factors multiply

radiation outside the body and its sensitive organs reduces health risks by orders of magnitude.

In particular, the work with accelerators highlights the difference between exposure of different organs and the importance of considering specific doses due to the many possible exposure scenarios. Manually working with a radioactive samples or substances might expose the hand in just a few mm distance to the source, while the torso remains at arm length ($\approx$0.5 m) distance. In this case already the distance scaling (2.49) leads to dose rate differences of a factor >1000 between hands and organs. The small organ factors of the parts constituting the hand make it robust and this exposure situation less dangerous. An example where this works against us: Checking the working of a X-ray tube target by eye through a vacuum window guides the beam-on Bremsstrahlung directly onto our eye. The eye takes the complete dose, but in the case of the eye, the range of the radiation becomes relevant. The outer part, the cornea has a similar resistance to radiation as the skin. The lens is the sensitive part, but about 3 mm of tissue have to be passed to reach it. While photons penetrate deep, lower energy electrons and ions cannot reach this depth. Dosimeters can measure these depth doses via corresponding absorption foils covering their radiation detector. The 10 mm depth dose H(10) represents the body depth dose, while 3 mm depth H(3) considers specifically the eye's lens and 70 μm depth H(0.07) the sensitive part of the skin beyond the dead cell layer.

The highest risk relates to the radiation source being inside the body. The incorporation of radioactive materials through ingestion, injection (e.g. through wounds), or inhalation directly exposes the most sensitive tissues to the radiation, independent of their range. Accelerators can produce many isotopes with incorporation risk, although generally the incorporation path has less relevance compared to fission produced isotopes. For the incorporation risk, the specific isotopes radiation (Table 2.5) and its resilience time become important, since in this case the exposure time depends on how fast the isotopes can be removed from the body. The received dose integrates over the time between incorporation and removal from the body. While an active removal of radioactive material remains possible from our skin (so-called decontamination), incorporated materials practically rely on the metabolism for removal. Table 2.7 compares the effective half-life as a measure of the resilience time for three isotopes, demonstrating the relevance of the biological half-life. Tritium (^{3}H) as a long-lived isotope forms radioactive water, which will be replaced by drinking regular water resulting in a short biological half-life. Drinking more water further

Table 2.7 Comparison of biological (human) and physical half-lifes for incorporation of three different isotopes

Nuclide	Biological half-life	Physical half-life	Effective half-life
^{3}H	10 days	12.3 years	10 days
^{131}I	80 days	8 days	7.3 days
^{137}Cs	110 days	30 years	109 days

Mathematics dictates the effective half-life to be shorter than its two contributions of biological and physical/decay half-life reduces the biological half-life through isotopic exchange. Iodine tablets distributed for protection in case of nuclear fission accidents saturate the body with iodine, preventing further uptake of the radioactive ^{131}I isotope produced in fission reactors by filling the body's reservoirs resulting in quick excretion of further iodine incorporation. Besides a few special cases, the possibilities are generally limited. Some elements such as plutonium even bind specifically to the bones resulting in practically life-long exposures due to the slow metabolism of bones.

This section may sound drastic and indeed the risks of radiation should not be underestimated, especially as modern accelerators potentially induce lethal doses to a person within seconds when standing in the wrong place or handling active materials wrongly. Due to the invisibility of the radiation the exposed person would not even notice this until it is too late. As a personal statement from the author who has frequent contact with such conditions and also responsibilities for others working under such conditions I have to make a reassuring statement, though. Dangers are part of our lives and the risks of radiation are statistically lower than those of driving a car. In fact more people die from the consequences of climate change every month than from radiation accidents in the whole history of nuclear applications. In the end, it has to be worth taking the risk. The many applications presented in this book hopefully motivate this. Just try to be as informed as possible about the device you are working on by taking part in its installation and maintenance, document what has been done to it and how procedures under active radiation have to be conducted, and last but not least think of what you are applying radiation for and how it helps people more than it endangers them.

So far we addressed only the impact of radiation on the human body, but it also affects technical devices. We saw in Sect. 2.5 the intended impact of radiation on particle detectors, but the same mechanisms apply of course to all devices. Unfortunately, a precise assessment of the impact of radiation on technical parts is as complicated as it is for the human body. The more complex the devices become, e.g. highly integrated circuits (ICs), the less predictable the dose limits are. If the ionising radiation can penetrate into electronic chips it will induce charge separation which interferes with the charges the chip requires for operation and data storage. False signals or crashes of programs are possible results as a kind of deterministic damage. These crashes can be recovered with a reboot. Also permanent damage or accelerated fatigue of chips and storage devices is possible through a change of the doping of the silicon by nuclear processes. This type of damage usually occurs in radiation detectors, resulting ion slowly increasing dark currents finally killing the detector. Alpha particles and low energy electrons are typically merely problematic, but high energy electrons, protons, neutrons, and photons penetrate deeply into the chip materials. Besides electronic devices also materials suffer from radiation damage, see Sect. 7.4. Plastics, with their complex molecular structure, suffer from all types of radiation by becoming opaque and brittle. Metals and ceramics are resistant to photons and electrons below about 1 MeV, but neutrons and ions strongly alter their thermo-mechanical properties in the irradiated volume.

2.7.2 Avoidance Strategies in Plant Conception

Radiation protection was a critical part of the plant conception since the beginning of nuclear industry, but in pressurized-water fission reactors the solutions somewhat differ from accelerator applications. Fission reactors make use of a large mass of fission fuel with enormous residual activity. The vessels are under pressure, at high temperatures, and feature significant heat production by the residual activity, even when switched off.

The intrinsically different construction and goals of accelerators lead to very different accident scenarios. Above a few MeV, an accelerator can produce radioactive isotopes, but the device itself and its parts do not require any active material. If the accelerator is switched off it does not produce any radiation, e.g. the X-ray tube at the dentist remains accessible if switched off. A fission reactor employs mostly isotopes of heavy elements decaying by α-decay (the emission of ^{4}He nuclei). The accidental emission of these isotopes represents a radiation and a chemical risk to the human body by incorporation, therefore the whole reactor is contained in a concrete housing. All fission reactors have the same physical concept, but the concepts of accelerator applications differ strongly in beam energy, power, and species. An accelerator potentially produces much more diverse isotopic spectrum since every element can serve as target (not only Uranium). In general, accelerator produced isotopes feature comparably higher γ-activity, less incorporation dose, and shorter half-lifes compared to fission isotopes.

After this short excursion to fission reactors we focus back on accelerators. In the last section we identified the four A's as the technical procedure/strategy for radiation safety. In this section, current approaches for translating this to technical designs will be discussed.

Activity

Activity or radiation dose rate, respectively, originates from the mostly constant **beam-on induced radiation** and the accumulating **inventory of radioactive isotopes**. Both are proportional to the beam current and in a more complex way connected to the beam energy. While the beam-induced radiation is independent of beam fluence, the inventory increases with fluence up to equilibrium with the radioactive decay. Higher beam energy always leads to higher radiation levels, but the relation can be everywhere from nearly zero increase with energy to an exponential increase, depending on the interaction physics.

The selection of beam energy represents the most important step towards the resulting activity and radiation types (photons, neutrons, charged particles) and produced radioactive isotopes. High-energy charged particle beams represent a strong activity by themselves and can penetrate even thick materials, but here we consider only secondary particles emitted by the interaction of these beams with matter. Electron beams produce X-rays with a broad spectrum (Bremsstrahlung, see Sect. 4.3) and certain element specific peaks resulting in photon energies up to the primary beam energy. Neutron emission requires particle energies above a few MeV, see Sect. 4.1.

Ions beams produce only little photon dose rate compared to electron beams, but above a few MeV neutron emission dominates the beam on dose rate. The strong biological factor of neutrons (Table 2.5) in the typical beam energy range further increases the biological relevance. Physics generates a strong connection between the beam related neutron emission and the accumulation of radioactive material inventory. For protons and deuterons, first nuclear reactions with light elements become possible above 1 MeV, while above 5 MeV are required for reactions with basically all of the periodic table. Heavier ions require substantially more energy of a few ten to hundred MeV due to the repulsive forces of the positive nuclear charges, which is of low application relevance. Generally, higher energies over-proportionally complicate the radiation protection (see also Sect. 2.7.3).

The nuclear reactions induce a strong relation of the activity to the ion beam species and the materials hit. The energy thresholds for nuclear reactions lead to significant differences of activity along the accelerator. At the particle source, we might be below the thresholds, while at the design energy we run into problems. The low energy part of the accelerators is generally less critical and we will focus on the high energy part in the following. The application defines the minimum required energy and species, but production efficiency and rate often demand a higher beam energy than physics (Sects. 3.4 and 5.1.4).

More technical freedom for reducing activity lies in the beam optics and components. Defining a low beam current/current density on the low energy side (e.g. with the particle source or an aperture) and conserving it on the high energy side via a high transmission (>90%) beam optical system (Sect. 2.3) yields lower activity than a high source output with a low transmission on the high energy side. A high transmission equals only little interaction with apertures and vacuum tubing, resulting in less beam-on and inventory radiation. Nevertheless, the system will not be a straight tube and the tube diameter in combination with the local acceptance (2.19) will tell us how many σ of the beam (the current to the wall) are lost at a specific position. Take a design as depicted in Sect. 2.3.2 Fig. 2.30 or Fig. 2.31 and think about where particles collide with the walls. This will be beam optical elements, apertures, dumps, and samples as critical points. The ultimate limit lies in the statistical nature of the beam resulting in a certain fraction of the beam hitting the vacuum walls as depicted in Fig. 2.26.

Let us consider an actual example. A 16 MeV DC accelerator is set up for continuous deuteron ion beam operation with 1 mA beam current. In the frame of the device planning, the beam tubes material needs to be selected. Commercially either the stainless steel 316L or the Aluminium-alloy 6082 are available for standard CF type vacuum tubing. Both alloys contain a set of elements and impurities. From the ion source emittance and beam-optics, we expect a flux of 1 μA (10^{-3} of the beam) lost to the vacuum tubing. A nuclear inventory code (Sect. 3.5) allows for calculating the expected nuclear inventory to optimize the device layout for minimum activity. Both materials reach significantly different activities after 1 year of assumed non-stop device operation at maximum power. 316L reaches an activity of $2.8 * 10^{13}$ Bq, while Al6082 only reaches $1.4 * 10^{13}$ Bq. In fact, the iron component in the aluminium alloy provides the largest contribution to the materials activity as it does also in 316L, due to the long half-life of ^{55}Fe of 2.74 years. The nuclides of lower importance differ in both materials. While for Al6082 the second highest activity related to tritium with a 31% contribution, this contributes only 2.7% to the 316L activity. ^{54}Mn originating from activation of Cr represents 27% of the activity in the stainless steel 316L, but only 1% in 6082. Both materials lose about 97–99% of their activity and dose rate after 1 year. The biggest difference is the radiation dose rate level, which lies for 6082 a factor 10 below the level of 316L, due to the different nuclides present in both materials.

The selection of proper materials (and their purity/quality) represents a major point of radiation protection in accelerator applications. The example shows Aluminium based vacuum components reduce the dose rates by a factor 10 in the investigated case, extending the safe range of beam energy and power and extending the possibility for workers to enter and maintain the device. Materials are important for radioactive isotope inventory, where the isotope specific dose rates come into play, but also for beam-on radiation in beam dumps, tubes, and analytical components such as Faraday cups. The emission of bremsstrahlung by electron beams increases for heavier elements, while for ion beams neutron emission, and with that the radioactive isotope producing nuclear reactions, can be avoided completely with heavy elements, at least up to some MeV (e.g. 5 MeV for proton beams on Ta). At higher energies, detailed calculations become necessary as demonstrated in the example above since the combination of energy, projectiles, and materials defines the possible products and activity (Sect. 5.1).

Distance

The key approach to increasing distance between source and staff is remote control. Remote control allows for practically infinite freedom in the position of the operator. With modern computer systems the operator could even be on the other side of the world. Nevertheless, regulations and common sense require staff to be on-site to be able to handle problems. Finding the balance in this interplay of **normal and off-normal operation** and understanding where and when human interaction and manual work is required is the key for efficient and safe distance concepts.

On the downside, increasing levels of remote control become increasingly costly. Above we discussed the difference between stainless steel and aluminium vacuum tubing. Vacuum technology is not designed for remote installation, but in applications, standard parts, designed for manual installation, are used. Standard parts offer lower cost at higher quality than custom parts, a difference in cost that adds up to the remote control costs. What if a pump breaks or a vacuum leak appears? These unforeseeable situations have to be foreseen in a plant conception. Paradoxical, but otherwise the whole system becomes, physically speaking, unstable.

Divide and rule: The concept of separating a larger task into independent subtasks helps in this respect. By concentrating different radioactive activities such as a cyclotron, the beam optics, the vacuum pumps, and the patient treatment or target, respectively, in separated rooms (often called bunkers due to the thick walls) of a building represents an ideal situation. Each room/laboratory layout can follow

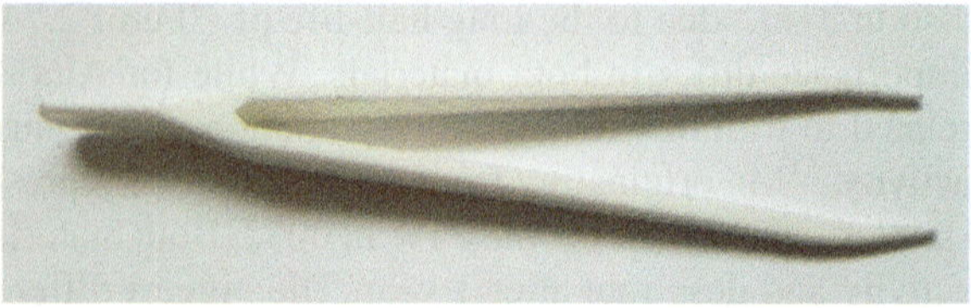

Fig. 2.55 Tweezers, a simple yet efficient way of increasing distance to radioactive material

the specific requirements, while nuclear inventory present in one room loses its importance for maintenance or operation (e.g. patient treatment) in another room.

The same applies for transport of radioactivity and the disposal of defective parts. Human handling of nuclear inventory always bears the risk of contamination (sticking of radioactive isotopes to body or equipment) and loss of material. Generally, the legislation rightfully distinguishes between enclosed radioactivity (in a rigid container so it cannot be touched) and open radioactivity (touchable with the potential for contamination). Depending on the form of the radioactive materials and the tasks different systems apply. For low levels of activity grippers and tweezers, Figure 2.55, allow for increasing the distance to radioactive samples by a few 100 mm, decreasing the radiation dose to the hands by orders of magnitude compared to finger handling. Isotopes with primarily α- and β-activity bear mostly incorporation risks. In this case a fume hood reduces the risk through the constant airflow directing released radioactivity away from the worker. With increasing activity glove-boxes increase this barrier efficiency via a hermetic sealing between radioactivity and worker. Isotopes with relevant photon emission require additional distance compared to the short-ranged contamination and incorporation risk. In this case, hot-cells with mechanical manipulators and lead-glass windows are required. Rabbit systems and conveyors inside the hot-cell allow dropping solid materials into shielded containers for further transport. In particular for medical applications, piping for gaseous and liquid products (e.g. ^{18}F for PET) directly connect the isotope production target with the chemical processing plant in the hot-cell, requiring no further human interaction with the accelerator exposed part.

Complex mechanical operations such as tightening a screw or aligning devices require complex robotic arms. Extreme examples of robotic remote handling arms are the systems for nuclear fusion reactors such as JET and the upcoming ITER. The arms feature many joints to move around the donut shaped vessel with total length of several ten metres in order to access functional components and replace them, see for Fig. 2.56. Operating these large scale arms requires new forms of man-machine interaction such as virtual reality to find, inspect, and hit the tiny parts to be replaced. The extreme costs and the technological difficulties to run such a system in a highly radioactive vacuum environment (considering e.g. greases for the joints) are justified by the reduction in down-time necessary to enter the device, or probably by being the only possibility to enter the devices at all.

Fig. 2.56 3D model of the MASCOT system designed at the JET fusion experiment. Two robotic arms allow replacement of complete wall modules inside the reactor. Similar systems are planned for future nuclear fusion reactors in order to reduce down-times in case of accidents and regular maintenance in spite of the extreme activity of the nuclear inventory. Reproduced from Imperial College London (wwwf.imperial.ac.uk/blog/student-blogs/2016/03/04/jet-is-cool)

Exposure time

Modern accelerator systems allow for a complete remote operation, reducing the exposure to maintenance periods with deactivated accelerator (besides the treatment of patients in medical applications). This situation avoids the exposure for the beam-on radiation via entrance control and safety interlocks. These systems switch-off the accelerator if doors to the accelerator room are opened or critical radiation dose rates are reached in certain locations.

The exposure to activation induced nuclear inventory remains. Many of the produced isotopes have only short half-life, therefore the waiting time before entering a facility with radioactive inventory becomes important. Figure 2.57 illustrates such a situation where already after minutes 90% of the initial activity disappears, while after 1 day 99% of the activity decayed. Knowledge of the produced isotopes and their properties becomes a critical point in avoiding relevant exposure. Upon entering a room of relevant nuclear inventory a planning and strict execution of work reduces the exposure time. A single person executing the work reduces the exposure to others. The German principle of supervising every working person with three non-working persons just multiplies the exposure time by four. Colleagues should be ready for help in case of problems but wait in safe distance or in a rotating duty scheme.

If waiting is not sufficient for avoiding exposure, or urgent interaction is required, control systems and detectors help identifying critical locations and assessing the possible exposure/working time before reaching critical dose levels at these locations. So-called Electronic Personal Dosimeters (EPD, Fig. 2.58) fit directly on the clothing of exposed staff. These devices measure the dose and dose rate and visualize them to the affected person. Specific dosimeters exist even for fingers and eyes to assess exposure time in manual work. The directional and radiation type sensitivity of these devices is not flat, leading to an uncertainty of the obtained results, but their local

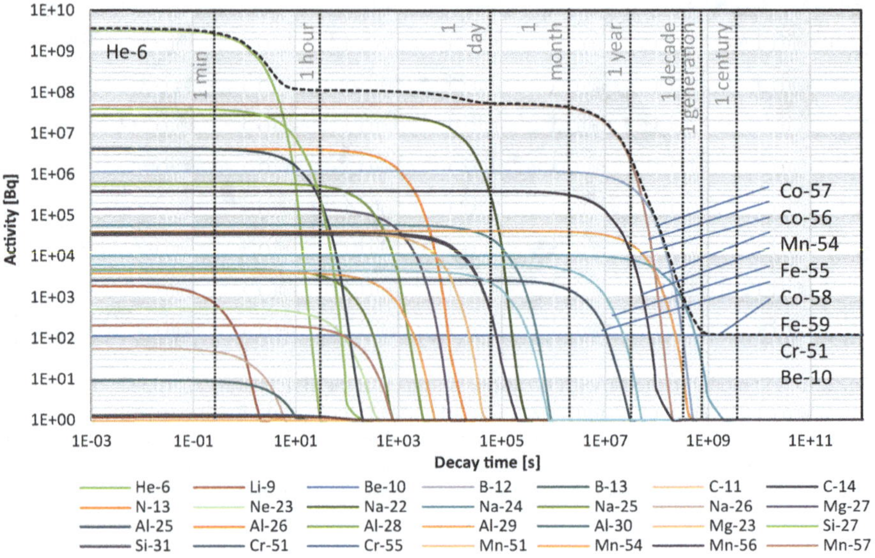

Fig. 2.57 Calculated activation and decay of the Nova-ERA neutron- source beryllium target disc (mass = 6.5 g) after 2000 h of irradiation with 1 mA, 10 MeV, and 4% duty cycle. Already after 1 min the activity decreases by a factor 10 due to the decay of ^{6}He. Over the years the dominant nuclide changes from ^{6}He over ^{56}Co to ^{10}Be. From Mauerhofer et al. (2017) published under CC BY 4.0

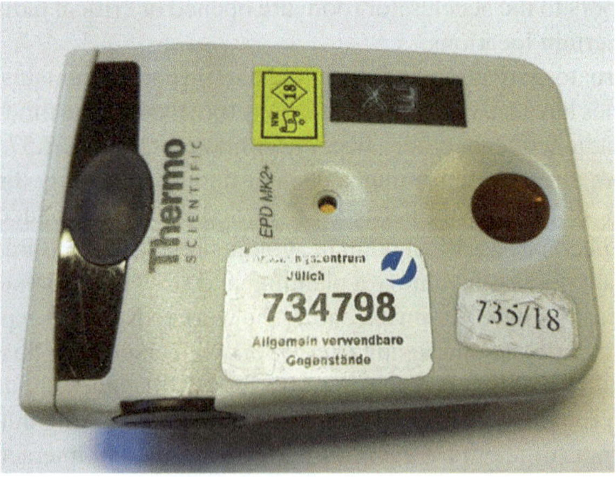

Fig. 2.58 An EPD used for direct personal dose rate estimation of photon (γ) and electron (β) doses. Correct placement for catching, technical properties and orientation towards the radiation source represent the main difficulties in application

information is the only mean to know the personal exposure time and the related risks.

2.7.3 *Shielding*

In the last section, general strategies for avoiding human exposure to radiation were discussed, but the last A was missing, so far. Among the four A's shielding (Abschirmung) represents the passive technical solution, while the others are rather of organisational nature. Shielding enables the use of standard equipment and non-radiation exposed staff by passively reducing the dose rates, hence the additional costs of shielding pay off even for technical devices. Shielding offers the further advantage of not only protecting humans but also important technical equipment such as electronics or detectors without excessive distance to the point of interest, since it further decreases the radiation dose rate beyond the pure distance scaling of (2.49). Figure 2.59 depicts such a shielding configuration for a sample observation system. The avoidance of a direct line between radioactive source and sensitive subject allows for placing shielding to further reduce the received dose.

The physical basis of shielding is the interaction of radiation with matter. We will discuss this in more detail in chapter 3, so the reader might want to come back to this section later. The four different types of radiation (neutron, ion, electron, and photon) interact physically very different with matter, hence we will discuss these four cases separately.

Photons

$$I(x) = I_0 * e^{-\mu x} \tag{2.50}$$

The shielding of photons emitted from beam-on or decay radiation follows simple physics. Photons interact with the electrons in the shielding material. The more electrons between you and the radiation source the stronger the reduction of photon dose rate. More electrons mean more atoms mean more material thickness and density

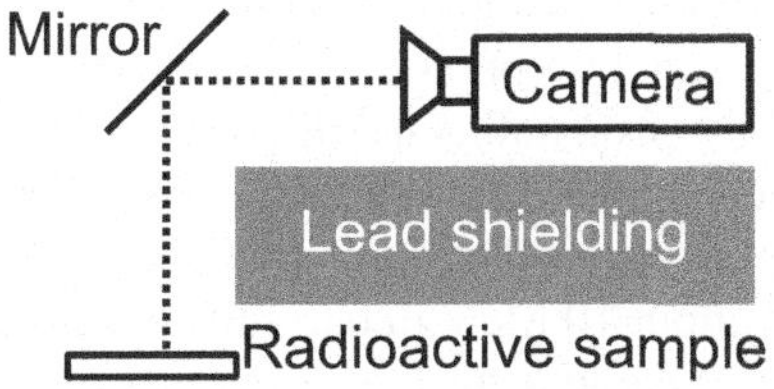

Fig. 2.59 Principle of a shielding setup for protecting an observation camera. A metal mirror does not suffer from photon radiation, hence it can be placed in high radiation areas. This 90° geometry typically reduces the received radiation dose by >90%. The principle can be stacked for further reduction. The maze-like entrance structures of radiation protected labs follow the same principle

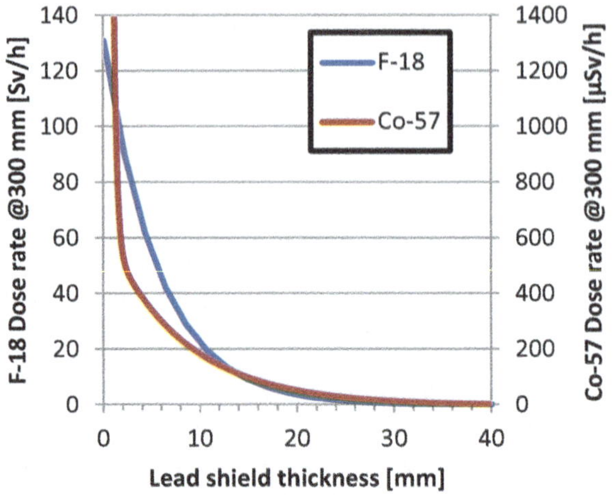

Fig. 2.60 Shielding thickness dependent dose rate of 1 mg of F-18 (medical tracer) and Co-57 (protons on iron) at 300 mm distance to a point source. 13 mm of lead absorb 90% of the F-18 radiation, while Co-57 requires only 0.1 mm. The Co-57 dose rate (μSv scale) with 0 mm shielding is even 39x higher than the F-18 dose rate (despite the Sv vs. μSv scale). The required shielding for F-18 is rather thick, due to the high energy (511 keV) photons emitted from the positron annihilation of its β^+ decay. Co-57 emits mostly photons and electrons below 140 keV, requiring less shielding thickness

equalling more shielding. This leads to the so-called mass attenuation coefficient μ and the exponential decay of photon intensity I with shielding thickness x according to (2.50). The mass attenuation coefficient depends on the shielding material and the photon energy, leading to different situations for different photon spectra as depicted in Fig. 2.60. The figure demonstrates the high importance of the photon energy, as the lower energy radiation from Co-57 requires significantly less shielding thickness compared to the one of F-18. On the downside, photons can never be absorbed completely, in contrast to all other radiation types. Every electron represents only a certain absorption probability for photons. The shielding cuts down a certain percentage of the dose, but a part of the radiation always remains, since (2.50) reaches zero only asymptotic. Therefore, often the tenth value (90% absorbed = radiation reduce by factor 10) is given for shielding materials.

Lead is the standard material for photon shielding as it features a high density at a relatively low price, but also iron and concrete are applied. Tungsten represents the best shielding per volume, but at much higher costs. We have to consider the photons discussed here (and generally considered in radiation protection) have at least a few keV of energy and up to about 10 MeV due to limits of nuclear decay physics. The transparency known from visible light has no relevance for these photons, as the absorption/attenuation coefficient strongly depends on photon energy. In the energy range considered here the absorption physics becomes simpler, combining only a few processes as depicted later in Fig. 3.3. Their efficiency drops by a factor 10^5 from 1 keV until about 1 MeV and stays mostly constant from there on.

The attenuating processes do not depend on the electronic structure of the material, therefore the shielding material can be transparent in the visible spectrum while opaque for high energy photons (and vice-versa). Transparent shieldings are typically oxides of heavy elements such as PbO (lead glass). Lead glass windows allow for manual operations with radioactive products for example for preparation of nuclear medicine products or scientific samples in hot-cells.

Electrons

Electrons, as charged massive particles, show a completely different behaviour than photons. Electrons feature a relatively well-defined range in matter. In contrast to photons they do not disappear one after the other, but like a driving car which ran out of gas they continuously lose kinetic energy on the way through the shielding until they stop completely. Unfortunately, they raise dust on their way in the form of Bremsstrahlung. This secondary radiation represents the complication of electron shielding as the photons feature a broad energy spectrum up to the electron energy with all the implications raised above. The use of light materials with low stopping power (Sect. 3.2) reduces the amount of photons raised. This requires a second shell of heavy elements for absorbing the emitted photons as stated above. Consequently, X-ray sources, as used for example in medical imaging (Sect. 4.3.1), do the exact opposite in order to produce intense radiation.

Ions

Due to the higher mass of ions, Bremsstrahlung hardly reaches relevant values, rendering shielding secondary photons unnecessary. Problems with the shielding of ions start above some MeV, when nuclear reactions (Coulomb barrier) and negative Q-value reactions become possible. Starting with (p, n) reactions, this results in the production of neutrons. These reactions produce high beam on neutron dose rates and nuclear inventory with its photon dominated dose rates at the same time. The shielding of ions themselves remains uncritical due to their relatively short range. Protons penetrate more than 1 mm of most materials only above 20 MeV and reach up to 69 mm in iron at 250 MeV. All heavier ions reach even shorter. This low range leads rather to high thermal loads, requiring special beam dumps/targets (Sect. 2.6) for handling the deposited power, than requiring radiation shielding.

Neutrons

Shielding of neutrons represents the most complex task among all particle types. It requires up to several metres of shield for neutrons of some 10 MeV. As an extreme example, the European Spallation neutron source ESS requires a 5 m concrete shielding for the neutron energies up to 2 GeV. Basically, neutrons combine all the above mentioned shielding issues at any neutron kinetic energy. We start our considerations at the upper range of neutron energies above about 10 MeV and go down from there on. The collisional stopping of neutrons is extremely inefficient at these energies compared to charged particles or lower energy neutrons. Nuclear reactions are relatively important, but their cross-sections are typically lower than for thermal neutrons, see Fig. 4.7 in Sect. 4.2.1. In this range nuclear reactions even produce additional neutrons via the dominant (n, xn) reactions, further increasing the neutron dose rate. Consequently, we cannot reduce their number, but only their energy via these reactions. Lead represents an efficient element for this task since it has good cross-sections and is still relatively cheap (shielding requires a lot of material). The multiplied MeV neutrons present after this shielding step bear even increased dose rates due to increased number and higher quality factor (Table 2.5). The lower energy on the other hand allows for more efficient neutron stopping (a.k.a. moderation). Hydrogen bound in water, concrete, or polyethylene (PE) is the ideal material as it is cheap and features a similar nuclear mass as neutrons resulting in efficient energy transfer (just like all billiard balls have the same mass). Only if the neutrons leave the MeV region to slower velocities, capture reactions (n, γ) become efficient for absorbing the neutrons, reducing their quantity. Unfortunately, at these energies only reactions with zero or positive energy output remain possible (positive Q, Sect. 3.3.2). This energy can only be released in the produced γ's, leading to a shower of secondary photons with high energies. The absorption of these photons requires another shielding layer as discussed above. Boron, namely its isotope ^{10}B, has a high (n, α) cross-section for thermal neutrons with $Q = 2.79$ MeV, making it one of the few exemptions from this rule since it emits α's instead of γ. This special behaviour makes Boron a common additive to neutron shielding materials, in particular concretes. In summary, the best neutrons shield features several layers of different materials mostly containing hydrogen and lead, but the complex multi-stage processes make an easy estimation difficult. Only sophisticated transport codes such as GEANT4, FLUKA, or MCNP allow for optimal shielding designs. Even better is avoiding high-energy neutrons from the beginning by choosing lower beam energies.

The multitude of radiation types, their combined occurrence, and the problems with high energy neutrons faced at accelerators above about 10 MeV scream for a unified shielding solution covering all radiation types in this case. Special radiation protection concretes deliver this solution by being a homogeneous mixture of hydrogen, light, and heavy elements. Additionally, they are easily cast into shapes, provide structural functions, and have a good price tag. Their elemental composition combines neutron/particle stopping/moderation and shielding. On the downside, concrete shielding require larger thickness and a higher overall weight as it is not as optimized as specific shields.

Legal Framework

Any radiation legislation has to consider the presence of natural radiation in our environment and enable the application of nuclear technologies. Over the years many common materials such as tungsten were found to be long-lived but radioactive with improving detection technology, but a sudden control of these materials just because of technological progress would be impractical and pointless. Furthermore, some landscapes feature higher natural radiation levels, for example the evaporation of radon from the soil of the black forest in Germany or the increased cosmic radiation at high elevation mountainous landscapes. Finally yet importantly, also humans carry around radioactivity, especially after nuclear medicine treatments, but you cannot restrict all of them to their private space. Due to personal rights, you even cannot imprison patients until they are decayed-off. A practical radiation protection law has to arrange all of these aspects with everyday life while still providing safety and freedom in the required situations. For these reasons and the strong impact of physics onto the legal implementation rather similar legal frameworks have developed all over the world. Radiation protection laws are definitely one of the few exceptions where physicists respectfully accept the lawyers' achievements and lawyers respectfully accept the physicists' accuracy of description.

In Germany a very strict and quantitative set of rules exist in two forms. The general legislation comes by the name Strahlenschutzverordnung (StrSchV) and Strahlenschutzgesetz (StrSchG) with some details on transportation regulated in the European Agreement concerning the International Carriage of Dangerous Goods by Road (ADR). These laws cover all usage cases and required documents and licenses, except for the operation with nuclear fission fuels for which a separate set, the Atomgesetz (AtG), exists. Interesting aspect which is somehow symptomatic for the understanding of nuclear physics in Germany: The nuclear power legislation and also nuclear power plants go by the term atom, although the technology relies on the nucleus not the chemical entity atom. The following will discuss legislation at the example of Germany during the write-up phase of this book in 2017.

The Free-handling limit forms the main solution developed to allow for a practical handling of irrelevant radioactive quantities and natural radioactivity. In the language of the physicist it is understood as a background level. In this case, legislation definitely had to adapt to the physical reality of measurements, which will not allow for the detection of arbitrarily small quantities, but will allow for the detection finite quantities. Free-handling limits also cover the different risk potentials of different isotopes by isotope specific limits. In the end (in German legislation) a large table for basically all existing isotopes develops, from which Table 2.8 shows a small extract.

The first two rows of Table 2.8 contain the activities allowed to be included in materials before a handling of these materials according to the radiation protection legislation becomes mandatory. The numbers in each row vary significantly, accounting for the individual risk and specific dose rate of each isotope. Free handling means, a license for handling is not required, but you still have to be aware of the radioactivity. Recycling and release limits are much lower, since the materials enter the regular

Table 2.8 List of selected nuclides with their corresponding free-handling and waste recycling limits within German legislation (Strahlenschutzverordnung 2001)

Nuclide	^{3}H (T)	^{14}C	^{18}F	^{40}K	^{55}Fe	^{99m}Tc	$^{137+}$Cs	^{185}W	$^{235+}$U
Free-handling (Bq)	10^9	10^7	10^6	10^6	10^6	10^7	10^4	10^7	10^4
Free-handling (Bq/g)	10^6	10^4	10	10^2	10^4	10^2	10	10^4	10
Recycling (Bq/g)	10^3	80	10	-	10^4	10^2	0.6	700	0.8

The superscript "+" indicates also daughter nuclides are included. The numbers roughly relate to the human health risks of the emitted radiation, but in fact they also represent a good compromise between practical, analytical, and economic considerations world where nobody is aware of the radiation. Let us consider the example of so-called tritium lamps which contain tritium and a scintillator to produce visible light over many years e.g. in clock faces. As long as the activity of the contained tritium stays below 10^9 Bq or 10^6 Bq/g (total device) the lamp could be freely handled in any radiation protected site without notification or license of the authorities. Using more tritium, for example for powering a fusion reactor, requires a specific license given only with reasoning. In order to leave the site with the tritium lamp its activity has to be proven to be $<10^3$ Bq/g and its dose rate <10 μSv/year. This procedure requires a qualified measurement (see Sect. 7.1.1) and documentation. After releasing the activity from the radiation protection area it will not play a legal role anymore. In contrast, an activity $>10^3$ Bq/g defines the watch as a nuclear product restricted to radiation protection zones. The last row shows numbers relevant for releasing devices, products, or samples from a radiation protection zone for recycling. These numbers become relevant for waste management procedures and lifetime costs of operating accelerators. To stay in the accelerator example: If we replace an aperture weighting 1 kg and containing, for simplicity, only ^{185}W as radioactive nuclide with an activity of $1.4*10^6$ Bq = 1400 Bq/g it would have to be released as (expensive) nuclear waste. Alternatively, the aperture could be stored inside the radiation protection zone until the activity decays below $7 * 10^5$ Bq = 700 Bq/g (one half-life of 75.1 days) or it could be separated for activated and non-activated parts in order to release a part as regular metal. An economical optimisation of materials and designs by considering the produced isotopes, their handling limits, the material price, the disposal costs, the dismantling, and possible storage costs requires the numbers within these tables.

The definition of relevant radiation doses separates source of natural and medical treatment origin from the technical ones. The radiation origin determines if it is accounted for or not. Luckily, radiation doses do not have a feeling of discrimination, and nobody bats an eye whether we call it natural, medical, or work related radiation. The first two are not accounted for as they are considered unavoidable or even beneficial. Civilians must be protected only from work related radiation. Only doses up to 1 mSv per year are allowed since this is considered negligible and a compromise between practical aspects (considering exponential attenuation of photon radiation, filtering efficiency of ventilations ...) and safety. Radiation exposed professionals are allowed doses up to 20 mSv per year in Germany. Two illustrative and complicated examples better explain this. Cosmic radiation leads to relevant doses (10–15 μSv/h) on long flights. For the crew this is work related, limiting their air time as they are in fact the employee group with the highest average yearly doses. For the passengers, the very same radiation is considered natural and not accounted for (no, not even if you are on a business trip). Second example: At the doctor, the reader is given an X-ray investigation. This medical exposure is not accounted for the reader. For the doctor this is accounted for as work related and allowed only within 20 mSv yearly limit. For this reason the assistants leave the patient X-ray room before acquisition. For the innocent pedestrian crossing the window of the X-ray laboratory, the radiation protection officer of the X-ray laboratory has to ensure under no circumstances a dose greater than 1 mSv per year can be received, even if the pedestrian camps in front of the window. Only with appropriate shielding and laboratory layout among calculations and technical specifications of the devices demonstrating this, the legislation will issue a license for operating the X-ray source.

Legislation distinguishes three different levels of radiation dose rate within such radiation protected *areas*. Declaration of areas represents the next important concept. Legislation considers doses and dose rates as restricting quantities for entrance. Normal areas not considered by radiation protection have to be limited to the above mentioned 1 mSv/year. Surveillance areas are company grounds where doses above 1 mSv/year are possible and consequently no one should enter them unintendedly for example by enclosing it with a fence. Controlled areas add an enclosure with limited entry, special warnings for radiation (Fig. 2.61), and the requirement for instructions and personal dosimetry for everyone entering it. With <3 mSv/h and >6 mSv/year the doses potentially received in these areas lie significantly above the surveillance area opening up possibilities for larger accelerators and nuclear inventory. Areas where dose rates >3 mSv/h can occur are called Closed/off-limits areas. These areas cannot serve as regular working space. Consequently, entrance must be avoided except for special emergencies. Off-limits areas require additional barriers and warnings to separate them from the rest of the controlled area. These areas can be temporary due to beam-on radiation for example in an X-ray imaging patient room (think of a CT scan applying 25 mSv in 30 min as presented in Fig. 6.7). These three classes of areas should not be confused with the concept of a room. Shielding around an X-ray tube, or a concrete wall around an accelerator target encapsulates the higher level area such that it can be potentially situated even in a normal laboratory room with a tape marking its borders.

The radiation protection officer (Strahlenschutzbeauftragter) is the captain of these areas. The legislation defines a set of ranks with increasing competence for handling of radioactive material and radiation emitting devices up to a full competence for handling, operation, and installation of devices. Higher ranks require higher levels of education and working experience in respective sites. The license for operation and installation of accelerators represents the highest level in the German system

Fig. 2.61 Warning sign found on controlled area entrances with a text stating "controlled area, caution: radiation"

requiring the highest levels of technical or academic education in addition to two years of practical experience at such devices. Independent of the rank, the officers always bear personal liability for all incidents under their supervision. Following the management law "No responsibility without competence" the officers also receive an unbreakable right of command (even before his/her boss) for all devices and persons in their site and a protection against dismissal.

This brings us to the concepts of licensing of devices and laboratories. Licensing of laboratories requires the staff and public doses to be within the limits for the requested type of area as mentioned above. For proving this to the authorities the device properties, the shielding, the total amount of nuclear inventory and the operational schedules have to match. Continuous monitoring ensures this during the later operation. Of course, the nuclear inventory changes with decay and possible production via charged particle irradiation, but licensing requires this to be considered in advance or the device operation needs to stop if the limits are reached by increasing activation. An example: We want to set up a laboratory using X-ray tubes for material analysis. The schedule foresees 50 working weeks with 40 working hours on a 5 days per week basis. This results in 2000 hours of potential exposure of the employees within the lab. Dividing the limit of radiation exposed workers of 20 mSv would result in a maximum local dose of 10 μSv/h. At this point the 4 A strategy of Sect. 2.7.2 starts by considering the required source strength of the X-ray devices. This value derives from the discussion of analysis methods in Sect. 7.1. Typically, a competitive analysis will require source strength exceeding these limits. The next level of thinking considers a shielding around the devices. Practical reasons will limit shielding thickness. Consequently, our lab layout will also consider keeping the employees at a distance to the sources. A credible distance concept requires electronic interlocks at doors or entrances, ensuring the staffs distance during beam-on to the legislation.

Many common devices, such as electron microscopes or X-ray scanners, feature only low-levels of radiation, not worth the costs of employing a radiation protection officer and maintaining dedicated closed rooms. These kind of established technologies have to stay within limits of beam energy and power to be operated without these requirements. Below 30 keV charged particle energy and with <1 μSv/h dose rates 0.1 m outside their boundaries licensing is not required since this range can for physical reasons only emit beam-on Bremsstrahlung photons. Devices with higher energies can receive design certification leading to the same freedom of operation, if the device features safety interlocks and appropriate shielding to maintain dose rate limits in the above stated range. These individual design certificates require a costly process and strongly limit the freedom for later changes or updates of the certified device. New generations require new certificates, a process common only for mass-market devices with beam energies well below the nuclear activation thresholds in the MeV range. Devices with higher dose rates at their outer boundaries must be licensed individually and operated in radiation protection areas.

Internationally slight differences in the details and numbers exist, but the general concepts discussed in this section can be found almost everywhere. A lot of international cooperation exists in the nuclear sector in the EU and beyond, but, so far, not a single legal space ranging beyond countries borders exists. As a result of this, the doses received in one country will not be legally relevant in another country. In the end, a wish for more international agreements and cooperation arises for the author from the current situation to increase cooperation and exchange in the nuclear field.

References

M. Abbrescia, V. Peskov, P. Fonte, *Resistive Gaseous Detectors* (Wiley, New York, 2018)

J. Badziak, Laser-driven ion acceleration: methods, challenges and prospects. J. Phys. Conf. Ser. **959**, S. 012001 (2018). dx.doi.org/https://doi.org/10.1088/1742-6596/959/1/012001

C. Braccoal., Beam studies and experimental facility for the AWAKE experimentat CERN. Nucl. Instrum. Methods Phys.Res. A **740**, 48–53 (2014)

T. Böhlen, F. Cerutti, M. Chin, A. Fassò, A. Ferrari, P. Ortega, V. Vlachoudis, The FLUKA code: developments and challenges for high energy and medical applications. Nucl. Data Sheets **120**, 211–214 (2014)

CERN, The official CERN FLUKA website. Von (2020). https://fluka.cern/abgerufen

T. Esirkepov, M. Yamagiwa, T. Tajima, Phys. Rev. Lett. **96**, 105001 (2006)

R. Faw, J. Shultis, *Radiological Assessment: Sources and Doses* (Amer Nuclear Society, 1999)

P. Forck, D. Liakin, P. Kowina, Beam position monitors (2009). cds.cern.ch, S. cds.cern.ch/record/1213277/files/p187.pdf.

GEANT Collaboration, *GEANT4: A simulation toolkit*. Von (2020). https://geant4.web.cern.ch/abg erufen

F. Hinterberger, *Physik Der Teilchenbeschleuniger Und Ionenoptik* (Springer, Berlin, 2008).

ICRP, Basic anatomical and physiological data for use in radiological protection reference values. ICRP Publication 89. Ann. ICRP **32** (3–4) (2002)

ICRP, The 2007 recommendations of the International Commission on radiological protection. ICRP Publication 103. Ann. ICRP 37 (2-4) (2007)

International Commission on Radiological Protection, The 2007 recommendations of the international commission on radiological protection. Ann. ICRP, S. (2007). ISBN 978-0-7020-3048-2

K.L. Jensen, P.G. O'Shea, D.W. Feldman, J.L. Shaw, Emittance of a field emission electron source. J. Appl. Phys. **107**, 014903 (2010)

G.F. Knoll, *Radiation Detection and Measurement* (John Wiley, New York, 2010).

A. Koning, et al., TENDL-2015: TALYS-based evaluated nuclear data library. Von (2015). https://tendl.web.psi.ch/tendl_2015/tendl2015.htmlabgerufen

T. Kuroda, Development of high current negative ion source. Fusion Eng. Des. **36**(1), 143–156 (1997)

C. Lejeune, Theoretical and experimental study of the duoplasmatron ion source. Nucl. Instrum. Methods **116**, 417–428 (1974)

Los Alamos National Laboratory, *MCNP Homepage*. Von (2019). https://laws.lanl.gov/vhosts/mcnp.lanl.gov/index.shtmlabgerufen

G. Lutz, *Semiconductor Radiation Detectors: Device Physics* (Springer, Berlin, 2008). ISBN-13: 978-3540716785

A. Macchi, A review of laser-plasma ion acceleration (2017). https://arxiv.org/abs/1712.06443

M. Mayer, S. Möller, M. Rubel, A. Widdowson, et al., Ion beam analysis of fusion plasma-facing materials and components: facilities and research challenges. Nucl. Fus. **60**(2), S. 025001 (2019); doi.org/https://doi.org/10.1088/1741-4326/ab5817

E. Mauerhofer, U. Rücker, T. Cronert, P. Zakalek, J. Baggemann, P. Doege, T. Brückel, *Conceptual Design Report NOVA ERA (Neutrons Obtained via Accelerator for Education and Research Activities) A Jülich High Brilliance Neutron Source Project* (Forschungszentrum Jülich GmbH, Jülich, 2017).

S. Möller, O. Kachko, M. Rasinski, A. Kreter, C. Linsmeier, In situ investigation of helium fuzz growth on tungsten in relation to ion flux, fluence, surface temperature and ion energy using infrared imaging in PSI-2. Physica Scripta T170, S. (2017). https://doi.org/10.1088/1402-4896/aa8a0a

M. Nastasi, J.W. Mayer, Y. Wang, *Ion Beam Analysis: Fundamentals and Applications* (CRC Press, 2014)

I. Obodovskiy, *Radiation: Fundamentals, Applications, Risks, and Safety* (Elsevier, 2019)

OECD Nuclear Energy Agency (NEA), JANIS. Von (2017). https://www.oecd-nea.org/janis/abgerufen

V.K. Olsen, E. Adli, P. Muggli, J.M. Vieira, Loading of wakefields in a plasma accelerator section-driven by a self-modulated proton beam, in *Proceedings of NAPAC 2016* (2016), S. ISBN 978-3-95450-180-9

J. Orloff, Handbook of Charged Particle Optics (Ttaylor & Francis, 2008)

Pfeiffer Vacuum GmbH, *The Vacuum Technology Book*, vol. 2 (Pfeiffer Vacuum GmbH, Asslar, 2013).

A. Popp, Dynamics of electron acceleration in laser-driven wakefields: Acceleration limits and asymmetric plasma waves. Fakultät für Physik der Ludwig-Maximilians-Universität in München, S. Ph.D. Thesis (2011)

A. Piel, *Plasma Physics: An Introduction to Laboratory, Space, and Fusion Plasmas* (Springer, Berlin, 2010)

G.W. Quax, A. Gottdang, D.J. Mous, A high-current light-ion injector for tandem accelerators. Rev. Sci. Instrum. **81**, S. 02A701 (2010). https://doi.org/10.1063/1.3259235

D. Reiter, *Hydkin - Reaction kinetics analysis online for Hydrocarbon catabolism in hydrogen plasmas*. Abgerufen am 23. 06 2018 von (2019). https://www.hydkin.de/plotform/index_hydhel.html

I.G. Segev, E. Yahel, I. Silverman G. M., Blister formation at subcritical doses in tungsten irradiated by MeV protons. J. Nucl. Mater. **496**, 77–84 (2017). https://doi.org/10.1016/j.jnucmat.2017.09.024

N. Tsoulfanidis, S. Landsberger, *Measurement and Detection of Radiation* (CRC Press, 2015)

UK Atomic Energy Authority, *FISPACT-II*. Von (12 2018). https://fispact.ukaea.uk/abgerufen

D.B. Williams, C.B. Carter, *Transmission Electron Microscopy: A Textbook for Materials Science* (Springer, Berlin, 2009)

H. Zhang, *Ion Sources* (Springer, Berlin, 1999).

Chapter 3
Interaction of Particle Beams and Matter

Abstract This chapter teaches the basic understanding of the beam-matter interaction physics of photons, electrons, and ions relevant for accelerator applications. The two main parts of energy-loss and nuclear interactions will be discussed on the basis of examples and practically relevant quantities. The mathematics of depth-dependent reactions combines these two aspects resulting in a 1D model. Depth dependent reactions explain about 98% of the interaction physics of the typical fixed thick target geometry of applications resulting in reaction probabilities and equations for practical efficiency optimisation and device layout. A few examples of established codes and practical implementation of the knowledge concludes the chapter.

This section will discuss the physical basics for understanding accelerator applications. By far most of them rely on the interaction of the accelerated beam with some kind of target, may this be a sample, a production target or a human being. The four fundamental forces of physics represent the basis of all beam-matter interactions. From daily life we know gravity, a force extremely weak when normalized to the number of particles required per unit strength. Its range is large, but its small strength makes it negligible in accelerator applications. The electro-magnetic force has a similar range, but a significantly stronger effect than gravity. It keeps the world together by using photons to let positive and negative charges interact. In contrast to gravity it has two different charge polarities which we call plus and minus, leading to a possibility of shielding it via neutralisation of opposing fields. We already learned about its importance for accelerator applications with respect to the electro-magnetic technology required, but it also results many interactions of charged particles with targets such as stopping or elastic scattering. The strongest force, the strong force, also features the lowest range. It holds the nucleus together, but its range restricts to nuclear dimensions. It requires six charge types/flavours usually named red, green, blue, and their anti-counterparts and is mediated by gluons. This force draws responsible for inelastic scattering in the form of fission and fusion reactions, allowing us to change the atomic nucleus, if we are able to bring two particles into their strong force range. Its strength draws responsible for the large specific energy content of nuclear fuels, exceeding the one of chemical (= electro-magnetically bound) fuels by about 10^6. Lastly, the so-called weak force is responsible for most radioactive decays and

S. Möller, *Accelerator Technology*, Particle Acceleration and Detection,
https://doi.org/10.1007/978-3-030-62308-1_3

neutrino interactions. Its mediators, the W and Z bosons are heavy, leading to a low range in the order of the atomic nucleus. Its weakness manifests in the extremely low interaction probability of neutrinos, which only interact via the weak force with matter (see later in Sect. 4.4).

In contrast to the typical situation in fundamental particle physics, where two beams counter-propagate and collide with each other in an interaction zone (fixed centre-of-mass), application targets consists of normal stationary matter (fixed target). Furthermore, the target comprises several different species, at least electrons and a set of more or less abundant elements (desired and impurities) leading to complex interactions. Besides this, also the kinematics and the chain of events of the interaction differ between colliding beam and fixed target situations.

The Rutherford experiment marks the original fixed target experiment, featuring already many of the physical aspects still relevant in accelerator applications today. Rutherford wanted to understand the nature of the atomic nucleus and its charge distribution. At that time, over 100 years ago, it was not quite clear whether the nucleus is a compact object or a cloud of positive charges mixed with a cloud of negative electron charges in the atom, since both would result in a neutral atom as seen from the outside. Rutherford wanted to falsify one of the models by a scattering experiment of an ion beam with a fixed target. Firstly, he calculated the kinematics and scattering probabilities (cross-sections) for each situation, yielding the famous Rutherford formula (3.1). With this knowledge, he designed an experiment for shooting some MeV α-particles (doubly ionised helium: $^4He^{++}$) from a nuclear decay through a thin gold foil with particle detectors around it. The experiment yielded an angular probability distribution of the scattered particles according to the compact nucleus model represented by his formula, constituting the current understanding of the atomic structure. In his experiment, Rutherford was lucky, because with the technology of his time he was already able to produce a foil thin enough (some 100 nm) to be passed by 5 MeV alpha particles, which was the maximum he was able to provide. We will see in the following sections that α-particles of that energy not even pass 10 μm of gold (assuming surrounding UHV) and hence the technological aspects were a key parameter in the success of Rutherford's ground-breaking experiment.

$$\frac{d\sigma_R}{d\Omega}(E) = \left(\frac{Z_1 Z_2 e^2}{4E_0}\right)^2 * \frac{1}{\sin^4(\theta/2)} \tag{3.1}$$

Each interaction has two sides, on the one hand its probability of occurrence called the cross-section (which was Rutherford's testing ground) and on the other hand the collision process of each individual instance of this reaction called the kinematics. To change from the view of the probability/cross-sections to the process/kinematics, we first define the situation. We call the accelerated particle shot into the situation the projectile. Definitely the projectile starts the situation, since it is initially the only moving particle (remember we are talking about fixed target situations). Please note this view is from the laboratory system, in the centre-of-mass system all particles are moving also in a fixed target reaction. The projectile of mass m_1 hits the target of

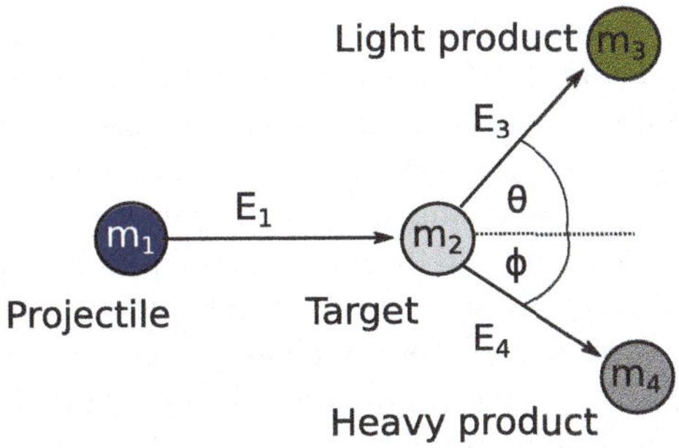

Fig. 3.1 Kinematics of a 2-body reaction with a stationary target in the laboratory frame (E_2= 0) has 9 parameters. The energies and angles of all particles have unique relations defined by mathematics. In contrast, a reaction with more products (*n*-body reaction) also conserves energy and momentum, but allows for more than one solution as will be explained later

mass m_2 with kinetic energy E_1. Both particles scatter and emit a light product m_3 and a heavy product m_4, each having a certain emission angle (Θ and ϕ) against the initial vector of the projectile. In 3D Θ and ϕ will describe circles when seen from the projectile movement direction due to the rotational symmetry of the process. The whole situation is depicted in Fig. 3.1.

This so-called two-body reaction is the standard situation we have to consider in accelerator applications. Single reactions with more than two input particles require extreme densities. Those conditions are technologically so far inaccessible or of minor importance and will not be considered in this edition (maybe in future ones). Reactions with more than two output particles commonly appear in nuclear decays, in particular β-decays feature 3 output particles (electrons for β^-, positrons for β^+, neutrinos and a heavy nucleus, see Chap. 5), and nuclear reactions can feature 3 (E_1 $>\approx$ 10 MeV) or more products with increasing projectile energy. These reactions add the complication of interconnected spectra for all outgoing particle properties, in contrast to the kinematics of the two-body reaction featuring only a single solution at each product angle. In any case, the whole situation has to fulfil momentum and energy conservation, which allows calculating the respective product parameters with the knowledge about the four masses and, in the case of two-body reactions, any four other parameters of the situation (leaving 1 unknown + 1 equation = unambiguous solution), see Sect. 3.3.2. The mathematical flexibility implied by these equations forms an important aspect of our physical understanding and also technological exploitation of beam-matter interactions. As we will see in the course of this chapter, everything interacts with everything, even with the vacuum, but the mathematical formulations allows us tailoring and identifying the reactions. This additional information compensates for the lack of information provided by detectors (Sect. 2.5). The same equations apply for material analysis, isotope production, or patient treatment, just with different unknowns in the equation system.

The interaction of beams and matter covers a wide range of specific physics. Not all of them are fully or even partially understood. In view of applications we divide the level of knowledge into three categories: Theoretical, semi-empirical, and empirical understanding. Full theoretical models requiring no external input, except for fundamental constants, so-called ab initio models, are the highest level of understanding. Think of a treasure quest. Theoretical understanding equals a situation where you have a full map containing all the information on what the treasure looks like, how much gold it contains, and the mm resolved path this would allow walking to the treasure blind with only your feet (or technology) limiting the amount of success. If you know there is a treasure somewhere, but you only have a plain path drawn on a handkerchief, without coordinates, scales, or the like you have a semi-empirical understanding. It tells you which turns to take and the dangers lurking on your path, but you do not know where to start or how long the way will be and which dangers wait on your path. The same applies for beam matter interaction. Some cross-sections, such as Rutherford's, were understood to their fundamental physics and a theory was found accurately describing them. Others have been investigated deeply and mathematical relations were found empirically, but certain constants, factors, or limiting cases cannot be covered by existing semi-empirical models. A few cases, such as the radioactive decay, were broadly investigated experimentally, but due to the lack of understanding no type of extra- or interpolation of data is possible. In this lowest level of understanding we only have an empirical qualitative estimate of the order of magnitude and the influence factors of the process, but we do not even know if this covers the full space of possible pathways of the process.

Rutherford understood the nature of the target structure in his gold foil experiment by the match of the cross-section calculated from his hard sphere model and the agreements with the experimental results. Interestingly, most of the α-particles actually passed the gold foil in Rutherford's experiment undisturbed. We can quantify the interaction probability w using (3.2) and the gold atomic density ρ by multiplying with the Rutherford cross-section σ and the foils thickness d.

$$w = \sigma(E) * \rho * d \tag{3.2}$$

The interaction probability remains in the percent range, even if we infinitely increase the gold foil thickness beyond Rutherford's thin foil. From common sense, but also from a mathematical limit consideration, it becomes obvious we didn't completely understand the situation: An infinite target thickness should yield a 100% reaction probability for each projectile, otherwise they would fly through the target, a situation empirically non-existent. By increasing the foil thickness beyond 10 μm we would come to know that the α-particles will already get stuck in the foil way before Rutherford's reaction probability even has a chance reaching 100%. Therefore, at least a second mechanism stopping the α-particles has to exist besides Rutherford-scattering. The question arises which interactions were missing in Rutherford's description, since a particle beam will not stop by itself, just like a spaceship will not stop by itself in the vacuum of the universe.

So far we skipped considering 98.7% of the particles in the gold foil, the electrons attached to the gold nuclei forming the atom. Gold has 79 times more electrons than nuclei (= nuclear charge Z), which can undergo the same 2-body reactions with the

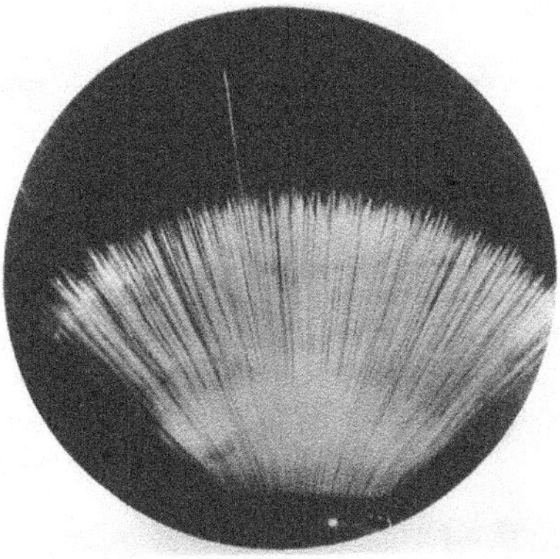

Fig. 3.2 An α-particle source (≈ 5 MeV α) placed at the bottom emits α's into a cloud chamber. The bright tracks indicate individual particle tracks. The particle range in the cloud chamber's alcohol mixture slightly differs for each particle due to statistical effects of the stopping, even a 50% longer spike is present. Reprinted from physicsopenlab.org CC-BY 4.0 license

projectiles as the nucleus, although with different kinematic parameters. Everything relies on the ratios of the interaction cross-sections and the results of the interaction. There are actually significant amounts of interactions with the electrons contained in the gold foil, but due to the strong mass difference of electrons and α-particles the energy transfer remains small in each interaction. In a reasonable approximation, the electrons act like an aether continuously slowing down the α-particles or, in general, charged particles, passing through them. Imagine it like walking through IKEA's Småland ball pool with the force required to push away the balls from your way slowing you down. The ratio of ball to human mass even approximately resembles the electron to α-particle mass.

Rutherford's experiment was designed in a way to minimize this slowing effect by staying in the limit of a thin target. Upon increasing the foil thickness we slowly leave the *thin target* regime and the energy-loss of the α-particles becomes visible/measureable. At a gold foil thickness of about 10 μm all α-particles will stop inside the foil and Rutherfords experiment would not yield any measurable quantity in the forward direction. In the backward direction the situation will also change, as the reactions from different depth will add up. The thickness related to this so-called *thick target* limit strongly depends on the beam and target properties. The situation gets nicely visualized in a cloud chamber in Fig. 3.2 with no α-particle reaching the top end of the cloud chamber.

Particle beams do not see distance when passing through matter. Of course, for the beam optical aspects of divergence and direction remain relevant as demonstrated in Fig. 3.2, but here we focus on the beam-matter interaction since the distances are relatively short (e.g. the 10 μm foil). In order to understand the way particle beams see matter let us consider the following three situations: A beam gets fired onto a solid metal, the same metal but as a metal foam with vacuum in the pores, and, last but not least, the same metal foam but with air inside the pores. The situations are

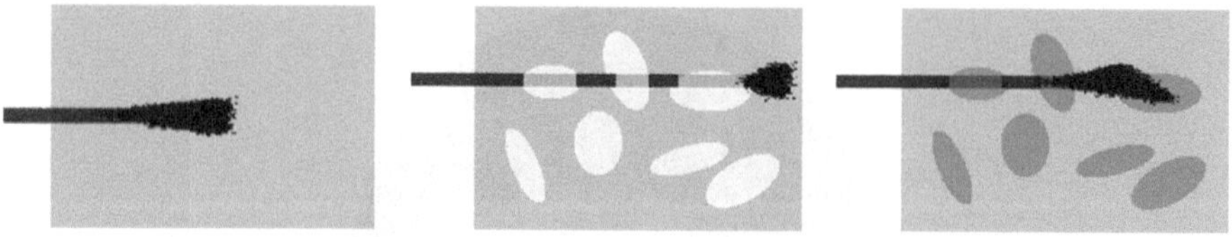

Fig. 3.3 Illustration of a beam (black line) in a solid (grey), a porous solid with vacuum inside the pores, and one with gas inside the pores. At the end of its range, the beam spreads out due to the statistical nature of scattering. In its interactions, a beam only sees matter, not distances. In vacuum, the beam travels undisturbed, like a space ship. In matter, the intensity of interaction depends on the matter density, hence gas volumes show lower interaction rates than solids

depicted in Fig. 3.3. In the first case, the beam travels a depth X into the material until its energy is dissipated. In the porous material the beam will lose energy when passing the metal, but in the pore's vacuum no energy is lost, hence the total range extends by the porosity aspect. In the third case, the pores also contain matter (gas), but the relatively low density of gas yields only a very small influence. The range in the pore exceeds the range in the metal, but gas still contributes to the energy-loss. Section 3.2 discusses the quantification of the beam stopping and its mathematical treatment.

Usually the projectile (photon, electrons, ions) beams density is insufficient for simultaneous reactions of several projectiles with single targets and reactions between projectiles are negligible due to the low relative speeds (= low emittance). Also target densities of normal matter are too low for reactions with multiple targets. This allows treating each beam particle individually in the so-called binary collision approximation (BCA). Therefore, the interaction of a beam with a target equals the sum over all the independent individual reactions. The BCA constitutes an important basis for our understanding and quantitative computer modelling of beam-matter interaction. This may sound trivial, but the exchangeability between the individual particle and the ensemble (beam) picture will become an important tool for understanding and mathematical treatment of beam-matter interactions.

In order to understand and work with something you have to give it a name. In addition to these energy-transfer reactions nuclear reactions become possible at higher projectile energies. Naming of nuclear reactions follows international conventions. The naming needs to include the target, projectile, and products. The reaction of a ^{12}C target with a ^{3}He projectile resulting in a proton and a ^{14}N product reads for example ^{12}C(^{3}He, p)^{14}N. For describing a class of reactions or shortening the naming, the same reaction could also be named (^{3}He, p) corresponding to a naming scheme (projectile, light product). (p,n) describes a typical reaction with two products, a 2-body reaction. Correspondingly (p,2n) and (p,n+^{4}He) describe reactions with three products, (p,3n) with four products and so on. If we want to describe a class of reactions, leading for example to the same element, we can introduce a variable x in the form (p,xn) with $x = 1$ to infinity to discuss reactions producing only neutrons.

3.1 Absorption and Reactions of Photons

We first take a step back away from the massive (mass > 0) charged particles produced in accelerators in order to improve our understanding of the interaction of particle beams with matter. Mass-free (or more correctly rest-mass-free) particles like photons necessarily have to travel with the speed of light, hence they cannot be slowed down. Logically for these particles, an energy-loss mechanism by friction is not possible, but energy can only be transferred via a reduction in quantity or intensity, respectively. This directly leads to a differential equation with an exponential decay solution of the photon beam intensity I in depending on the distance d passed in matter, equivalent to a constant absorption probability for each individual photon per passed matter particle.

$$\frac{I}{I_0} = e^{-d\mu} \tag{3.3}$$

Due to the lack of the friction mechanism, photons typically achieve the longest attenuation length μ and hence range (distance d) for a given kinetic energy of the considered species (e^-, ions, neutrons). The attenuation length increases with increasing photon energy. We already saw the technical effect of this fundamental physics in the detector Sect. 2.5 in Fig. 2.44 with the required detector thicknesses being largest for photon absorption. Also in radiation protection, Sect. 2.7.3, this and the exponential decay law lead to thick shielding requirements for photons.

Having said photon beams only lose energy by a reduction of intensity is actually not entirely accurate. For lower energies, scattering dominates the interaction of photons and matter. Scattered photons are absorbed and instantly reemitted in a different direction and hence cannot be considered as the same particle or part of the same beam population. A set of mechanisms exists for the interaction of photons with electrons. Scattering processes (approximately) conserving the photon energy dominate the photon matter interaction for lower energies up to the binding energies of electrons to atoms (e.g. 13.6 eV for H). This class of elastic process retains coherence (= phase relation) with the original beam. A prominent example among this is the Rayleigh scattering which leads to the blue sky, since the processes cross-section is inversely proportional to the fourth power of the wavelength, scattering more blue than red or green light in the observer's direction. With increasing photon energies or shorter wavelength, respectively, the elastic scattering energy transfer increases. With sufficient energy transfer, remember our scattering partner binds to atoms, the electrons gain enough energy to leave their binding state. Starting with this energy, the elastic scattering becomes incoherent, since the electron receives part of the energy. This so-called Compton scattering follows a probability distribution given by a cross-section called the Klein-Nishina formula, (3.4).

$$\frac{\mathrm{d}\sigma}{\mathrm{d}\Omega} = \frac{1}{2}\left(\frac{Z^2 e^2}{4\pi m}\frac{E'}{E}\right)^2 \left[\frac{E'}{E} + \frac{E}{E'} - \sin^2(\theta)\right] \tag{3.4}$$

This differential cross-section describes photon scattering from free resting point charges (electrons or ions), a situation only approximately true for electrons in atoms, with mass m, charge Ze, and the scattering angle Θ between incoming (energy E) and outgoing photon (energy E'). This formula only approximately describes the situation, but allows understanding the basic trends with an analytical description. The scattering cross-section decreases with the photon energy, hence Compton scattering becomes less efficient for high energy photons. The cross-section also decreases with increasing energy transfer and scattering angle. For photon energies small compared to the electron rest-mass $m_e c^2$ only negligible energy transfer occurs; the low energy limit of Compton scattering yields the coherent elastic scattering process discussed above. In all cases, (3.4) leads to a continuum of scattered photon energies, similar to Bremsstrahlung where electrons penetrate matter.

In parallel to the incoherent scattering Einstein's photoelectric-effect, the ionisation of atoms or the freeing of bound electrons, respectively, takes place. Equation (3.5) describes the cross-section of this inelastic process. It requires photon energies above the binding energy of the electrons to their atoms. Energy in excess of the binding energy will end up as kinetic energy of the released electron. All atoms above hydrogen (H) feature several electrons, each with different binding energies. The ionisation of the electrons from the innermost shell, called the K-shell, requires the highest energy in the order of a few 10 keV for heavy elements. The higher a binding level in the atomic shell, the lower its binding energy due to the core charge shielding of the inner electrons. Each binding state represents an independent instance of the photoelectric-effect, leading to so-called absorption edges at the given binding energies.

$$\frac{d\sigma}{d\Omega} = \text{Constant} * Z^5 * E^{-3.5} \tag{3.5}$$

The free spot of the released electron will quickly be reoccupied by another electron. The involved binding energy remains the same, but now has to be released in the form of a photon. In particular for the inner binding shells, also bound electrons from other higher shells can reoccupy the free position. These inner conversions emit photons with an energy given by the difference between initial and final binding state. A table of possible conversions arises, from which the innermost shells (K and L) are given in Fig. 3.4. The absorption of photons with an energy equal to the binding energy or higher can only lead to a complete release of the electron. The process can be triggered not only by photons, but also by charged particles as we will see later.

At the highest photon energies, new inelastic scattering processes add up to the interaction list. Up to here all photon interactions involved scattering with more or less free electrons. The quantum nature of the bindings implies certain specific energy limits. With photon energies above 1022 keV a new interaction process with the nucleus becomes possible. This interaction converts the photon energy to matter/mass. All physical processes have to conserve energy, momentum and quantum numbers. Consequently, for producing massive particles only matter anti-matter pairs can be produced. With a rest-mass of 511 keV/c^2, 1022 keV and more

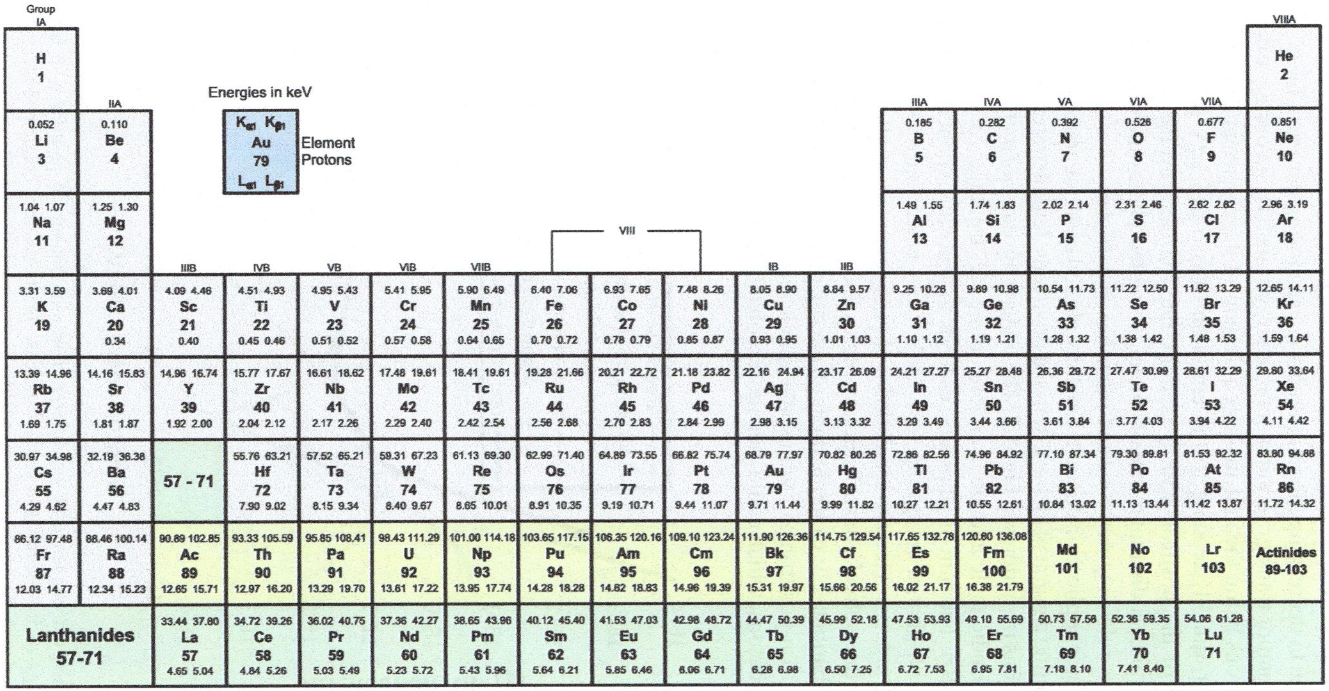

Fig. 3.4 Chart of the first four X-ray absorption/emission energies along the periodic table of elements. Generally emission and absorption of photons are connected in physics, hence all of these energies relate to absorption edge for photons passing through the elements and also emission lines when exciting the atoms via particle beams

allows for the production of electron-positron pairs. The process requires a nucleus to balance the momentum. Balancing the momentum with an electron requires slightly higher energy due to decreasing momentum per energy for lighter particles (= same amount of momentum transfer requires more energy transfer), favouring a nucleus as partner. Positrons being the anti-matter equivalent of the electron cannot survive in normal matter, quickly leading to the emission of two 511 keV photons from the annihilation of the positron with a random electron (the reverse process). At even higher energies, a disintegration of nuclei, so-called photo-nuclear reactions, becomes possible, releasing neutrons and other heavy particles.

In the accelerator application context, typically photon energies are between 1 keV and 10 MeV. Plotting a graph of all the discussed processes demonstrates the diversity of effects of photons in matter, see Fig. 3.5. The underlying data are well documented for the whole periodic table and are available online e.g. from the XCOM database (Berger et al. 2010). The sum of all processes leads to a mostly exponential decrease of attenuation with photon energy up to 1 MeV, from where on it stays constant. We have seen all processes tend to break down photon energy to smaller and smaller chunks. Some of these chunks have discrete energies due to quantum effects and some continuum distributions. For example the high energy processes convert a 2000 keV photon to an electron-positron pair with some kinetic energy. The particles annihilate to two 511 keV photons, which could then most probably Compton scatter to photons

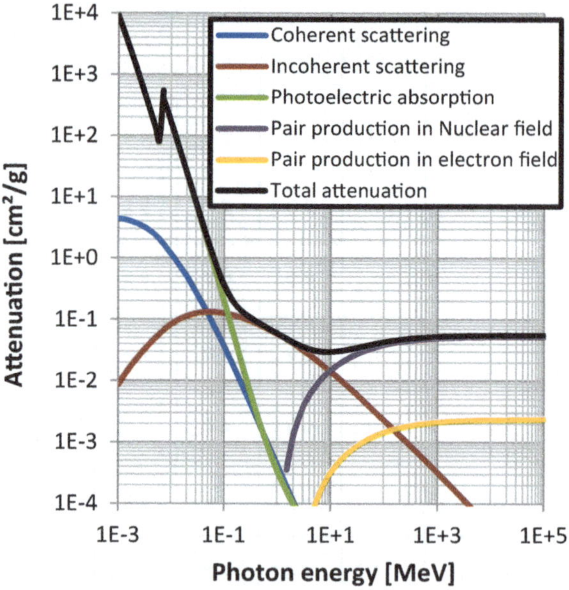

Fig. 3.5 Photon energy dependence of the different reaction channels of photons with iron. At 7.06 keV the K edge of iron leads to a sudden increase in attenuation, compare Fig. 3.4. The attenuation multiplied with the material density yields the exponential fall-off length μ, (3.3). Data from NIST XCOM Database (Berger et al. 2010)

and electrons of even lower energy, which then induce photoelectric electrons and photons, and so on. This kind of salami tactics consumes higher energy photons in a chain of events with many secondary particles involved finally ending up at particles with negligible energy.

Accelerators cannot directly influence or produce photons, but they arise from several processes of accelerated charged particles as secondary particles. We already came to know the Bremsstrahlung and synchrotron radiation, which describe spectra of photons produced by deceleration of charged particles, e.g. when passing matter. The decay of radioactive inventory and numerous nuclear reactions represents the second relevant source of high energy photons. These photons originate from the atomic nucleus and were given the name gamma-ray (γ) in contrast to X-rays originating from the atomic electron shell. Therefore, in the accelerator context photons potentially represent a problem, since they contribute to the radiation dose rates, but originate from fundamental processes. On the other hand many applications rely on using the production of photons. For example, the interaction of charged particles with bound electrons produces known photon energies, Fig. 3.4, allowing for elemental identification.

3.2 Range and Stopping of Charged Particles

Beams of massive charged particles (electrons, ions) follow different physics than photons, as their number is conserved by fundamental laws (like the conservation of velocity for photons), but their velocity is variable. For deep insight into the physics and mathematics of particle beam stopping and interaction with matter the reader is referred in particular to the book (Sigmund 2006) and also the accompanying book to the famous ion stopping software SRIM (Stopping and Range of Ions in Matter) which includes many examples and numbers of ion stopping (Ziegler et al. 2008).

Stopping of charged particles predominantly arises from the interaction of the beam with the electrons in matter (IKEA's Småland ball pool friction effect), similar to photons. The terms stopping, stopping power, specific energy loss, dE/dx, and friction are used more or less as synonyms describing an effect of an energy loss per length (e.g. MeV/mm), normalised to mass density (e.g. MeV cm^2/g), or per passed atoms [e.g. keV/(10^{15} atoms/cm^2)]. In fact, since the collisions with electrons induce stopping, distances cannot play a role for energy loss (as discussed in Fig. 3.3), making the energy lost per passed atom/area the most fundamental quantity. The Bethe-Bloch formula (3.6) for stopping power S in energy E lost per travelled length x describes this for ions in matter. Stopping depends on ion velocity not energy, but energy is usually the interesting quantity for other aspects of beam-matter interaction.

$$S_B(E) \equiv \frac{\mathrm{d}E}{\mathrm{d}x} = -\frac{n_e z^2 e^4}{4\pi m_e v^2 \epsilon_0^2}\left(\ln\left(\frac{2m_e v^2}{I\left(1-\frac{v^2}{c^2}\right)}\right) - \frac{v^2}{c^2}\right) \tag{3.6}$$

With electron density n_e of the material, projectile elemental charges z (e.g. $z = 1$ for electrons and hydrogen ions), projectile velocity v, a mean excitation energy of the target material $I(\approx 10\,\text{eV})$, and m_e, ε_0, c, and e fundamental constants. At energies below a few 100 keV/amu (atomic mass unit) ions additionally lose relevant amounts of energy due to collisions with nuclei (the billiard table effect). This nuclear aspect of stopping transfers energy to the target nuclei leading to cascades where individual particles hit nuclei and transfer enough energy for the hit nucleus to hit further nuclei, see Sect. 7.4.

Electrons also lose energy due to collisions with target electrons, but with their correspondingly higher velocity at a given energy the relevance of several effects changes. Relativistic effects and Bremsstrahlung reach a relevant level at significantly lower energy compared to ions. In particular ions have to be beyond our considered energy range of 250 MeV for this, while electrons require only about 1 MeV. The Berger-Seltzer-Formula describes the stopping power of electrons consisting of the collisional and the Bremsstrahlung part. Bremsstrahlung dominates the energy-loss of electrons for example above 10 MeV for Pb or 400 MeV for H targets:

$$S_E(x) = \frac{e^4 n_e}{8\pi \epsilon_0^2 m_e c^2} * \frac{1}{1 - \frac{1}{\gamma^2}} * \left[\ln\left(\frac{2(\gamma + 1)}{(I m_e c^2)^2} \right) + F \right] \tag{3.7}$$

Equation (3.7) follows a different trend with similar input parameters and the Lorentz factor γ as a measure for the electron energy. The function F adds a term which depends on the target material properties and the electron velocity, typically small in the energy range considered here. Figure 3.6 compares the total stopping powers of electrons and ions. The behaviour for different ions is similar with a clear maximum in stopping power around a few 100 keV/amu and a minimum in the GeV range. Electrons in contrast have a minimum at around 1 MeV with higher values for

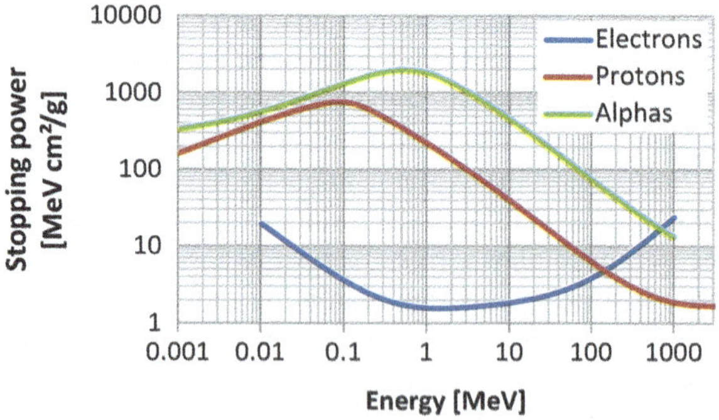

Fig. 3.6 Stopping power of charge particles in carbon. The stopping per length is calculated by multiplication of the values with the material density. Up to about 100 MeV the stopping power of electrons is about 100 times smaller compared to ions. Furthermore electrons show a different behavior, while different ions have similar but shifted stopping functions. Data from National Institute of Standards and Technology (2019)

small and large energies. In all cases, we see the graphs do not strictly follow (3.6) and (3.7) in particular in the low energy half. The equations describe an important part of physics, but additional higher order effects add up to the collisional stopping in the high and low energy limits.

Unfortunately, physics becomes very complicated in particular at lower energies. The electrons in the solid also move with the so-called Fermi velocity, changing their stopping effect if the projectiles have comparable velocity. Electrons on the other hand suffer additional energy losses by Bremsstrahlung emitted due to their strong deceleration in matter above a few MeV. We will not discuss the details of these higher order corrections in detail, but refer the reader to the given literature mentioned earlier. To summarise the findings: In particular at low energies stopping cannot be completely described analytically, but rather semi-empirically with fits to experimental data. This implies a limited accuracy in these energy ranges, compared to a full theoretical understanding. Uncertainties of the best known stopping powers range up to 6% (Ziegler et al. 2008), with the accuracy of full stopping models improving towards 1–2% at a few MeV/amu (the sweet spot).

For calculating the stopping S_{Mix} of elemental mixtures such as stainless steel or human skin, Bragg's rule applies. Within this rule, the total stopping power is given by the atomic fraction ρ_i of the weighted sum over the individual stopping powers S_i of each pure element, (3.8). Deviations from Bragg's rule occur in materials with strong chemical interactions between their constituents, in particular with light elements. While deviations <2% occur for metals and heavy elements, a value of 6% was found e.g. for SiO_2 and H_2O and values up to 20% for other special cases. Due to the practically infinite number of compounds, only a few common compounds were experimentally investigated for the correction factor. Some of them are available in the SRIM code (Ziegler et al. 2008).

$$S_{\mathrm{Mix}} = \sum S_i * \rho_i \tag{3.8}$$

The stopping power translates to a particle range. After this range the projectile has lost its energy and neutralizes with the target. In particular with ion projectiles shot into solids, this process is called implantation. The variations in stopping power lead to a typical energy deposition curve for ions. This so-called Bragg-curve, shown exemplary in Fig. 3.7, not only tells us the average depth of implantation of ions, but also demonstrates the strongly inhomogeneous deposition of collisional damage and beam power with its peak. This inhomogeneity increases with ion beam energy for energies above the stopping power maximum due to the monotonous decrease of stopping power with increasing energy, see Fig. 3.6. For technical materials the Bragg peak induces thermo-mechanical problems due to the combination of collisional damage, maximum power deposition, and implantation of ions. The particle range x calculates by integrating the stopping power S of the target of density ρ_S from the primary beam energy E_0 (= projectile energy E_1) till zero beam energy according to (3.9).

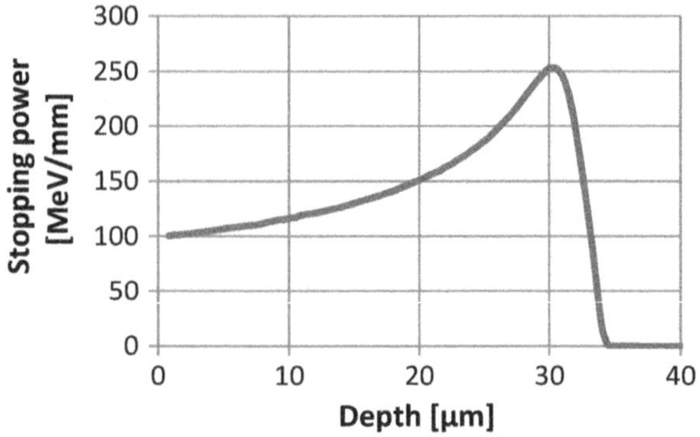

Fig. 3.7 The energy-loss curve of 5 MeV α-particles penetrating an elemental mixture representative for human skin (mostly H, C, O with $\rho_S = 1$ g/cm^3), a case discussed in 2.7. The particles feature an average range of 33.5 µm with a Bragg-Peak (maximum energy-loss) at 30 µm depth. At the right end of the peak the ions fully stop (= implantation)

$$x = \frac{1}{\rho_S} \int_0^{E_0} \frac{1}{S(E)} \mathrm{d}E \Rightarrow \mathrm{d}x = \frac{\mathrm{d}E}{\rho_S * S(E)}, \tag{3.9}$$

The underlying effects of stopping involve the collision between the projectile and single particles, leading to a statistical fluctuation of individual energy loss events. Figure 3.2 showed such an example and its result on the range. This statistical broadening of range and energy-loss is called straggling. Straggling broadens the particle distribution in energy and space, resulting in a beam energy distribution broadening at a fixed depth and a range broadening, respectively. The collisional nature of stopping induces not only a longitudinal variation of energy and range, but also a transversal component broadening the beam diameter. These statistical effects mostly follow a normal distribution (Fig. 2.22). The particles above and below the maximum see different $S(E)$. Due to the variations of S with energy the energy distribution becomes skewed. Figure 3.8 depicts the impact of straggling in an exemplary case at different depth. Practically effects of roughness and thickness tolerances over the finite beam spot sizes add up to the physical straggling. Due to their mass equalling the collision partner's mass electrons experience orders of magnitude intenser straggling than ions.

In contrast to electrons, ions can change their effective charge during passage of matter by picking up or stripping-off electrons. Equation (3.6) highlights the importance of the charge z for S. The change of charge state of ions mostly leads to fewer electrons attached to the nucleus for energies above some hundred keV. Because of this it is often called stripping (of electrons). The effective charge increases with increasing beam energy. Atoms with many electrons require energies even exceeding our 250 MeV limit for stripping off all electrons. The effect makes the initial projectile charge state practically irrelevant for stopping power. No matter whether a He atom, He^+, or He^{++} projectile, the stopping power remains the same, only the projectile

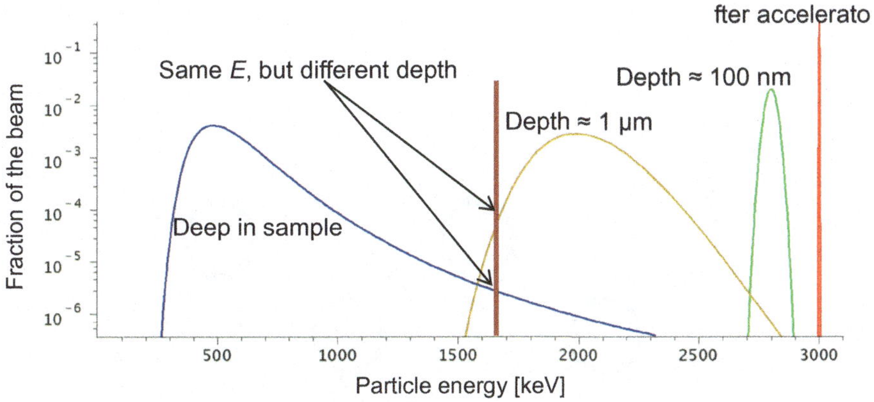

Fig. 3.8 Sketch of the beam energy distributions of 3 MeV ^{3}He ions in carbon at three different depth x. The deeper the particles reach, the broader and more skewed their energy distribution. As a result particles of the same energy can be found in different depth (vertical bar)

energy matters. The statistical nature of collisions also affects the effective charge, resulting in an energy-dependent statistical distribution of the charge states of the individual ions constituting a beam. Besides its impact on stopping power, stripping has a high practical relevance for ion beam generation and control. Examples such as the Tandem accelerator Sect. 2.2.1 and the negative ion extraction in cyclotrons Sect. 2.2.2 rely on stripping.

The effects discussed so far assume an evenly distributed electron cloud inducing the stopping. Gases and liquids fulfil this condition, but solids with crystallographic structure feature different, inhomogeneous electron distributions. Figure 3.9 illustrates such a distribution with several hotspots of electron density n_e and large areas of nearly zero electron occupation probability. The picture strongly depends on the crystallographic axis. Crystals appear as a stack of channels to the projectiles. When hitting a channel, the stopping significantly reduces due to the lower n_e, if the solid is sufficiently ordered and the particles hit in the crystallographic direction (practically that means normal incidence). In polycrystalline materials, the random grain orientation hardly leads to fundamental directions (the crystallographic indices with only ones and zeros) aligned to the surface normal, suppressing the effect. Only single crystals feature channelling. Channel acceptance angles increase with nuclear charges Z_1 and Z_2 of projectile and target, and decrease with projectile energy, since a pure coulombic interaction potential between the positive projectile and target nuclei charges forms the basis of this effect (Nastasi et al. 2014). In practice channelling yields information in crystallographic analysis of silicon wafers by MeV helium ions, achieving maximum acceptance angles in the order of 1° or by analysing the backscattering of focused electron beams from the individual grains in polycrystalline samples, see Sect. 7.1. The contrast derives from the fact that a backscattering into the acceptance angle (see Rutherford cross-section (3.1) has low probability, hence the projectile faces much lower stopping power than the backscattered products detected outside the acceptance angle.

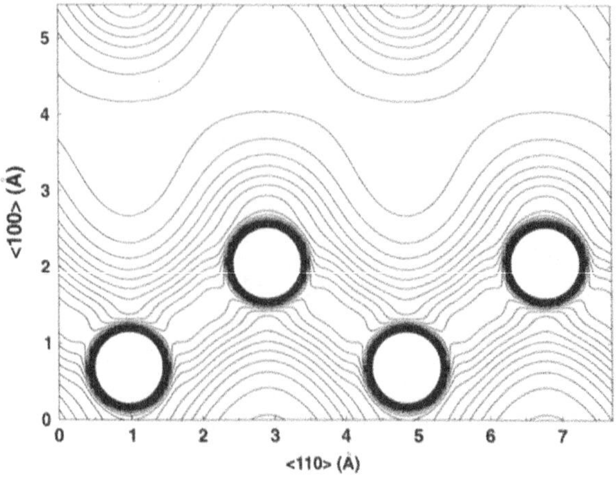

Fig. 3.9 The electron distribution of silicon in a < 110> †plane. The silicon atoms are clearly visible, as is the bond between the nearest neighbours. The contour interval is 0.05 e with contours going from 0.05 to 1.5. The electron density is high near the nuclei and low in between. Reprinted with permission from Elsevier from Sillanpää (2000)

3.3 Nuclear Reactions

– *Stupidity identifies itself by repeating the same action and expecting different results in each instance.*

This sentence, as true it is for real life, completely fails for the sub-atomic level of physics. The basic principle of quantum mechanics is uncertainty. Therefore it is indeed logical to expect different outcomes for every single beam particle impacting onto the same target. This fact should not be over interpreted, though. Summing up the outcomes over numerous interactions will lead to statistically solid probabilities, but it's just probabilities so everything remains possible.

Nuclear reaction cross-sections, or just cross-sections, describe the probability for a certain interaction pathway to happen. Graphically we can imagine them as a target disc which we try to hit with a projectile. In accelerator applications, usually the projectile, as a part of the charged particle beam, moves towards numerous targets atoms which are at rest. Every projectile sees the target discs of the many atoms it approaches, but still most of the space is empty making it improbable to hit a target. As a rule of thumb usually only up to a few percent of the projectiles undergo nuclear reactions. As with stopping nuclear reactions probe matter, independent of its arrangement in the form of density or porosity only the amount of passed atoms/area counts. In fact, generally many different nuclear reactions can take place between given projectile and target, each with its individual cross-section.

In order to describe the cross-section of a specific interaction, the unit barn (= 10^{-28} m^2) was defined. We could think of it as the area of the imaginary target disc of each target particle. Its dimension is extremely small, but of course our projectile will face many targets on its way, since the atomic density of matter is large. The probability P of interaction is given by the cross-section σ times the number of target particles N_T per target area A.

$$P = \sigma * \frac{N_T}{A} \tag{3.10}$$

Equation (3.10) applies independently to every constituent (element and isotope) of the target with their individual cross-section. Different types of reactions can occur with the same type of target. We already came to know the elastic nucleus scattering as relevant contribution to stopping at low energies or the elastic scattering with the electrons as primary contribution to stopping at higher energies. In contrast to elastic scattering, which conserves the total kinetic energy E, nuclear reactions are inelastic. Inelastic reactions enable a transfer between mass and energy, generally not conserving the kinetic energy (but of course the total energy). Due to the equivalence of mass and energy [Einsteins famous (2.6)], the physical law of conservation of energy remains intact. The Q-value expresses this energy redistribution in the quantity of energy transferred in the nuclear reaction (usually in keV). Consequently, elastic scattering reactions feature $Q = 0$ while inelastic reactions identify by $Q > 0$ (mass consumption) or $Q < 0$ (mass generation) defined by (3.11) for a 2-body reaction as depicted in Fig. 3.1.

$$Q = \Delta E = (m_1 + m_2 - m_3 - m_4)c^2 \tag{3.11}$$

Physically, the Q-value results from a difference in the sum of the rest masses, related to different nuclear binding strength, between the particles before and after the nuclear reaction. This so-called mass defect calculates from the difference in the actual nuclear mass from the sum of the masses of an equivalent amount of isolated protons and neutrons. The highest Q-values are found in reactions of light elements with ^{6}Li(^{2}H,^{4}He)^{4}He marking the summit with $Q = 22.38$ MeV. Very thorough studies with sub-keV precise nuclear mass data exist, summarised in the ongoing Atomic Mass Evaluation project(Huang et al. 2017) and made publicly available on numerous websites (e.g. http://oecd-nea.org/dbdata/data/structure.htm or http://nrv.jinr.ru/nrv/webnrv/qcalc). A few examples were compiled in Table 3.1.

The particles produced in a nuclear reaction can enter excited nuclear states, similar to excited atomic states, further reducing Q below the ground state mass difference. The nucleus features a shell structure of protons and neutrons similar to the atomic shell and its excitations as shown for three configurations of nucleon number $A = 14$ (^{14}C,^{14}N,^{14}O) in Fig. 3.10 (left). With enough energy provided by $Q + E$, nuclear reactions can produce ^{14}N in its excited state as the measurements shown on the right demonstrates. Here three different proton energies corresponding to three different Q-values of the ^{12}C(^{3}He,p)^{14}N reaction are detected. Most excited states quickly decay by emission of a photon with the corresponding energy, but even if the de-excitation is energetically possible, it can be hindered by the conservation of angular momentum. Every nuclear level features not only an energy, but also an angular

Table 3.1 Examples of mass defects (= binding energy), first 3 excited states, and coulomb barriers with protons. Protons naturally do not have excited states due to their single nucleon nature

Nucleus	P	4He	12 C	18O	18F	180Ta	181Ta
Mass defect (keV)	0	28,296	92,162	139,808	137,370	1,444,662	1,452,239
Mass defect (keV/nucleon)	0	7074	7680	7767	7632	8026	8023
Nuclear states (keV); (spin) (Parity)	0; 1/2+	0; 0+ 20,210; 0+ 21,010; 0−	0; 0+ 4438.9; 2+ 7654.2; 0+	0;0+ 1982.07; 2+ 3554.84; 4+	0; 1+ 937.2; 3+ 1041.5; 0+	0; 1+ 39.54;2+ **77.2; 9−**	0; 7/2+ 6.237; 9/2− 136.26; 9/2+
Proton Proximity barrier (keV)	181	364	1027	1338	1523	9555	9545

State transitions require spin differences of 1, otherwise the transition is forbidden and long-lived such as the 3rd state of ^{180}Ta also known as ^{180m}Ta. Proximity barriers from (Blocki et al. 1977). Nuclear levels from NuDat 2.7 [https://www.nndc.bnl.gov/nudat2] and Nubase2016 (Audi et al. 2017)

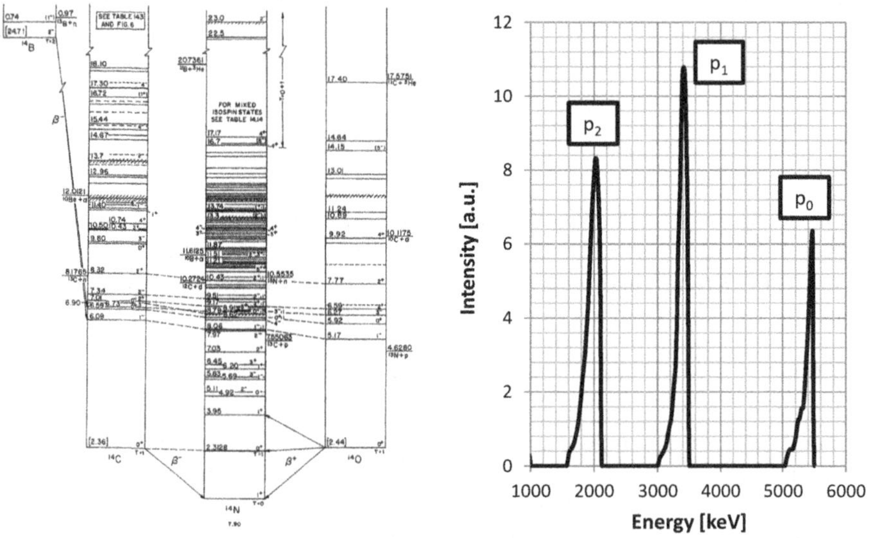

Fig. 3.10 Left: Nuclear levels of the nuclei with 14 protons + neutrons (A = 14). ^{14}N represents the only stable nucleus. All nuclei feature numerous excited states. From every state different pathways towards the lowest energy state exist via emission of particles and photons. In the ground state ^{14}O decays via β^+, while the first few excited states prefer emitting a proton forming ^{13}N. From(TUNL Nuclear Data Evaluation Project). Right: ^{12}C(^{3}He,p)^{14}N reaction at E = 2.5 MeV and 160° producing ^{14}N in the ground and the first two excited states. The emitted proton energy reduces corresponding to the excitation energy leading to three distinct proton peaks

momentum and parity. Physicists call transitions which have a difference of more than 1 unit of spin between initial and final level forbidden, since 1 corresponds to the spin carried (away) by a photon. Nuclear physics apparently makes it increasingly improbable to carry away more spin. The longest lived example for a forbidden excited state transition is ^{180m}Ta with a difference of 9–1 spin units between excited and ground state (Table 3.1) resulting in 1.8 * 10^{15} years half-life, but also other isotopes such as ^{99m}Tc feature practically relevant half-lives due to this effect.

Having fixed target situations in accelerator applications also requires a heavy moving particle after the reaction in order to conserve momentum. This prohibits a simple consumption of projectiles with only one product particle, but requires at least two products (2 variables need 2 equations). But nuclear physics further limits the possibilities. Nuclear reactions have to follow additional nuclear conservation rules, most importantly the conservation of particle number and electric charge. The amount of protons, neutrons, and leptons (namely electrons + neutrinos, see Sect. 4.4) will not change during the reaction, only a transfer between projectile and target is possible. The situation changes for β decays since these involve the weak force. In the β decay the conservation of lepton number becomes important. The conversion of a proton to a neutron, or vice versa, changes the nuclear charge which has to be compensated by emitting a positron or an electron, respectively. This violates the conservation of lepton count, consequently a neutrino or anti-neutrino, respectively, has to be emitted additionally. Physicists invented the so-called Feynman diagram to cover all possible reaction and decay routes, but this theoretical construct goes too far for applications. The actual probability and branching between different possible reactions is described by the individual reaction cross-sections or decay probabilities.

Not every reaction allowed by the conservation laws will also occur. Reactions with $Q < 0$ have a threshold since this missing energy has to be provided by kinetic energy of the projectile (conservation of energy). A chemist would call these endothermic and reactions with $Q > 0$ exothermic, but of course thermal energies have no meaning for the MeV energies involved in nuclear reactions. Since this book discusses charged particle beams and target nuclei also consist of similar charged particles, the electro-magnetic Lorentz force produces a barrier potential for reaching a nuclear proximity required for nuclear reactions. We can understand the nucleus as an armoured tank with its electrical charge building some kind of Coulomb armour. With projectiles of low kinetic energy fired for example from a handgun or a rifle we cannot penetrate its armour, but the projectile will bounce off. The more punch we have the higher the probability to penetrate the armour instead of bouncing off. The same applies to nuclear interactions, the higher the projectile energy, the lower the cross-section for elastic scattering and the higher the nuclear reaction cross-section. Table 3.1 lists a few examples of barrier potentials towards proton projectiles derived from analytical calculations. These barriers scale-up roughly with the number of protons involved in the reaction. The barrier energies only indicate at which projectile energies nuclear reactions become possible, but any barrier can be tunnelled in quantum systems. For this reason nuclear reaction cross-sections usually start with an exponential increase from low energies towards higher energies as the next section will elaborate.

3.3.1 Cross-Sections

In applications, nuclear reactions primarily involve ion beams. Nuclear reactions with electron projectiles require similar energies as with ions in the 10 MeV region, implying certain application drawbacks connected with the high electron velocity as discussed in Chap. 2. Furthermore, electron-nucleus reactions are somewhat limited by the fact that, in contrast to the constituents of ions, electrons are not present in the nucleus, limiting the possible reactions and products. For these reasons nuclear reactions with electrons have little application relevance and will not be discussed in this edition. Anyways, many physical aspects are independent of the projectile species.

The magnitude of the cross-sections strongly depends on the projectile energy and the projectile-target combination. In some cases also the nuclear polarisation state significantly influences the reaction cross-section (Ciullo et al. 2016). Nuclear reactions require higher energies than the elastic reactions underlying stopping due to the proximity of projectile and target required by the short-ranged nuclear forces responsible for nuclear reactions (Coulomb barrier effect). The evolution of the cross-section σ with the projectile energy E is described by the total cross-section $\sigma(E)$ (sometimes also just called cross-section). The example in Fig. 3.11 demonstrates the variation of the total cross-section of the ^{18}O(p,n)^{18}F reaction over five orders of magnitude between 2 and 200 MeV projectile energy. The cross-section first increases, then reaches a maximum of about 300 mbarn at $\approx$ 6 MeV, and then decreases towards higher energies again by five orders of magnitude towards 200 MeV. Empirically, many cross-sections follow a qualitatively similar behaviour with varying peak cross-section, peak width, and projectile energy at the maximum.

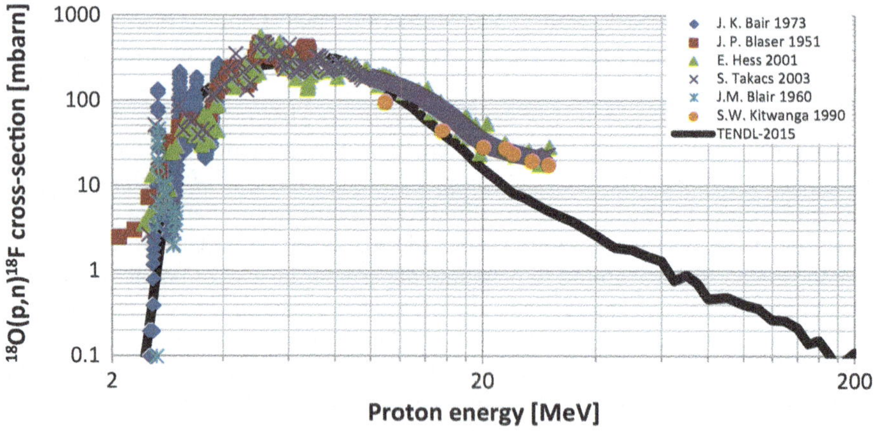

Fig. 3.11 An extract from the JANIS OECD Nuclear Energy Agency (NEA) (2017) database comparing experimental and theoretical differential cross-section for the ^{18}O(p,n)^{18}F reaction. Symbols mark experimental data, while the line shows calculated cross-sections from TENDL-2015 (Koning et al. 2015)

Figure 3.11 compares the semi-empirical cross-section TENDL-2015 (Koning et al. 2015) with various experimental data. All datasets roughly agree within a factor three to each other, but TENDL-2015 is unable to reproduce some of the features, for example the resonances in the 2–4 MeV region, which are, on the other hand, also ambiguous in the experimental data. The cross-section shows a typical threshold reaction, in this case $Q = -2438$ keV as displayed in the mass defect difference between ^{18}O and ^{18}F in Table 3.1. From the threshold on, the cross-section increases exponentially due to tunnelling of the proximity potential. For $Q \geq 0$ reactions, the threshold with its exponential increase will adapt to the order of the barrier potential. For ^{18}O(p,n)^{18}F the Coulomb barrier of 1338 keV is lower than the energy threshold (see Table 3.1) and therefore not relevant. The existence of resonances arises from the quantum mechanical particle-wave duality. Each particle has its individual wave and as soon as projectile and target wave come into contact, resonant overlaps, similar to the interference of light waves, can occur. The resonances change the cross-section at specific energies by orders of magnitude, either increasing or decreasing it. Towards higher energies, generally the cross-section decrease due to reducing particle wave-function overlap.

A moving projectile defines a unique direction/vector with its direction of movement. This vector represents a symmetry axis for the reaction, leading to a non-isotropic emission of reaction products. In other words, the cross-section changes with the angle towards the direction of movement, leading to differential cross-sections $d\sigma/d\Omega$ depending on the exit angle of the products, also called reaction angle, and projectile energy. Figure 3.12 shows an example. The quantity Ω describes the solid angle into which the given cross-section can be measured at the given energy E and reaction angle. Equation (3.12) allows calculating the solid angle from a given area A of a sphere of radius r. We can see it as a detector of area A with a distance of r to the point of reaction.

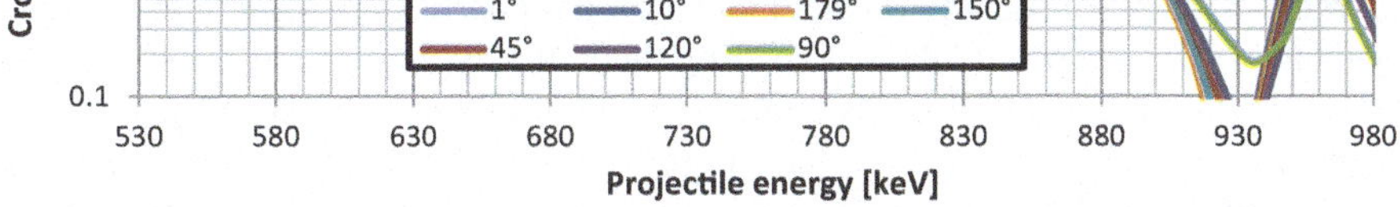

Fig. 3.12 Differential cross-section of the ^{18}O(p, ^{4}He)^{15}N reaction for detecting the ^{4}He at the stated laboratory reaction angles. The resonance at 629 keV is weakest at 90° and increases up to a factor 4 towards higher and lower angles. At 828 keV the cross-section decreases with increasing angle by about 20%. Data from Sigmacalc (Gurbich 2016) *R*-Matrix fits to experimental values

$$\Omega = A/r^2 \tag{3.12}$$

Differential cross-sections have a considerably lower level of available data in the literature due to the fact that determining them requires measuring the moving products in-situ and not only the amount of present isotopes at some point in time after the process (ex situ). Total cross-sections rather find use in isotope production and activation applications, while differential cross-sections find use in analytical methods due to their connection to detection and reaction kinematics. As in Fig. 3.11, the differential cross-section in Fig. 3.12 features an exponential growth from the low energy side and several resonances. In the differential form, a clear dependence of the resonance height on the reaction angle becomes visible. In $\sigma(E)$ only the angular average would be visible, but each resonance varies differently with reaction angle.

The determination of precise reaction cross-sections represents an integral part of many technological advances in accelerator applications, but also a challenging one. Physicists try to reduce the effort by determining only the data specifically required, leaving many holes in the data landscape. In many application cases discussed later in this book a certain element with several stable isotopes is used to produce a specific isotope via nuclear reactions. In this case, potentially several reaction types, usually sets of multi-product reactions such as (p,xn) can lead from different isotopes to the same nuclide. In this case the individual isotopes reaction cross-sections were summed up to a so-called production cross-section. Production cross-sections are a simplification with several drawbacks, but the advantage of easy experimental determination using natural isotopic composition. As an example ^{182}Re via protons could occur from natural tungsten from reactions with its various isotopes via ^{182}W(p,n)^{182}Re, ^{183}W(p,2n)^{182}Re, ^{184}W(p,3n)^{182}Re, or ^{186}W(p,5n)^{182}Re reactions. If we measure only the final ^{182}Re activity we will not be able to distinguish between the isotope specific reactions. In contrast, the reaction cross-sections could be determined only with an isotopically purified target of a single (tungsten) isotope.

Unfortunately, theoretical physics has not proceeded to a point where mathematical descriptions for all nuclear scattering reactions exist. We already saw the Rutherford cross-section in (3.1) with its accurate theoretical description of elastic scatterning up to a few MeV as an example of such a theoretical description, but the underlying reaction does not involve a nuclear interaction in the sense of everything beyond the electrical charge visible to the outside (the billiard ball model). Existing approaches based on quantum calculations on the quark level (Quantum-chromodynamics) offer the potential for delivering theoretically derived cross-sections, but the computational effort strongly scales with the number of involved quarks and brings current supercomputers to their limits, even for hydrogen isotope reactions. A full solution of the problems probably requires the next level of computer technologies, more complete physical understanding of the strong nuclear force or of the four fundamental forces in general.

Nuclear physicists found a set of models describing nuclear reaction cross-sections to inter- and extrapolate from given experimental data. These semi-empirical equations combine an adequate theoretical model description of the overall process and require fitting of (theoretically) unknown parameters to the experimental data. This procedure yields a formula for calculating cross-sections over a certain energy and angular range with a statistical uncertainty given by the data. The problem is, the range of validity is not known and systematic errors in the experimental data or model could be unwillingly absorbed in the result. Hence semi-empirical models are a useful tool for data analysis and computations, but the results must be handled with care, especially in the extrapolation region where the risk of systematic errors of the model increases.

$$\sigma_{\text{Res}}(E) \sim \frac{1}{(E - E_{\text{Res}})^2 + \Gamma^2/4} \tag{3.13}$$

In general, reaction cross-sections combine resonant and non-resonant regions. We already learned the start of nuclear reactions follows an exponential increase due to the tunnelling of the Coulomb barrier. The Breit-Wigner or also Lorentz resonance function (3.13) describes the probability distribution of resonant interactions. This cross-section depends on the central energy of the resonance E_{Res} and a resonance width Γ. Interferences with the non-resonant part of the reaction can lead to the typical down-up or up-down resonance where both cross-sections subtract and add up (or vice versa) before and after the resonance energy. Figure 3.13 shows three examples of this process with different Γ in all cases.

The next level beyond this simple analytical fitting requires more complex models implemented in several codes. Implementing the above idea of resonances and the interaction of particle waves in the sense of the Schrödinger equation leads to the so-called R-Matrix algorithms such as the code SigmaCalc (Gurbich 2016). These fitting algorithms combine the features of several states (3.13) known from experimental data to a matrix which then allows solving the Schrödinger equation, reproducing cross-sections in the resonant and non-resonant region over an energy and angle

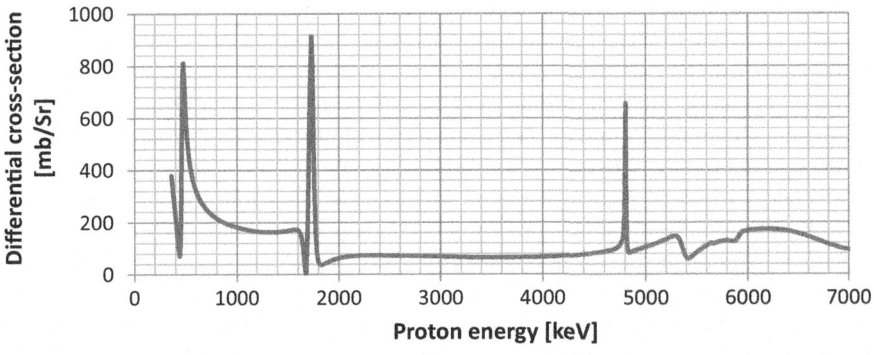

Fig. 3.13 Differential cross-section of $^{12}C(p,p_0)^{12}C$ at 165°. Already at 360 keV this elastic reaction deviates from the Rutherford behaviour, in spite of the 1027 keV barrier. The data show several types of different resonances in the given energy range. Data from *R*-Matrix fit of SigmaCalc (Gurbich, 2016) to a set of experimental data

range given by the experimental data. Nuclear models such as the codes Talys (this is behind the frequently updated TENDL cross-section database) or Empire (Herman et al. 2007) follow a different approach by combining a multitude of limited physical models of specific nuclear interactions. These codes produce reasonable total cross-sections as we saw in Fig. 3.11, although resonant features might be missing. For differential cross-sections the agreement significantly worsens. Nevertheless, these codes often provide the only available data for new developments and over the years progress on the underlying nuclear data continuously improves the result quality of these codes.

Knowing what cross-sections are, the applicant will ask how to determine them experimentally. Cross-sections represent a probability at a given energy. Consequently, we required a defined known energy for the reaction in the target, an identification of the reaction, and a detector for counting the amount of reaction products in relation to the fluence of projectiles. For detecting the occurrence of a given nuclear reaction we have to detect the products of this reaction. This could be the fast light products (mostly p, α, and n) or the heavy product, which usually remains in the target due to its limited range. Stopping reduces the initial beam energy, leading to a mixing of different energies upon passing a target. This requires either a local measurement, an energy resolved measurement, or a restriction of stopping by using thin targets and large detector solid angles for compensation.

Differential cross-sections are mostly measured via thin targets. The exact quantity depends on the element and beam energy, but typically lies in the order of 1 μm thickness. The target thickness induces a contradiction due to the well-defined energy with thinner samples (lower stopping), but the higher statistics with thicker samples (3.2). The detection of the heavy products follows the beam irradiation via ex situ spectroscopy of the decay radiation together with isotope identification via characteristic spectral libraries available for most isotopes, e.g. (Nucleonica GmbH 2014). This pathway yields differential cross-sections. For non-monoisotopic target elements, similar reactions potentially lead to the same product isotope. (p,xn) reactions are the prime example, but also other reactions, decays, or impurities lead to the same problem. In this situation only isotopically enriched targets allow unfolding the problem with an ex situ analysis.

In-situ detecting the light product via particle detectors suffers from the problem of catching all angles around the target via physical placement of detectors. A differential cross-section results from a single detector angle. The in-situ detection allows for an energy resolved measurement of the products. This additional information in principle allows unfolding the stopping for 2-body reactions and using thick samples to generate complete spectra in one measurement. The backward calculation of the kinematics (see next section) has its drawbacks, but offers the potential for accelerated determination of cross-sections via stopping induced beam energy scanning (Möller S., Analytical continuous slowing down model for nuclear reaction cross-section measurements by exploitation of stopping for projectile energy scanning and results for 13C(3He,α)12C and 13C(3He,p)15N, 2017). In the end the best method derives from the energy range, the required accuracy, and the specific reaction to be investigated.

3.3.2 Kinematics

The cross-section describes which reactions occur with which frequency/probability when a given beam impinges on a target. Kinematics describes how the particles move into, and, in particular, out of this reaction. The most important choice for understanding kinematics relates to the inertial system considered. In other words, what do we see as resting and what as moving? As always in this book, we have a moving beam particle and a stationary target defining our initial conditions. This is called the laboratory (Lab) system, the interial system we also reside in. Nuclear decays feature only a single stationary particle in the beginning, significantly simplifying the situation as no preferential direction or relative velocity exists. Due to this symmetry, decays simply emit their products isotropically (equally probable in all directions) and will not be considered primarily in this section.

The Centre-of-Mass (CM) system represents the true physics of each interaction. In this system both projectile and target move, but with exact opposite momentum. In collider experiments, two beams move towards each other, making the CM system identical to the Lab system. All other cases require re-calculation of energies, angles, and cross-sections to switch between both. The Lab system represents our, and in the case of stationary targets, the targets point of view. For applications we are interested in the Lab system, since it describes the results we see. Switching from Lab to CM system reduces the kinetic energy available for reactions. In the Lab system, the centre of mass, as a point in between projectile and target, has to move towards the target. A kinetic energy and momentum not present in the CM system. This CM movement could be understood as a virtual particle carrying the remaining kinetic energy. The recalculation between CM and Lab and the view of the CM system belong to the fundamental particle physics or nuclear physics and will therefore be omitted in this book. For more details on the calculations related to this, the reader is referred to any standard nuclear physics book.

The physics of nuclear reactions and decay kinematics depends on the number of involved particles. Each particle existing before and after the reaction has a set of kinematic properties, namely mass (m), kinetic energy (E), and movement vector (v) or in other notation energy and momentum vector. In applications only very few reactions involve more than two particles on the input side, due to the small probability of coincidence for the usual beam densities. This means we consider reactions of a projectile with a target. Already in Sect. 3.3.1 we learned about the existence of reactions with two products such as ^{181}Ta(p, n)^{181}W with a light product, the neutron, and a heavy product. The wording of light and a heavy product originates from nuclear reactions favouring to form the strongest bound products (see Fig. 8.1), which are typically heavy elements. Correspondingly nuclear reactions tend to release neutrons or protons or if the Q-values are attractive ^{4}He. These so-called 2-body reactions yield a unique solution for the kinematics of the products. Figure 3.1 depicts the 2-body situation. The application aspect of the kinematic theory starts when not all kinematic properties are known. Solving 2-body kinematics relies on the mathematical logic of requiring as many equations as we have unknowns, see (Zagrebaev et al. 2019) for an online tool. The conservation of energy and momentum vector, see (3.14), provides the equations. Particle 1 (bearing E_1, p_1) represents the projectile, particle 2 the target and particles 3 and 4 the products. Particle 3 represents the light product and particle 4 the heavy product particle. Consequently, if we know all kinematic properties of an interaction except for two (remember each particle bears two properties), the kinematic equations will yield unique results for the two unknowns.

$$\begin{aligned}\vec{p_1} + \vec{p_2} &= \vec{p_3} + \vec{p_4}\\ E_1 + E_2 &= E_3 + E_4\end{aligned} \tag{3.14}$$

Here the momenta are given as vectors. The pure forward momentum of the initial situation can receive an additional transversal momentum component balanced between the two product particles. Since two particles necessarily lie on the same plane, the whole situation becomes rotationally symmetric about the axis defined by the movement vector of the projectile towards the target. This rotational symmetry is the same as discussed with the reaction angle in Sect. 3.3.1. With the target (particle 2) initially at rest, p_2 and E_2 become zero. Reactions with $Q \neq 0$ break the conservation of mass by opening an exchange channel between mass and kinetic energy. We need to take the transfer between mass and energy via the Q-value into account. This modifies the conservation laws of (3.14) to

$$\begin{aligned}p_1 &= p_3 * \cos(\theta) + p_4 * \cos(\phi)\\ 0 &= p_3 * \sin(\theta) - p_4 * \sin(\phi)\\ E_1 &= E_3 + E_4 - Q\\ m_1 + m_2 &= m_3 + m_4 + Q/c^2\end{aligned} \tag{3.15}$$

In production applications we usually know the projectile and target properties, while in analytical applications projectile and the properties of one or two products are known from detectors. The solution for a missing quantity, such as a product energy, is anything but straightforward. According to (Nastasi et al. 2014) Appendix 4, the product energies in the Lab system read in the non-relativistic case with $E_2 = 0$:

$$E_3 = \frac{E_1 m_1 m_3}{(m_1 + m_2)(m_3 + m_4)} \left[\cos(\theta) \pm \sqrt{\frac{m_2 m_4}{m_1 m_3}\left(1 + \frac{Q}{E_1} + \frac{m_1 Q}{m_2 E_1}\right) - \sin^2(\theta)}\right]^2 \tag{3.16}$$

$$E_4 = \frac{E_1 m_1 m_4}{(m_1 + m_2)(m_3 + m_4)} \left[\cos(\phi) \pm \sqrt{\frac{m_2 m_3}{m_1 m_4}\left(1 + \frac{Q}{E_1} + \frac{m_1 Q}{m_2 E_1}\right) - \sin^2(\phi)}\right]^2 \tag{3.17}$$

$$\theta_{\max} = \arcsin\left(\frac{m_2 m_4 (E_1 + Q)}{m_1 m_3 E_1}\left(1 + \frac{m_1 Q}{m_2 (E_1 + Q)}\right)\right)^{0.5} \tag{3.18}$$

$$\sin(\phi) = \left(\frac{m_3 E_3}{m_4 E_4}\right)^{\frac{1}{2}} \sin(\theta) \tag{3.19}$$

The book (Nastasi et al. 2014) provides a comprehensive list of kinematic equations beyond (3.16)–(3.19) in its appendix. The equations include the possibility of elastic ($Q = 0$) and inelastic ($Q \neq 0$) nuclear reactions by allowing for different masses of incoming (m_1, m_2) and outgoing particles/products (m_3, m_4) and a kinetic energy production or consumption via the Q-value. Equations (3.16) and (3.17) offer two possible solutions ($\pm$) depending on the mass ratio of projectile and target. For a projectile lighter than the target, both products could be scattered in the forward direction (think of a grazing impact) or in opposite directions (think of a frontal impact). If the projectile mass equals the target mass or exceeds it, both products have to move in the forward direction due to conservation of CM momentum. Only one solution remains.

In other words, for a given set of properties of one product, the properties of the second product are strictly defined by the conservation laws. Practically this means, if we analyse one product for its energy and mass at a certain scattering angle, e.g. Θ, and we know the projectile mass, the 2-body kinematics, e.g. in the form of (3.16), can tell us for example which mass the target had. In this case m_2 and m_4 are two unknowns, but we can add the conservation of mass from (3.15) as the second equation required for a unique solution with 2 unknowns. In some cases logical exclusion principles yield extra information on the involved particles via conservation of nuclear numbers (proton, neutron, electron count) or known Q-values. This recalculation represents an important fact for accelerator based analytics by making the measurement of different quantities physically equivalent for understanding the whole reaction.

3-body or n-body reactions such as (p,2n) or (p,xn), respectively, follow different rules than 2-body reactions. The definition of the kinematic properties of one product does not define the kinematic properties of the other two products, since these two have no mathematical rule how to share the remaining energy and momentum. Energy, momentum, and mass are still conserved, but the additional degree of freedom yields distribution functions instead of singular values as depicted in Fig. 3.14. Imagine dropping a bag of food onto a bunch of dogs: All the food will be consumed for sure, but each time you do it every single dog will receive a different amount of food. This n-body situation is independent of whether a projectile-target situation or a decay such as the β^- decay emitting a heavy decay product, an electron, and an anti-neutrino is considered. A prominent example of this problem is the KATRIN experiment searching for the neutrino mass via detection of the energy distribution function of the electron emitted in the decay of tritium. The problem of this analytical experiment lies in the decreasing counting statistics towards the high-energy end of the electron energy distribution.

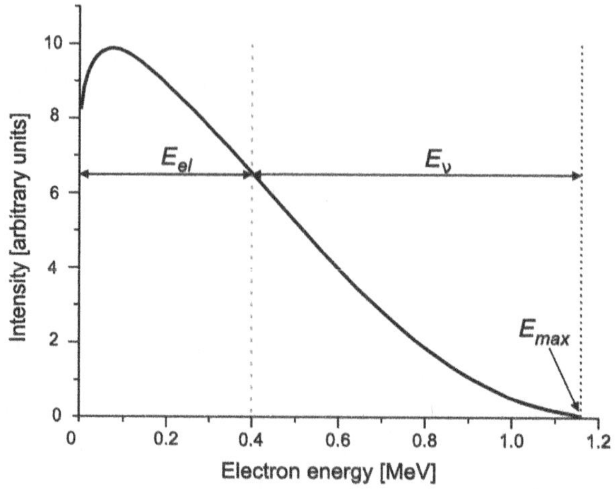

Fig. 3.14 Energy spectrum of the electrons escaping from the β-decay of ^{210}Bi, which is a 3-body decay (heavy nucleus, electron, muon). Due to their kinematic freedom n-body decays always feature product energy spectra. The example shows an electron energy $E_{el} = 0.4$ MeV would correspond to a neutrino energy $E_\nu = 763$ keV, but all other combinations are also possible

In the distribution function lower energies occur more often, since more allowed combinations of properties for the other particles exist in this region. Physicists call this the density of states. Considering the limiting case of one particle receiving the maximum possible energy (E_{Max}), the other two particles have to feature an exactly opposite momentum, due to the limited available energy, and only one state of the whole system remains possible. For simplifying the situation we can reformulate the problem to a 2-body reaction with an imaginary box combining two of the three products to one. In any case the physics of ambiguous solutions remains the same, making n-body reactions an at least unattractive situation for analytics.

3.4 Depth- and Stopping Dependent Reactions

Considering the individual particle picture, a charged particle traveling through matter constantly loses energy, but it also has a probability to scatter depending on energy and projectile-target combination. Combining these effects yields the reaction probability of a nuclear reaction described by the cross-section σ in a depth interval of z to z_1 or energy interval E_0 (= beam energy) to E_1, respectively. In the energy picture, the ratio of the reacting ρ_R to the stopping ρ_S matter comes into play when considering compounds consisting of several elements. In these mixtures, all target particles induce stopping, constituting ρ_S. Only specific target species interact via σ, constituting ρ_R.

$$W \equiv \int_{z_1}^{z} \frac{\Delta W(E)}{\Delta x} \mathrm{d}x = \int_{z_1}^{z} \sigma(E(x)) * \rho_{\mathrm{R}} \mathrm{d}x = \frac{\rho_{\mathrm{R}}}{\rho_{\mathrm{S}}} \int_{E_1}^{E_0} \frac{\sigma(E')}{S(E')} \mathrm{d}E' \tag{3.20}$$

Equation (3.20) describes the reaction probability W in 3 different forms. First the reaction in an infinitesimal slice of the target of thickness Δx integrated over the considered depth, second the energy (E) dependent cross-section times the reaction partner density integrated over depth and lastly the depth is solved for the depth dependent energy via the means of the stopping power S. Setting $E_1 = 0$ assumes full stopping of the projectile in the target, a situation typical for accelerator applications. For analytical purposes the reaction depth will be important to resolve further information such as depth and local concentrations of elements and isotopes ρ from the analysis. For production purposes, the integral over the whole accessible depth, the range, or the integral over the energy from the initial accelerator energy to zero, respectively, is more relevant. In either case, the relevant quantities change over depth and this location/depth of reactions happening forms a quantity of interest.

Figure 3.15 explains the different parts where stopping and depth define the reactions and whether products remain in the sample or leave it. Projectiles entering the target at a shallow angle, i.e. α close to 90°, induce reactions closer to the surface. The depth of the reaction scales according to (3.21). Variation of the impact angle enables virtually increasing the stopping power of the material for the projectile. The same scaling exists for the exit angle of the products, enabling differentially changing the effective stopping of projectile and products, a trick often used to increase for example the depth resolution of analytical methods. Unintended impact and exit angle variations relate to surface roughness and porosity.

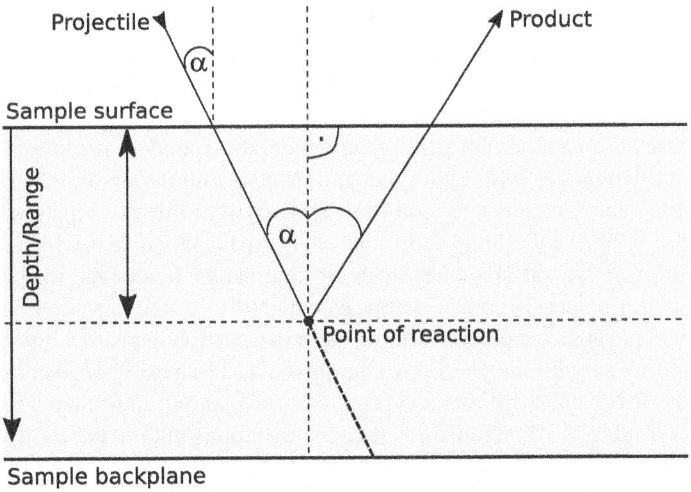

Fig. 3.15 An exemplary depth dependent scattering situation. The way in and out of the sample differ depending on impact and exit angles. Leaving the sample requires sufficient product energy, depending on depth and exit angle of the products

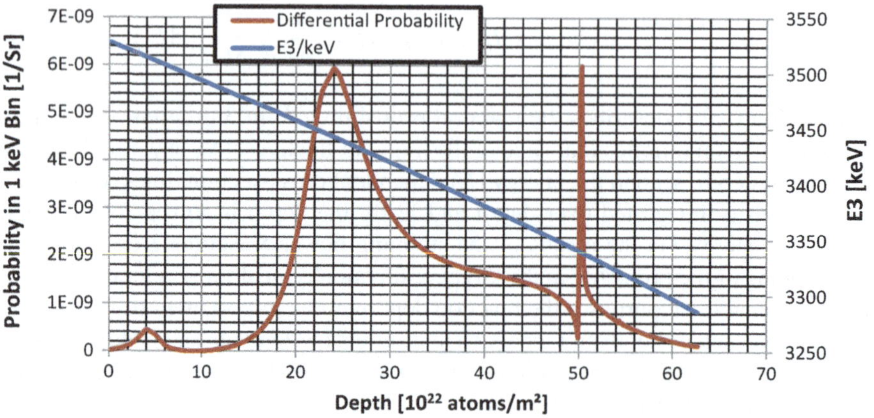

Fig. 3.16 Depth calculation of the $^{18}O(p, {}^4He){}^{15}N$ reaction of 988 keV protons in SiO_2 showing the depth profile of reaction probability together with the light product (^4He) energy E_3 at 150° scattering angle. The reaction has a range of about 6.2 * 10^{23} atoms/m^2 (recalculate with density to have it per length) with a lower energy limit of 520 keV due to the onset of the reaction barrier, see Fig. 3.12

$$\text{Depth} = \text{Range} * \cos(\alpha) \tag{3.21}$$

Considering a perpendicular incidence, $\alpha = 0$, of a proton probing for ^{18}O in a SiO_2 (silicon dioxide) and combining Fig. 3.15 with the (3.19), (3.20), and (3.16) yields Fig. 3.16. Here we calculated depth in the sample in units of atoms passed vs. the light product energy E_3 at 150° scattering angle for the reaction $^{18}O(p, {}^4He){}^{15}N$ (cross-section in Fig. 3.12) with 988 keV projectile energy E_1. For recalculation of this two-body reaction kinematics take care of the units and magnitudes. Note that the calculated E_3 is given at the depth of the reaction, when leaving the sample additional energy will be lost on the way out. Neither was straggling included.

The figure demonstrates several aspects of depth dependent reactions. The light product energy changes with depth, but in this case only by about 243 keV, while the projectile energy changes by 468 keV. The origin of this discrepancy lies in the reaction $Q = 3979.8$ keV, adding additional energy to the kinematics which dominates the momentum conservation in the two-body kinematics. In the reaction probability, the cross-section clearly becomes visible, as stated by (3.20). If this reaction was used for analytical purposes, the behaviour of the cross-section leads to a low sensitivity to ^{18}O close to the surface (<15 * 10^{22}atoms/m^2). The two resonances at 25 and 50 * 10^{22} atoms/m^2 on the other hand produce strong signals in limited depth ranges. At depth beyond 62 * 10^{22} atoms/m^2, the sensitive range but not the projectile range ends, since the cross-section exponentially approaches the reaction Coulomb-barrier.

Applying this setup for example to a heterostructure, which features a significant ^{18}O concentration only at the surface, e.g. due to a ^{18}O tracer gas exposure, only negligible amounts of detectable products will be produced. The method will not be sensitive to the sample structure. The method is right, but the projectile energy was wrongly configured. Changing the projectile energy E_1 from 988 to the 830 keV (cross-section resonance) optimizes the detection properties via the understanding of depth dependent reactions.

In most applications, efficient usage of the expensive accelerator technology is a central aspect. The term *efficiency* always requires specification of the quantity the process makes optimal use of. In this case we are looking for energy efficiency, since it not only determines a part of the application cost structure, but it also defines target heat load and radiation safety aspects. Here in particular the stopping dependency of reactions influences the result. Applications usually require a defined amount of specific reactions, therefore the optimal beam energy and current have to be found before installing and setting up the accelerator. Considering the full (average) track of the projectiles via depth dependent calculations allows us understanding this efficiency. Instead of calculating the beam energy with maximum reaction probability W, we have to add the input projectile energy E to (3.20) to obtain a quantity representing production efficiency H:

$$H(E) = W/E \tag{3.22}$$

The maximum of (3.22) yields the point of maximum beam energy efficiency. In other words: For a required reaction rate, corresponding for example to a production rate of isotope X, the maximum in H represents the minimum in required input beam power. Technically spoken this energy marks the point where increasing the beam current yields the smarter choice than increasing the beam energy. The evolution of H depends on the fundamental physics of S and σ and their evolution. Mathematically, the maximum in H corresponds to a zero-point in its derivative. Of course several local maxima can be present, e.g. when the reaction cross-section σ features several maxima/resonances.

$$\frac{\mathrm{d}H(E)}{\mathrm{d}E} = 0 \tag{3.23}$$

Equation (3.23) tells us the shortest and most efficient route to our production goal by minimizing the costs for the accelerator (Beam energy E_0), electrical power input ($E_0 \times$current I), and target heat loads (also $E_0 \times I$). Will every reaction have this zero-point? A zero-point in the slope of H corresponds to the reaction cross-section decreasing faster than the stopping power.

Let us consider a specific case. For inducing nuclear reactions typically some MeV of projectile energy are required. For light ions this value lies beyond the maximum in stopping power (typically between 0.1 and 1 MeV, see Fig. 3.6), hence the stopping power monotonically decreases for higher projectile energies for any target. The nuclear reaction cross-section on the other hand can be very dynamic with resonances or at least a single broad resonance at a few MeV. Figure 3.17 compares S and σ with the ^{18}O(p,n)^{18}F reaction used for producing ^{18}F-PET tracers (see Sect. 6.1). The reaction features a simple single resonance cross-section and a threshold around 2.5 MeV, qualitatively similar to many other reactions, check

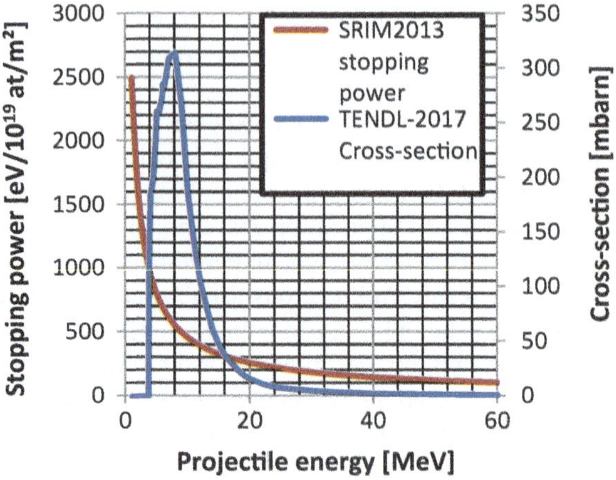

Fig. 3.17 Plot of the evolution of stopping power and nuclear reaction cross-section of $^{18}O(p,n)^{18}F$ used for the production of ^{18}F labelled medical products in ^{18}O enriched water ($H_2^{18}O$)

(OECD Nuclear Energy Agency (NEA) 2017). At low energies, S decreases while σ increases, but above the maximum in σ at 8 MeV both decrease.

Ok, if both values decrease we have to dig in deeper. The ion stopping power of (3.6) demonstrated a decrease of stopping power with something like $\ln(E)/E$, while the resonance of cross-section in the form of (3.13) decrease with $1/E^2$, a faster decay. The ratio σ/S then scales with $1/(\ln(E)E)$ resulting in a monotonic increase of the integral over this ratio. The integral of this function yields $\ln(\ln(E))$. Taking into account the proportionally to the beam energy increasing invested energy of (3.22) we end up with $\ln(\ln(E))/E$. These functional shapes we will also see in Fig. 3.18.

What can we do for increasing the output of a given reaction when beam energy is fixed? In numbers this efficiency means: At 12 MeV the reaction probability reaches 0.33% compared to, for example, 0.46% at 24 MeV. With 24 kW beam power (12 MeV/2 mA or 24 MeV/1 mA) we obtain $4.1 * 10^{16}$ reactions/s at 12 MeV, but only $2.8 * 10^{16}$ reactions/s at 24 MeV. The 46% higher efficiency at 12 MeV comes along with less unwanted activation, but also about 3 times increased power load on the target window required for the water target.

Figure 3.18 compares the energy efficiency of the $^{18}O(p,n)^{18}F$ reaction by production per projectile and per MeV invested energy. The figure shows a strong increase of efficiency starting at a few MeV, originating from the simultaneous decrease of the stopping power and increase of the reaction cross-section as displayed in Fig. 3.17. Between 20 and 30 MeV this increase levels off at a reaction probability in the typical order of 1%, here 0.5–0.6%. While the reaction probability still increases slowly, the efficiency factor H stagnates, its derivative approaches zero. At this point the accelerator layout would suggest rather increasing the accelerator current than its energy, since both deliver an identical benefit for the production rate. Increasing the beam

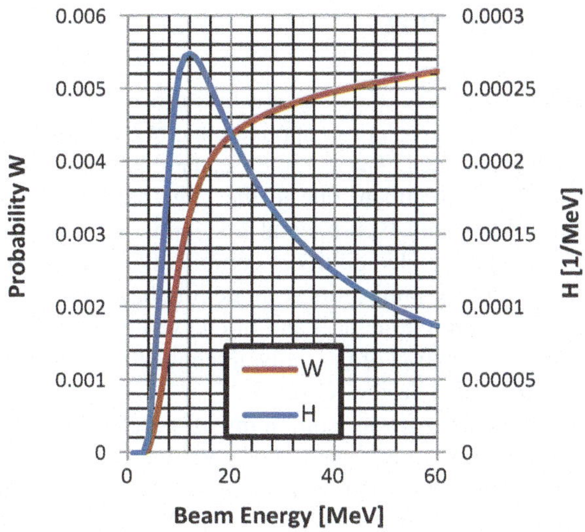

Fig. 3.18 The integration of (3.20) from the energy given on the x-axis down to zero represents the reaction probability for the $^{18}O(p,n)^{18}F$ on the whole path of the proton projectile till its stop in the H_2O target. Above the cross-section maximum at 8 MeV the slope of both curves strictly reduces with a maximum in H at 12 MeV

energy yield more problems of radiation protection and device cost, while increasing the beam current does not.

For electrons, the situation is upside-down and we see a minimum of the stopping power in the order of 1 MeV with increasing stopping power towards higher energies. When considering the energy efficiency of a reaction this induces a generally bad situation, since the minimal stopping power coincides with the lower limit for nuclear reactions. To be fair to the electrons: Electron stopping powers are lower than ion stopping-powers in the considered range, see Fig. 3.6.

Practical and economic aspects limit the maximum beam energy in applications. Physical limits require given minimum energies in order to induce the foreseen reactions as well as reach required penetration depth'. In this section, we learned how to navigate in the grayscale between these extremes via the physics of depth dependent reactions. As an example consider an electron microscope: At higher electron beam energy additional elements become visible by enabling more X-ray transitions (Fig. 3.4), but if the measurement aims at analysing a thin layer, e.g. a thin coating on a thick substrate, higher beam energy would finally lead to information mixing of layer and substrate, since increasing electron energy increases the projectile range, penetrating the thin layer.

3.5 Computer Modelling

Computer modelling of particle-matter interaction processes has become the main tool to derive quantities for accelerator applications, radiation protection purposes, and the layout of new devices. The complex equations presented in this chapter require a computer based treatment due to the large amount of considered reactions, individual properties, and experimental data input. In real world problems, the geometry and time-evolution add additional complexity to the housekeeping of the processes. In spite all these complications, modelling based on the best knowledge usually hits the reality within a factor 3 and often even better. Therefore computer models enable a full construction of basically all applications presented in this book in the idea of computer aided design (CAD).

Unfortunately, no single software combining all accelerator application aspects exists so far, but different programs need to be combined as modules for modelling an application setup. This starts with beam optics (see Sect. 2.3), thermo-mechanical modelling (e.g. Sect. 2.6), radiation protection (Sect. 2.7), and ends with beam matter interaction, namely stopping, nuclear reactions, detectors, and decays.

Let us first consider the peculiarities of a practical situation in a qualitative manner using an example featuring the main points of many applications. The considered example is a spallation neutron source with a proton beam of some hundred MeV and a tungsten target where the beam releases neutrons via (p,xn) reactions. These neutrons supply several experiments including a medical patient treatment. The beam produced in the accelerator part travels towards the neutron production target through stainless steel 316 vacuum tubes. The beam has a finite emittance; hence at least a minute fraction of the beam continuously hits the beam tube, considering a normal distribution. For the machine layout, we need to know if it will be possible to exchange steel parts for maintenance (radioactive inventory limits) and how far away vacuum pumps and electronics have to be placed in order to survive the radiation. At the target, the beam impacts the tungsten metal and suffers the continuous stopping. As we learned in the last section, relevant changes of beam energy and nuclear reaction probability occur already on the μm scale. Furthermore, tungsten consists of five natural isotopes (+impurities) each having around 20 different reactions above 100 MeV. For the treatment of patients this primary spectrum, originating from the proton reactions, is too broad. The primary neutrons interact with their surroundings, leading to thermalisation, reactions, and broadened neutron spectra. The produced neutrons on the other hand interact only weakly, with relevant changes on the 10 mm scale. A clever placement and material selection of structures and coolant flows significantly affects this spectral broadening and reduces the dose to the patient for a given treatment result. Finally, the neutrons interact with the patient. To allow the patient to survive the treatment only limited radiation doses/exposure times can be applied in order to avoid introducing more problems than we solve. This knowledge needs to be present before treating the first patient, but it strongly couples with the building and accelerator and target layout. Both can hardly be altered after installation. Simulations and also practical testing of components become necessary.

Following the nuclear interactions of potentially thousands of particle species (currently over 3000 isotopes are known) in 4D space-time is a mathematical and computational challenge. Two main mathematical approaches allow for a solution of these systems. In the first approach, large systems of analytical equations such as (3.24) enable an exact and computationally inexpensive solution on the time scale of minutes. The equation calculates the density N of all n involved isotopes/nuclides over the application time t via losses by exponential decays with half-lifes λ and the production from the decay of mother nuclides. Unfortunately, these advantages are countered by the difficulty to include complex geometries in the equations. Reasonable approximation of real situations by idealised physics allows an application of the analytical formalism to real situations. In our spallation example, the activation of the beam tube forms such an example. The produced quantities of activity are small, do not couple back to the beam or its surroundings and the distribution is homogeneous along the beam, since only very few beam particles get lost to the tube wall. The coupling of geometry to the nuclear interaction remains small, allowing neglecting it and assuming the beam tube as a point object receiving an input flux of ions. Even if additional interaction hot-spots such as apertures exist, these can be treated independently with an individual equation set.

$$N_n(t) = \sum_{i=1}^{n} \left[N_i(0) * \left(\prod_{j=i}^{n-1} \lambda_j \right) * \left(\sum_{j=i}^{n} \left(\frac{e^{-\lambda_j t}}{\prod_{p=i, p \neq j}^{n} (\lambda_p - \lambda_j)} \right) \right) \right] \tag{3.24}$$

In the second, so-called Monte-Carlo approach, codes simulate the hypothetical life of single particles through the given application geometry. This includes its movement, reactions and daughter particles. When interacting with matter, the particles have a certain probability of having one or another reaction and it is just decided by the throw of a virtual dice which of those marks the particles destiny. Naturally, less common situations become statistically underrepresented but yet may be important. Consequently, an accurate result requires a large amount of test particles to catch all possible situations, but the statistical question mark will always remain relevant. Furthermore, the time domain becomes difficult to access, since every point in time requires a simulation of the full geometry. In the spallation example, all site and construction related aspects are best solved using Monte-Carlo methods. Running several million test particles through the building will produce as many different particle histories, but most of the building volume will also get probed by these test particles. The simulation approximates the building and the beam conditions to be constant.

For calculating the inventory of daughter nuclides following a nuclear reaction process, the individual decay pathways of each nuclide have to be followed and the quantity of each daughter depends on the quantity of all its ancestors due to the connection by decay. The problem arises not only after a beam irradiation, but also during the beam irradiation since these daughters can also react with the beam forming new nuclides otherwise not accessible. This further complicates the full process to a complex coupled equation system. The Bateman-equation, (3.24), describes the situation without source term (production by beam interaction) with the half-life time constant λ_n, the quantity N_n of a certain nuclide n. The decay chain starts with the mother nuclide $n = 0$ at time t and evolves for all the daughters i. Adding a source term and a physical library of nuclide data to this equation leads to the basis of a first application code such as FISPACT-II (Culham Centre for Fusion Energy 2018).

FISPACT-II implements the mathematical solution of (3.24). Equation (3.24) also demonstrates the strength of this approach, since changing the decay time t is straightforward in this approach and enables a direct calculation of the isotope specific inventory over time. Monte-Carlo approaches such as the freely available GEANT4 from CERN (geant4.web.cern.ch), FLUKA (www.fluka.org), or MCNP6 (laws.lanl.gov/vhosts/mcnp.lanl.gov) of the large-scale beam-matter interaction have their strength on particle transport and interactions in complex geometries. Whenever time dependence is of minor importance, for example in shielding calculations, detector sensitivity studies, or planning of medical irradiations, the Monte-Carlo approach reveals its strength in geometrical calculations.

These large-scale codes include vast amount of physics coupling all possible radiation fields (photons, ions, neutrons, electrons), but in particular for analytical applications the physics of these code packages remains too general. These applications require specific codes including only the physics and technical aspects relevant to a single task, while trying to maintain a certain level of productivity in the form of device control, data frameworks, user interfaces, and computational speed. Atomic scale effects such as displacement damage, collision cascades, and surface sputtering exploiting the Monte-Carlo approach in the so-called binary collision approximation are SRIM (srim.org) and SDtrim.SP (www2.ipp.mpg.de/~stel/SDTrimSP.html). Analysis codes for analytical methods integrate specific material models with differences connected to the sensitivity of the corresponding method. The small angle X-ray scattering code SASfit (sourceforge.net/projects/sasfit) for example includes >200 structural models with physical relevance and separable impact on the scattering spectra in connection with a fitting algorithm adapted to consider the relevant parts of the spectra. The MeV ion-beam analysis codes SimNRA (Mayer, SimNRA User's Guide IPP Report Number: IPP 9/113, 1997) and NDF (www.surrey.ac.uk/ion-beam-centre/research-areas/ion-beam-analysis) on the other hand apply a layered sample structure with elemental concentrations individual to each layer, since the method features a depth resolution with elemental sensitivity. For methods with industrial maturity such as electron induced X-ray emission (a.k.a. energy-dispersive X-ray spectroscopy) instrument manufacturers provide their own codes.

All of these codes and packages represent application specific compendia selected and optimized for a more or less specific task. Adding a source term to the equation system or simulating a nuclear reaction spectrum or whatever these code packages do relies on codes describing the basic physics of ion-matter interactions. This chapter demonstrated the complex physics behind beam-matter interactions, namely stopping, reactions/interactions, and kinematics. Starting already at the stopping power complex physics and numerous experimental data have to come together. For this topic several codes exist such as A-star and P-star for protons and α-particles and e-star for electrons (National Institute of Standards and Technology 2019). The most common code for arbitrary ions is the program SRIM(Ziegler et al. 2008) with the latest version of 2013 being available freely on www.srim.org. SRIM allows also for integration in user programs with the help of its sr.module outputing stopping powers if given projectile and target properties. The SRIM software features a large set of examples and input data comparisons proving its accuracy in the few percent range down to a few keV/amu. For electrons the program CASP (http://www.casp-program.org/) provides stopping powers, while CASINO (http://www.gel.usherbrooke.ca/casino) already extends beyond pure stopping and straggling. The Nucleonica database (Nucleonica GmbH 2014) provides a compendium of codes for electrons, ions, and also positrons and muons.

Calculation of nuclear reaction cross-sections remains a major challenge to date. It has to be distinguished between ab initio codes calculating the nuclei on the quark level and nuclear model codes based on a semi-empirical approach. The former include extreme amounts of interactions, since all quarks of a nucleus interact with all other quarks and gluons via the physics of quantum chromodynamics (QCD). These ab initio codes do not require input other than fundamental physical constants. Unfortunately, current computational technology limits the capabilities of QCD calculations rendering it currently irrelevant for the complicated resonance cross-sections of elements such as lithium and beyond. In contrast, nuclear model codes allow for calculating reaction cross-sections for any projectile-target combinations, but they require input about the structure of the involved nuclei and physical models of the involved interactions. These structural data include experimentally determined energy level schemes and state lifetimes peppered with the typical experimental limits and accuracies. The most famous codes are Talys (www.talys.eu) and EMPIRE (www.nds.iaea.org/empire). The Talys code supplies the TENDL database, often cited here, which contains total and differential cross-sections for nearly all possible reactions.

Many of the above mentioned software packages do not feature full graphical user interfaces (GUIs), but rather rely on a scripting language for user input. Programming becomes a central skill of a least higher level accelerator application experts and researchers. This relates not only to the operation of codes, but also post-processing of data such as fitting and extraction of results. The simple adjustment of a beam position requires only three points for a skilled operator able to apply a polynomial fit of 2nd grade to the data, while the simple non-programming skilled approach might require acquiring a dozen data points, stepwise approaching a minimum value. Programming or data science, respectively, nowadays often termed artificial intelligence "AI" in business presentations, becomes the accelerator experts sixth sense. The author recommends to any reader from personal experience to acquire solid programming skills when working in this field. Even human resources departments value the programming skills of accelerator physicists due to this natural connection. The programming language Python has proven to be maybe the most valuable option in the recent years. Not only is the syntax quite flexible and straightforward and its whole is based on open source, but most importantly it features numerous packages with pre-programmed complex functions for mathematics, data handling, and so on. Some of the graphs presented in this book also found on Python codes.

```
import numpy as np
from scipy.interpolate import interp1d
import scipy.integrate as integrate
import matplotlib.pyplot as plt

def integral(z):   # generate a function for integration
    x = np.genfromtxt("C:/pn cross-sec.txt", delimiter='\t', skip_header=1, usecols=0)
    y = np.genfromtxt("C:/pn cross-sec.txt", delimiter='\t', skip_header=1, usecols=1)

    xs  = np.genfromtxt("C:/p stopping h2o.txt", delimiter='\t', skip_header=1, usecols=0)
    stop= np.genfromtxt("C:/p stopping h2o.txt", delimiter='\t', skip_header=1, usecols=1)

    fsig = interp1d(x, y)
    fstop = interp1d(xs, stop)
    return integrate.quad(lambda x: fsig(x)/fstop(x), 1, z, limit=5000)

def integralOverE(z):   # generate a function with normalisation
    x = np.genfromtxt("C:/pn cross-sec.txt", delimiter='\t', skip_header=1, usecols=0)
    y = np.genfromtxt("C:/pn cross-sec.txt", delimiter='\t', skip_header=1, usecols=1)
    xs  = np.genfromtxt("C:/p stopping h2o.txt", delimiter='\t', skip_header=1, usecols=0)
    stop= np.genfromtxt("C:/p stopping h2o.txt", delimiter='\t', skip_header=1, usecols=1)

    fsig = interp1d(x, y)
    fstop = interp1d(xs, stop)
    return (integrate.quad(lambda x: (fsig(x)/fstop(x)), 1, z, limit=5000)/z)

xnew = np.linspace(1, 60, num=60, endpoint=True)
Integ= np.array([integral(xi)[0] for xi in xnew])  #slow but ok, thats not the problem
IntegE= np.array([integralOverE(xi)[0] for xi in xnew])  #slow but ok, thats not the problem

plt.plot(xnew, IntegE*10, 'o', xnew, Integ, '-')
plt.legend(["H x10", 'W'], loc='best')
plt.show()
```

The above code generates Fig. 3.18 from two files containing plain text lists of discrete points of the corresponding projectile energy dependent reaction cross-section and stopping power. The program interpolates these discrete lists forming integrable functions. The integrals are evaluated in a range of 1–60 MeV and plotted. Packages such as SciPy (Jones et al. 2001) and Numpy (Walt et al. 2011) ease this task by providing ready to use functions for most mathematical tasks. Packages for the export and complex plotting of the data exist allowing for fully automated parameter studies and acceleration of repeated analysis tasks. Even nuclear data packages exist for python integration.

References

G. Audi, et al., The NUBASE2016 evaluation of nuclear properties. Chin. Phys. C **41**, 030001, S (2017). https://doi.org/10.1088/1674-1137/41/3/030001

M. Berger, J. Hubbell, S. Seltzer, J. Chang, J. Coursey, R. Sukumar, D.S. Zucker, K. Olsen, *NIST XCOM: Photon Cross Sections Database* (Von abgerufen, 2010). https://www.nist.gov/pml/xcom-photon-cross-sections-database (2010)

J. Blocki, J. Randrup, W. Swiatecki, C. Tsang, Proximity forces. Ann. Phys. **105**, 427–462, S. (1977). https://doi.org/10.1016/0003-4916(77)90249-4

G. Ciullo, R. Engels, M. Büscher, A. Vasilyev, *Nuclear Fusion with Polarized Fuel* (Springer 2016)
Culham Centre for Fusion Energy, *FISPACT-II Material Handbooks*. (Abgerufen am 05 2018 von, 2018). https://www.ccfe.ac.uk/fispact_handbooks.aspx
A.F. Gurbich, SigmaCalc recent development and present status of the evaluated cross-sections for IBA. Nucl. Instrum. Methods Phys. Res., Sect. B **371**, 27–32 (2016). https://doi.org/10.1016/j.nimb.2015.09.035
M. Herman, R. Capote, B. Carlson, P. Oblozinsky, M. Sin, A. Trkov, V. Zerkin, EMPIRE: nuclear reaction model code system for data evaluation. Nucl. Data Sheets **108**, 2655–2715 (2007)
W. Huang et al., The AME2016 atomic mass evaluation (I). Evaluation of input data; and adjustment procedures. Chin. Phys. C **41** 030002, S (2017). https://doi.org/10.1088/1674-1137/41/3/030002
E. Jones, E. Oliphant, P. Peterson, et al., *SciPy: Open Source Scientific Tools for Python* (Von abgerufen, 2001). http://www.scipy.org/
A. Koning, et al., *TENDL-2015: TALYS-Based Evaluated Nuclear Data Library* (Von abgerufen, 2015). https://tendl.web.psi.ch/tendl_2015/tendl2015.html
M. Mayer, *SimNRA User's Guide IPP Report Number: IPP 9/113* (Max-Planck-Institut für Plasmaphysik, Garching). www.simnra.com
S. Möller, Analytical continuous slowing down model for nuclear reaction cross-section measurements by exploitation of stopping for projectile energy scanning and results for 13C(3He,α)12C and 13C(3He,p)15N. Nucl. Instrum. Methods Phys. Res. Sect. B **394**, 134–140 (2017). https://doi.org/10.1016/j.nimb.2017.01.017
National Institute of Standards and Technology, *NIST: Introduction of e-star, p-star, and a-star*. (Von abgerufen, 2019). https://physics.nist.gov/PhysRefData/Star/Text/intro.html
M. Nastasi, J.W. Mayer, Y. Wang, *Ion Beam Analysis: Fundamentals and Applications*. CRC Press (2014)
Nucleonica GmbH., *Nucleonica Nuclear Science Portal*. (www.nucleonica.com) *Version 3.0.49* (Nucleonica GmbH Karlsruhe, Germany, 2014)
OECD Nuclear Energy Agency (NEA), *JANIS* (Von abgerufen, 2017). http://www.oecd-nea.org/janis/
P. Sigmund, *Particle Penetration and Radiation Effects: General Aspects and Stopping of Swift Point Charges* (Springer, Berlin, Heidelberg, New York, 2006). ISBN 13978-3-540-31713-5
J. Sillanpää, Electronic stopping of silicon from a 3D charge distribution. Nuclear Instrum. Methods Phys. Res. Sect. B: Beam Interact. Mater. Atoms Vol. **164–165**, 302–309 (2000)
S.V. Walt, Colbert, S. C., G. Varoquaux, The NumPy array: a structure for efficient numerical computation. Comput. Sci. Eng. **13**, S. 22–30 (2011). https://doi.org/10.1109/mcse.2011.37.a
V. Zagrebaev, A. Denikin, A. Karpov, A. Alekseev, M. Naumenko, V. Rachkov, M.A. Naumenko, V. Saiko, *NRV Web Knowledge Base on Low-Energy Nuclear Physics—2 Body Kinematics* (Von abgerufen, 2019). http://nrv.jinr.ru/nrv/mobilenrv/kinematics/Kinematics2Body.html
J.F. Ziegler, J.P. Biersack, M.D. Ziegler, *SRIM—The Stopping and Range of Ions in Matter* (Chester, 2008)

Chapter 4
Secondary Particle Generation with Accelerators

Abstract The beam-matter interaction results in a multitude of secondary particles. These particles bear information useful for analytical method and the potential for applications through their controlled generation. The knowledge of their emission physics results in technological options for the control of their properties. The production of secondary ions, electrons, photons, neutrons, and exotic particles will be discussed based on relevant application examples and connections to future chapters. In particular, neutron and photon sources based on accelerated induced secondary particle physics already have numerous applications. Different source options will be discussed together with layout optimisations.

> You know the best thing about a foursome? If one quits you still have a threesome (House M.D.)

In the last chapter, we investigated basic physics of particle-beam matter interactions and we already noticed the physical aspects to have different relevance for the applications. The terms *production* and *analytical* application form a rough frame of the directions of accelerator applications. There is an overlap extending beyond both being accelerator applications, but also several distinctions. Production applications rely on terms such as energy efficiency, rates and reaction pathways, while analytical applications more focus on separation/resolution of particles, ratios of processes and energy dependencies. What they have in common is a need to control the secondary particles released during the interaction.

The scientific method requires three steps for controlling a process: First describe it to be sure what exactly you are talking about; second analyse it to reveal its patterns; third rehearse your analysis via feedback in order to verify it to a maximum level of confidence (the reason why physicsts determine fundamental constants with 18 digit accuracy).

In general, accelerators allow for the production/release of all types of fundamental particles from the beam targets. Usually it is even unavoidable to produce or release certain particle types due to the physics of the interaction processes. Firing a particle beam onto a target yields a situation similar to throwing a stone into water, see Fig. 4.1: Upon contact the water deforms, some of it is released in all directions, backwards we see splashes and in forward direction bubbles of air. Just like this, the

S. Möller, *Accelerator Technology*, Particle Acceleration and Detection,
https://doi.org/10.1007/978-3-030-62308-1_4

Fig. 4.1 A stone thrown into water releases a burst of matter from its target similar to a particle hitting a target

projectile impact on target atoms will release splashes of matter and energy in all directions (see Fig. 1.2). Water, being a liquid, will quickly recover to its original state just as matter will. The analogy lacks in the interaction of the stone with the water on the nuclear level. Furthermore it is a deterministic situation, while on the atomic scale processes follow statistical rules.

So what are we talking about with the term secondary particles? It denotes all particles (ions, electrons, neutrons, photons ...) produced in the reaction, or accelerated by a reaction collision, leaving the reaction zone/target. This definition is in fact a bit vague since for example the same product can leave the target or not, depending on it being directed into the target or directed outwards. The particle leaving the target will be detectable, though, a major difference for the application. Applications can make use of secondary particles. The energy/energy distribution of the product particles becomes a crucial part, since low energy particles hardly leave the target. Generally, the higher the reaction energy, the more secondary particles are emitted. In case of surface near low energy products this logic can flip over due to a local reduction of the interaction probability with increasing energy.

Why should we want to produce particle from a particle beam? Ion and electron beams are easily produced, see Sect. 2.4, and these beams can be tailored due to their charged nature by beam optics, while in particular uncharged particles (photons and neutrons) or even strange fundamental particles are not easily emitted or influenced by our electro-magnetic technology. Secondary particles extend the range of accessible particles, if we can transfer our knowledge and ability to control the charged particles to the secondary particles. Furthermore, the secondary particles carry information about target and beam-matter interaction due to connecting physics such as kinematics and energy-loss. In spite of their useful aspect, secondary particles also represent a burden due to radiation safety.

The origin of secondary particles lies in the non-deterministic nature of nuclear physics. It behaves a bit like the British people in the 2019 Brexit discussion: It's unclear what it wants but it's totally clear what it does not want. The conservation laws (momentum, energy ...) make clear statements of what's not possible. Anything possible will also happen with a certain probability. The following sections will explain our understanding of the underlying processes generating the different particle types.

$$E = mc^2 \tag{4.1}$$

Last but not least, Einstein's famous (4.1) connects to secondary particle generation with accelerators. It connects the electrical energy we put into the accelerator with the particles we release. Although in the application range of energies, secondary particles are mostly released from bound states, the E in Einstein's equation can stand for the electrical energy from a power plug (with correct handling of units) which the beam-matter interaction converts into free particles.

4.1 Electrons, Atoms, Molecules, and Ions

Generating secondary electrons and ions requires the transfer of a sufficient amount of energy from impacting projectiles to the target atoms to break their bonds with the surrounding matter and release target particles as free secondary particles, see e.g. (Gnaser 1999) as a more detailed reference. In this section we will consider secondary particles as particles already present in the target before the reaction, not particle generated for example in a nuclear reaction or decay due to the different physics connected with their emission. The equations of energy transfer based on conservation of momentum were discussed in the last chapter. An important physical conclusion from kinematics for secondary particles lies in the mechanism generating backward oriented secondary particles. Secondary particles typically feature low kinetic energies due to the kinematic contradiction of high energy transfer and momentum reversal. A high energy transfer from projectile to target implies a head on collision which conserves the forward momentum. The low energy limits the secondary particle range to a few ten nm. This limits the escape of the secondary particles to surface near depth. The closest surface is the backward surface, making this the direction relevant for applications. From kinematics, the primary particles (projectiles) can be deflected backwards in a single collision, if they are lighter than the target. A target particle, in contrast, requires several subsequent interactions with other target particles in order be redirected to the backscattering direction. These multiple scattering events depicted in Fig. 4.2 reverse the initial momentum directed into the target towards the backscattering direction.

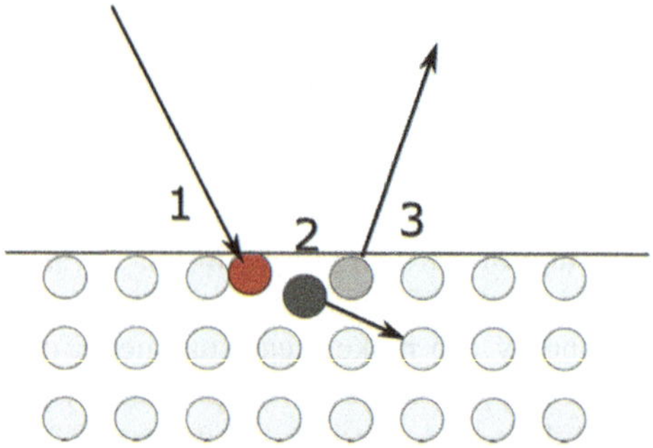

Fig. 4.2 The chain of events of secondary particle release. The red incident projectile hits a surface near atom (1). This primary knock-on atom (PKA) or electron receives a part of the projectile energy freeing it from its lattice binding. The primary hits another surface near particle (2). This secondary particle receives a fraction of the primaries energy, but again changes the direction of momentum (arrows). With sufficient energy it passes the surface barrier (3) being released as free, sputtered atom/electron

In addition to kinematic aspects energy-loss plays an important role for both the projectile and the secondary electrons and ions (in contrast to photons). On its way out of the sample, the secondary particle experiences energy-loss. Consequently, only the particles starting close to the surface feature enough energy to actually reach the surface, particles generated deeper will stop within the target. The projectile type, its energy, in this context called the impact energy, and its angle of incidence on the target represent the defining quantities for the generation of secondary particles and their properties from a given target. Heavier projectiles at energies around the stopping power maximum generally produce the highest secondary particle yields (yield = secondaries/projectiles). Shallow impact angles tend to shorten the chain of events shown in Fig. 4.2, shifting it closer to the surface and therefore increasing the yield.

The target surface represents an important boundary for secondary electrons and atoms/ions. Electrons and ions constitute matter, therefore binding energy keeping matter in its state are equivalent to threshold energies for releasing free secondary particles. The surface represents a special situation for this binding, since basically half the binding partners are missing for the surface atoms. The corresponding binding energies are the so-called work function for electrons (known from thermal electron sources) and the surface binding energy for ions (known as sublimation energy). For example polycrystalline tungsten features a work function of 4.5 eV and a surface binding energy of 8.79 eV. The projectile has to displace the secondary particle from its initial binding, then it moves towards the surface where it finally surpasses the surface barrier. Interestingly this chain of events results in a secondary particle energy only weakly depending on the projectile impact energy (Behrisch and Eckstein 2007; Seiler 1983).

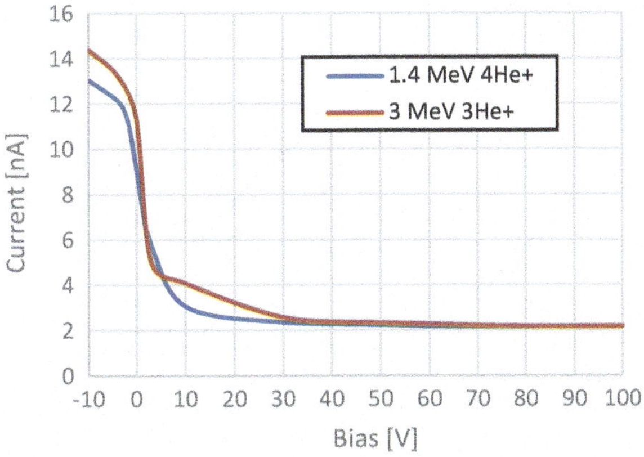

Fig. 4.3 Apparent beam current versus target bias for two He projectile types impinging at 0° on a biased polished tungsten sample. Positive biasing draws back secondary electrons to the target, if their kinetic energy is below the biasing voltage. The secondary electron energies are mostly identical for both impact energies. The current at +100 V bias represents the impacting ion current, hence each projectile releases up to 7 electrons at −10 V biasing (14/2)

Figure 4.3 shows an example of the apparent current composed of the impacting ions and the backscattered secondary electrons for two different impacting ion energies. Most of the secondary electrons have energies below 10 eV, in spite of the MeV projectile energy. The kinetic energy E distribution function f_E of the emitted particles typically follows the Thompson distribution (Eksaeva et al. 2017) (4.2). The only parameters are the distribution parameter a and the surface binding energy E_{SB}. Consequently, the model states an impact energy independent emission energy, only material dependent values influence the emission energy. In reality, minor parts at the high energy fall-off show an impact energy dependence.

$$f_E(E) = \frac{a(a-1)E * E_{\mathrm{SB}}^{a-1}}{(E + E_{\mathrm{SB}})^{a+1}} \tag{4.2}$$

The angular emission distribution of secondary particles typically follows a cosine function of the angle against the surface normal, see Fig. 4.4. This cosine typically features a non-one exponent which weakly depends on impact energy for a given target. For secondary electrons emitted by electron impact this holds true independent of the electron impact angle (Seiler 1983), but for secondary ions and atoms, the angle of maximum emission points towards the specular reflection direction of the projectiles as we would expect from the billiard like collision kinematics. For very low ion impact energies and materials of special order such as single crystals and certain metals the maximum deviates from the specular direction, following for example the crystallographic planes or resembling the so-called butterfly shape.

Whether a target atom is released as secondary positive ion, negative ion, or neutral depends on the surface. The initial binding energy of an atom with its electrons strongly depend on its chemical surrounding, leading to the so-called matrix effect.

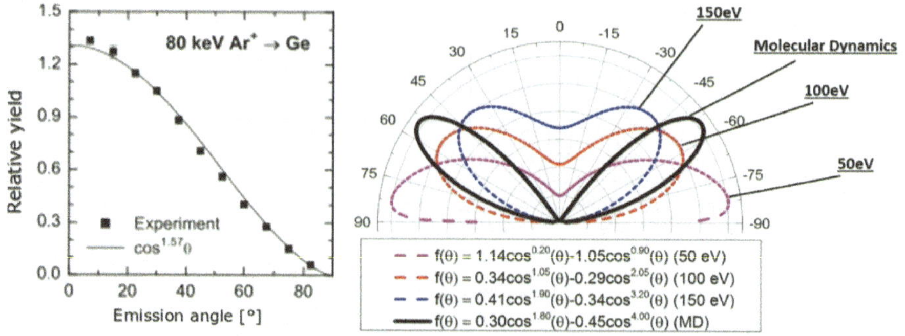

Fig. 4.4 Left: Cosine angular distributions of sputtered Ge particles emitted into the backscattering half-space against the surface normal. Reproduced from (Behrisch and Eckstein 2007). Right: W particles sputtered by low energy Ar ions with 50–150 eV impact energy are emitted in a butterfly shaped distribution. The dashed coloured lines represent experimental data, the black solid line a molecular dynamics simulation. Reproduced from (Eksaeva et al. 2017) published under CC-BY 4.0

Practically roughness, oxidation, and impurities further change the surface bindings from the bulk. In any case, the neutral emission mode is energetically favourable, but not inevitable. Sputtering with Alkali projectiles, e.g. Cs, strongly enhance the emission of negative ions due to their loosely bound outermost electron which they readily contribute to the surface upon implantation. Electron attracting elements, in particular oxygen, on the other hand increase the amount of positive ions while suppressing negative ion emission. Consequently, the release type of ions strongly depends on the target elements in the same manner. An oxygen atom would be emitted rather as negative ion than as positive ion, while a Cs atom prefers the positive over the negative state.

The process of secondary ion release is called sputtering. Due to their low mass and stopping power electrons hardly enable the emission of atoms and ions, but it requires other heavy particles to release them. Sputtering consumes the surface via the release of material, making it useful in applications of microscopic drilling/machining. The binary collision approximation in the form of the semi-empirical Eckstein formula (Behrisch and Eckstein 2007) (equation 4.3) or the SRIM (Ziegler et al. 2008) and SD.Trim.Sp codes describes sputtering with high accuracy. Equation (4.3) describes the total sputtering yield Y_S as a function of the ion impact energy E_0, the stopping power S, and the empirical fitting constants *A, B, C*, E_{thres} (= sputtering threshold) which are available in (Behrisch and Eckstein 2007). The binary collision model works well in the typical accelerator context, but reaches limits in the impact energy region below a few 100 eV typical for plasmas. With lower energies multi-body interactions gain increased importance for sputtering, requiring more complex molecular dynamics simulations. Energies above MeV and heavy projectiles mark the other limit of the binary collision approximation. Here the majority of energy deposits via electronic losses, the relevance of nuclear (billiard) collisions decreases. For these so-called swift heavy ions the physics of sputtering changes to a collective process local thermal spikes and evaporation.

$$
\begin{aligned}
Y_S(E_0) &= A * S(\lambda E_0) * \frac{\left(\frac{E_0}{E_{\text{thres}}}-1\right)^C}{B*\left(\frac{E_0}{E_{\text{thres}}}-1\right)^C} \\
\text{with } \lambda &= \frac{m_2}{m_1+m_2} * \frac{\left(\frac{9}{128\pi}\right)^{\frac{1}{3}} * \frac{\varepsilon_0 h^2}{m_e e^2}}{Z_1 Z_2 e^2 \sqrt{Z_1^{\frac{2}{3}} + Z_2^{\frac{2}{3}}}}
\end{aligned}
\tag{4.3}
$$

For special combinations of projectile and target, where volatile molecules can be formed, chemical interactions induce a loss of target material, so-called chemical sputtering. Unless sputtering this process is not a direct result of collision cascades, but relates to the accumulation of atoms loosely bound to the surface, so-called ad-atoms, produced by the projectile impact. Chemical reactions on the surface form volatiles, e.g. H ions on C form CH_4, which subsequently evaporated. The balance of surface concentration of projectile ad-atoms and the evaporation induces a temperature-dependence as demonstrated in Fig. 4.5. Higher temperatures open up more chemical sputtering channels, for example W sputtering by O ions features a chemical component due to the relevant vapour pressure of WO_2 above 1100 K.

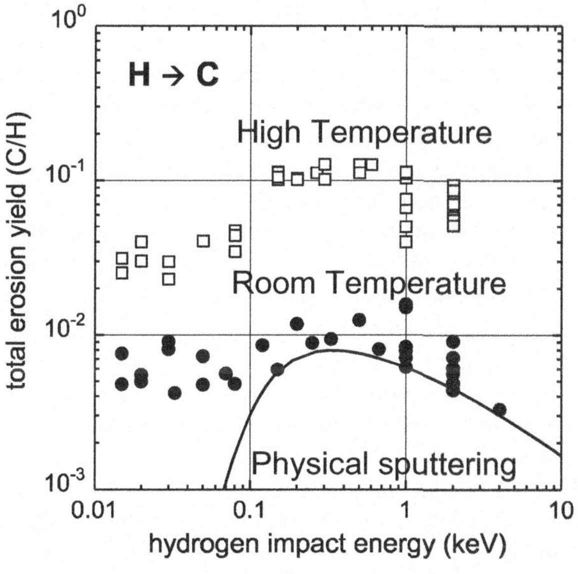

Fig. 4.5 Sputtering/Erosion yield of graphite (carbon) against H impact energy. Open symbols represent surface temperatures of 570–920 K, filled symbols are around room temperature (≈300 K). Collision cascades induce physical sputtering. The formation of volatiles such as CH_4 induces chemical sputtering for chemically interaction projectiles and targets only. The chemical sputtering yield mostly depends on the surface temperature, while physical sputtering depends mostly on the impact energy. Consequently, physical sputtering (solid line) dominates at lower temperatures and higher impact energies, while chemical sputtering dominates at low impact energies. Reproduced from (Behrisch and Eckstein 2007) with permission by Springer

4.2 Neutrons

I've Been Looking for Freedom (David Hasselhoff)

Neutrons are naturally bound to nuclei. Free neutrons have a limited half-life of about 610 s. Their rest mass exceeds the proton mass by 1.3 MeV, consequently they decay to a proton, an electron, and an electron anti-neutrino. Yet their specific properties make it worth setting them free. Neutrons are the only heavy particle without electric charge (except for some special particles). This implies several important differences to charged particles: The electronic energy-loss mechanism does not apply to neutrons. As a consequence, neutrons have orders of magnitude longer ranges than charged particles and they can penetrate matter down to thermal energies in the meV range. In contrast to ions, neutrons cannot be bound to matter, therefore remaining free even at zero kinetic energy. No charge also means no coulomb barrier hindering nuclear reactions. While ions require several MeV, neutrons react with nuclei at any energy (respecting conservation of energy). This decouples the beam energy and reaction rates from the reaction Q-value. Neutrons enable production of other isotopes than ions, an important degree of freedom for isotope applications.

The absence of nuclear charge implies important application drawbacks. So far we have no means of accelerating neutrons. We can only produce free neutrons via nuclear reactions releasing them from the atomic nucleus. Thereafter we have to live with the release energy. Collisions with light nuclei, in particular the hydrogen bound in water, enable reducing the neutron energy (aka. Moderation), but these collisional processes cannot produce sharp beam energies as we know from charged particle beams, see Fig. 4.6. Collisional processes will produce statistical energy distributions, preventing for example non-ambiguous kinematic relations. Most neutron releasing reactions anyways produce broad energy spectra due to their multi-body decay nature. Even for sharp primary spectra the high range of neutrons enables interactions with the surroundings generating mixed spectra as displayed in Fig. 4.6.

The absence of nuclear charge also prevents the application of beam optical devices. For neutron sources producing an isotropic emission the flux density will decay with the square of the distance to the neutron source. It will not be possible to focus the complete source flux onto a single target like with a charged particle beam or amplify the neutron beam using external energy input as with optical Lasers. Neutrons guide tubes provide a small relief from this practical problem. Seeing neutrons in the wave picture enables total reflection, as known from light optics, from the guide tube walls. Total reflection prevents absorption and transmission of the incident neutrons when incident in a shallow angle onto the wall. The incident angle must be smaller than a critical angle defined by the arcsin of the ratio of refractive index of two adjacent materials (e.g. air vs. road for light). For neutrons with their little interaction with matter only minute differences in the refractive index to the vacuum value exist. Traditionally nickel coatings on glass tubes were used, but newly applied layered materials feature slightly higher refractive indices, enabling the reflection of neutrons from larger incident angles, further reducing the losses. The

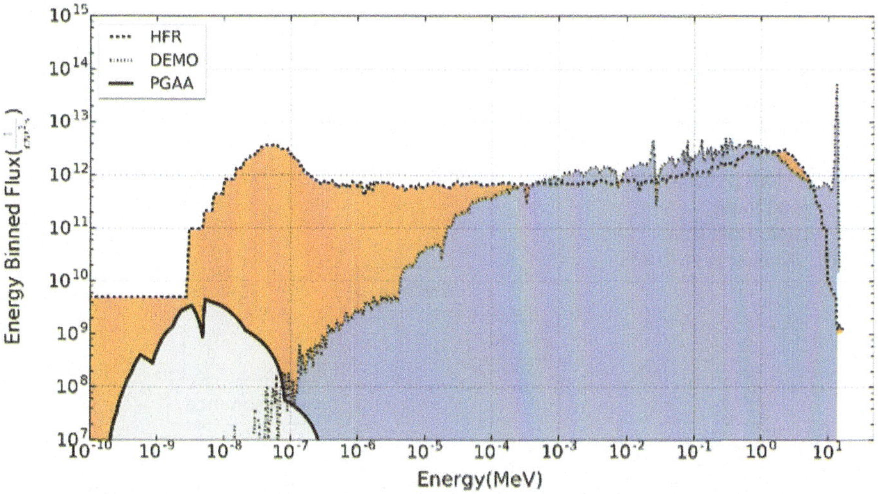

Fig. 4.6 Binned neutron flux spectrum for sample exposure in the HFR research fission reactor in Petten. DEMO is a planned nuclear fusion power plant, but in spite of it producing only 14.1 MeV and 2.45 MeV neutrons from the D-T and D-D reactions a broad energy spectrum can be observed in its inside. The PGAA analytical instrument at the FRM-2 reactor works with a cooled neutron spectrum. Data extracted from DEMO from https://fispact.ukaea.uk/wp-content/uploads/2016/10/CCFE-R1636.pdf; HFR from https://fispact.ukaea.uk/wp-content/uploads/2016/10/Handbook_HFR_UKAEA1532.pdf; PGAA from (Kudejova et al. 2008)

neutron guides enable fitting more end-stations to a given neutron source without excessive flux density losses associated with the distance law. At the same time they can be tailored to add functionality via filtering of energy, direction, or spin of the neutrons. Unfortunately, the effect only works for relatively long neutron wavelength connected to cold neutrons (meV).

The nuclear reaction cross-sections of neutrons typically features three different behaviours along the μeV to MeV range of neutron energies. At low energies up to about 1 keV the cross-section decreases exponentially from high values. In the keV to MeV range multiple sharp and strong resonances occur from a relatively constant baseline. Above MeV again a region of decreasing cross-section starts, but at a lower level compared to the low energy range. This lead to the definition of three main neutron energy ranges, the thermal, the resonance (also epithermal), and the fast neutron energy range as depicted in Fig. 4.7. Thermal neutrons originate from collisions with matter at a certain temperature, therefore the region further separates according to the thermal temperature into room temperature (=thermal) and cryogenic temperatures typically in the liquid hydrogen range (=cold).

Traditionally so-called research (fission) reactors produce application relevant fluxes of neutrons from uranium fission. These reactors produce a thermal neutron output flux of a few W up to a few 10 MW (e.g. the FRM-II reactor) with maximum flux densities in the order of 10^{15} neutrons/cm^2 at the core and some 10^{10} n/cm^2 at far end-stations. The source strength of the best research reactors lies in the order of 10^{18}

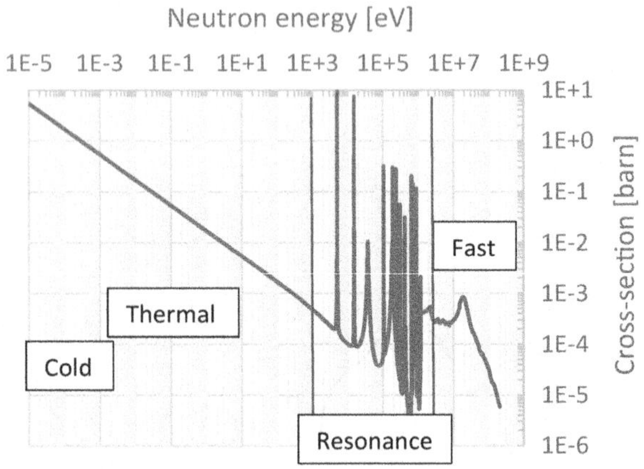

Fig. 4.7 Representative neutron reaction cross-section of the $^{30}Si(n, \gamma)^{31}Si$ reaction used for neutron doping of semiconductors. Three distinct regions of behaviour can be found in the cross-section. Data from TENDL (Koning et al. 2015)

neutrons/s. The radioactive waste and the highly enriched uranium with >20% ^{235}U (weapon usable) required for maximum neutron fluxes started a political debate about possible alternatives for these reactor types. A replacement for the regular 3–5% ^{235}U enriched fuel is possible at the expense of lower neutron fluxes and service life. In most cases lower neutron flux limits the productivity of the applications (flux and runtime are connected), therefore lower fluxes devalue the expensive reactor and end-stations. Together with the decaying acceptance of fission reactors this strengthens the search for possible high flux alternatives not relying on fissile fuels.

Neutrons are applied in a wide range with a certain overlap with charged particles, but often in a complementary way. Industrial applications include the doping of silicon via neutron capture of ^{30}Si forming ^{31}Si which decays to ^{31}P or the high sensitivity analysis of geological samples for resources. In the scientific context, neutrons have a special meaning for soft and biological matter due to the low energy deposition of thermal neutrons, but also numerous analytical applications in materials research exist. In the medical context neutrons aid in particular the production of certain medically relevant radioisotopes, but also direct treatment of cancer with neutrons irradiation is established.

4.2.1 Accelerator Neutron Sources

Accelerators with their possibilities for extreme beam power density and flexibility provide the main alternative for fission reactions for producing intense neutron beams. Beam focussing leads to small neutron source extend, enabling high fluxes even at lower powers. In contrast to fission reactors, beam optics enables close distances between neutron source and application resulting in increased usage ratios of produced neutrons. Bringing source and application closer together improves the geometric efficiency or in other words the ratio of flux density to source strength. Some, not all, concepts work completely without heavy metals and result in only little activation, important aspects for a new decentralised application concept of neutrons. The flexibility of accelerators and the many option in connection with the complicated technical developments of this new type of neutron source only recently resulted in the first competitive devices.

Many nuclear reactions produce neutrons so lets first sort them to get a start on the technological concepts. Historically the $^{9}Be(^{4}He, n)^{12}C$ reaction was used for producing neutrons from α-decaying radioactive isotopes resulting in a radioisotope neutron source. Mixtures of for example ^{241}Am produced in fission reactors with beryllium metal are still in use as reference sources with neutron fluxes up to about 10^6 neutrons/s. Practically we would like to see at least ten orders of magnitude higher fluxes/source strength. The reaction rate depends on the ratio of reaction cross-section to stopping power (see Sect. 3.4). Consequently, protons and deuterons have advantages over helium projectiles since they offer lower stopping power combined with technically favourable ion source properties (see Sect. 2.4). Independent of the projectiles two classes of reactions exist: 2-body reactions with light isotopes (practically hydrogen isotopes, lithium, and beryllium) such as $^{2}H(d,n)^{3}He$ or $^{7}Li(p, n)^{7}Be$ and multi-body (p, xn) reactions with heavy elements such as tungsten. Light isotopes offer small stopping power and low reaction thresholds. For example the reaction $^{7}Li(p, n)^{7}Be$ has $Q = -1.64$ MeV. Due to the limited amount of not super-short-lived light isotopes and the requirement of only a few MeV projectile energy only little problematic radioisotopes are produced with light targets. Every reaction produces a single neutron with fixed energy. In contrast, reactions with heavy targets require at least a few 10 MeV projectile energy for surpassing the Coulomb-barrier. A typical target is W due to its good thermo-mechanical properties and acceptable costs. Here $^{nat}W(p, xn)Re$ reactions produce several neutrons per reaction, but the multi-body reaction results in broad neutron energy spectra. Above about 100 MeV projectile energy a fission of the target nucleus becomes possible, releasing even more neutrons. The efficiency of an individual reaction decreases above a few 10 MeV, but as more reactions releasing more neutrons per reaction become available, the efficiency improves with beam energy. The spallation process can even be boosted with neutron induced fission reactions. These so-called accelerator driven systems multiply the spallation neutron count by a factor 10 or more by using them for inducing sub-critical fission reactions e.g. with uranium. This special neutron source type will be discussed in Sect. 8.1.

MeV photons allow releasing neutrons from the atomic nucleus via the nuclear variant of the photo-effect. Electron based neutron sources exploit the same mechanism by first producing high energy Bremsstrahlung photons from several 10 MeV electrons which subsequently induce the nuclear photo-effect. In both cases the (γ, n) process becomes efficient in the so-called giant dipole resonance region typically found in the region of 13–25 MeV photons (Herman and Fultz 1975). Basically, the photons bring the nucleus to a high energy excited state as depicted in Fig. 3.10. This state then decays via emission of nuclear constituents. In spite of the resonance being "giant" the overall process is not as efficient as ion induced processes and generally results in very high photon radiation levels, forming a practical limit for radiation shielding.

The previous discussion already sorted out a few concepts. Let us start with a small-scale neutron source, a so-called electronic neutron generator (ENG), with $\approx$100 keV deuterons inducing (d, n) on tritium targets. Small DC accelerators with metal targets loaded with tritium such as Titaniumhydride as depicted in Fig. 4.8. Most of the beam energy converts to heat due to energy-loss in matter. For detailed calculations see Sect. 8.3.1. The reader can calculate the required beam current and power for a given neutron output using the reaction probability. The heat load induced in a shallow layer due to the small ion range limits the neutron output. A technical variant uses slanted targets for increasing the receiving target area as depicted in Fig. 4.9, at the trade-off of a larger target. Target size and the minimum distance strongly influences the neutron flux received by an application. Neutrons distribute isotropically, therefore a larger working distance results in lower flux. With a source-size in the order of 1 cm^3 but a minimum working distance of about 3 cm the source flux distributes over 113.1 cm^2. A source strength of a typical commercial product of 10^9 neutrons/s results in a maximum flux density of 8.8×10^6 neutrons/(cm^2s).

The high stopping power at the low projectile energy results in a low yield of neutrons per MeV of projectile energy. According to Fig. 8.10 the reaction probability in a Titaniumhydride lies around 2×10^{-5}, multiplied with 120 keV ion energy we

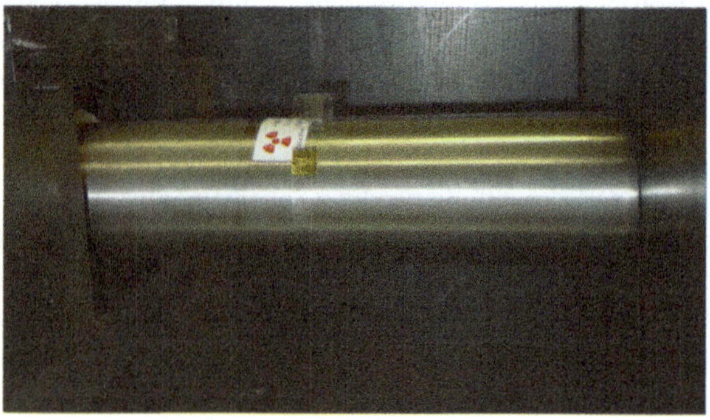

Fig. 4.8 An electronic D-T neutron generator encased in black polyethylene blocks. The tube encasing has a diameter of roughly 60 mm and emits up to 10^9 n/s. The device weights about 15 kg in total. The output decays over operating time due to the consumption of the tritium, lasting about 10,000 h. Copyright Forschungszentrum Jülich/Rahul Rayaprolu

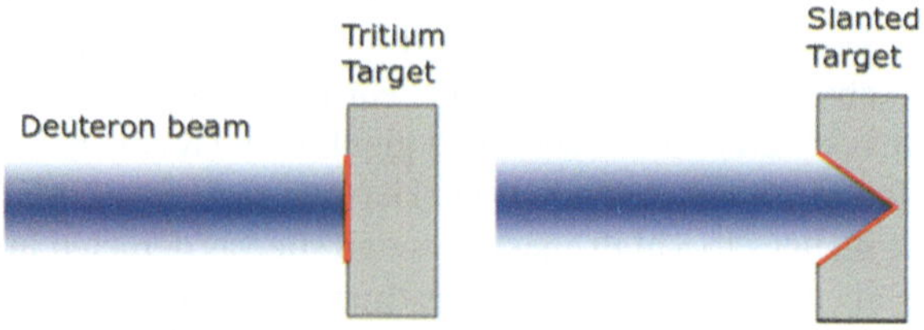

Fig. 4.9 Technical approach for reducing the areal heat flux in a beam target by increasing the impact area through inclination

end up with 6 GeV of energy invested per emitted neutron. This low efficiency limits the neutron output. Consequently, increasing the projectile energy will extend our technological limit of source strength.

Unfortunately the T(d, n)^{4}He reaction cross-section drops towards higher energies, encouraging us to consider other reactions when aiming at higher efficiency in the MeV region. Competitive concepts of such neutron sources appeared together with the shut-down plans for fission based sources, e.g. (Mauerhofer et al. 2017). This concept achieves a high flux density via a 10 MeV proton beam incident on a compact and flexible beryllium neutron source with close positioning of moderators and end-stations. The MeV beam energy leads to a limited localized nuclear activation of components as depicted in Fig. 2.57, activation starts becoming an issue in this energy range. At the required ion fluxes of >1 mA on a 40 mm diameter target implantation of gaseous species imposes a threat for mechanical stability of the target. Bubbles can form at the end of the ion range, rapidly destroying the target via gas damage (Sect. 7.4). Either the target has to be hot enough to outgas the implanted species or it has to be thinner than the ion range in order to implant the ions into the cooling fluid behind the target. In this case, a 0.7 mm thin target was chosen due to the ion range of 0.8 ± 0.04 mm. The target acts as a beam window as discussed in Sect. 2.6 between vacuum and cooling water in this case. Mechanical stability represents an issue, defining a lower beam energy limit connected to a minimum thickness. Due to the bad thermo-mechanical properties of lithium designs with lithium targets rather choose liquid lithium as a flowing curtain, see Fig. 4.10. The lithium flow combines target and coolant functions. Lastly, in particular deuteron, gas targets enable the direct use of the lightest target element. Reasonable ion ranges require pressures in the bar range, too much for the vacuum system of the accelerator. Either a solid window, limitations for the beam power, or clever and powerful vacuum systems in combination with beam optics can be applied.

With the technical concepts ready we can take a look if the MeV region actually brings a gain. Figure 4.11 compares a beryllium with a tungsten target. At 3 MeV the beryllium cross-section reaches relevant values, reaching a maximum at 10 MeV. A further increase of beam energy above 10 MeV brings practically no increase in cross-section, only slight gain due to decreasing stopping power. Heavy elements on the

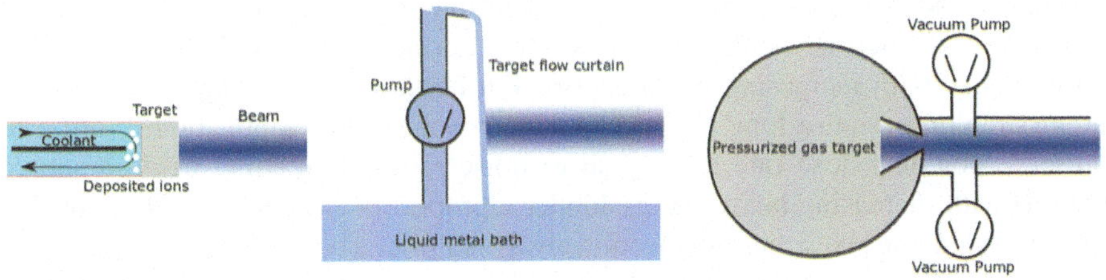

Fig. 4.10 Sketch of the three basic target options for neutron sources. Solid window like target, liquid flow target, and gas chamber target. The gas target chamber applies a cascade of gas apertures and pumps in order to reduce the vacuum pressure inspite of the high influx from the gas target

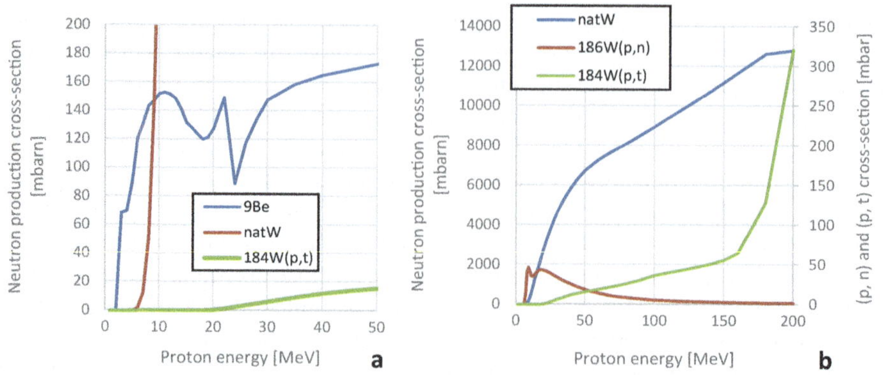

Fig. 4.11 Neutron production cross-section for p+^{nat}W and p+Be in comparison to an individual (p, n) cross-section and the tritium production cross-section relevant for radiation safety. Additional reactions open up at higher proton energies resulting in a monotonous increase of the neutron production cross-section with energy for the heavy tungsten, but a constant value for Be above 50 meV. At the same time radiation safety issues strongly increase with relevant tritium production above about 13 meV. Data from TENDL (Koning et al. 2015)

other hand allow for a multitude of (p, xn) reactions, producing several neutrons per reaction, which sum up with increasing energy. The comparison of the (p, n) reaction cross-section with the neutron production cross-section in Fig. 4.11b shows (p, n) dominates the neutron production cross-section up to 9 MeV, but thereafter other reactions take over resulting in a different behaviour than with the light target. The cross-section monotonically increases and the stopping power decreases, the process becomes increasingly efficient reaching a value of 120 MeV/neutron at 200 MeV with on average 1.66 neutrons produced per proton. At the end of this books energy range of 250 MeV the spallation regime starts. Here the proton energy suffices for fission reactions of the target nucleus, breaking the nucleus into two smaller nuclei and a set of neutrons. The neutron production cross-section continues to increase, but the numerous isotopes produced further increase nuclear waste issues.

Kinematics connects the maximum neutron energy of accelerator neutron sources to the incident beam energy. In most cases high neutron energies are rather a complication than an advantage, see for example the neutron induced cross-section in Fig. 4.7 screaming for thermal neutrons. The moderation and shielding (see Sect. 2.7.3) grow in size with increasing beam energy. Furthermore, increasing activation and bremsstrahlung require long distances between source and application in order to preserve practical dose rate and signal to noise ratios. Extending neutron source regions with increasing beam energy hinder a proportional increase in flux density. In conclusion, the beam energy selection always implies a compromise between flux, cost, size, and practical problems.

A practical setup requires maintenance and end-of-service provisions. This converts nuclear activation products to radioactive waste. Naturally, activation strongly scales up with beam energy since more reactions open up. For a 10

MeV proton source with beryllium target considerations were already presented in Fig. 2.57. In this case target activation resulted mainly from Iron alloyed to the Beryllium for processing reasons. Be and other light ions produce short-lived isotopes from proton irradiation. The only critical product from light ions is tritium as long-lived yet hardly detectable radioactive product, accompanied by secondary products due to the neutron radiation. In spallation targets, the reaction results directly in radioactive isotopes and a broad range of elements, leading to structural activation with spallation energies. Concerning other secondary particles, with increasing energy more species arise as reactions with more negative Q-values become possible.

Currently several scientific accelerator neutron sources are in operation or under realisation. The lab-scale NOVA-ERA source was already discussed. Numerous commercial systems based on proton and deuteron ion beams with handheld to laboratory sizes providing similar source strength up to several 10^{14} neutrons/s are available (IAEA 2000). The currently most extreme example is the European Spallation Source (ESS) under construction in Lund. This device applies a 2 GeV proton beam to a rotating tungsten target producing up to 10^{18} neutrons/s ($\approx$FRM-II). It employs a 400 m long proton LinAC with 5 MW beam power operated in pulsed mode for neutron time-of-flight analysis methods. The beam impacts on a 2.6 m diameter rotating tungsten target. Up to 42 beamlines enable connecting a wide range of instruments and experiments.

4.2.2 The Specific Energy Efficiency

Generally, energy efficiency and costs currently becomes increasingly important in technical concepts and this also affects accelerator applications. Definitely it would be a waste of ink printing all the zeroes typically present in the value of the energy efficiency of scientific experiments. For applications, the energy efficiency constitutes a central quantity. This is not only about costs of electricity, as of course consuming less energy to achieve the same goal is always cheaper, but mainly about technical limits as we saw in Sects. 2.2, 2.6 and others. If a system is properly optimized, the energy efficiency starts to limit its output due to thermal, legislation, or other limits. At this point, a target cannot produce more isotopes, a neutron source cannot generate higher brilliance, and the available laboratory cannot fit a larger machine. Therefore, the energy efficiency has to be considered for both production and analytical tasks.

As discussed in the last section, the production of neutrons can either exploit light targets with low stopping power S and a single neutron per reaction or heavy targets with high stopping power and several released neutrons per reaction via (p, xn). For the calculation of energy efficiency, we make use of the equations from Sect. 3.4. We select two exemplary reactions, namely the bombardment of beryllium with protons as discussed in (Mauerhofer et al. 2017) and the bombardment of tungsten with protons. Figure 4.12 depicts the calculation results based on integration of (3.20). In the energy region below 21 MeV the beryllium target has significant advantages with a factor 18 higher reaction probability at 10 MeV. Furthermore, the ion range exceeds

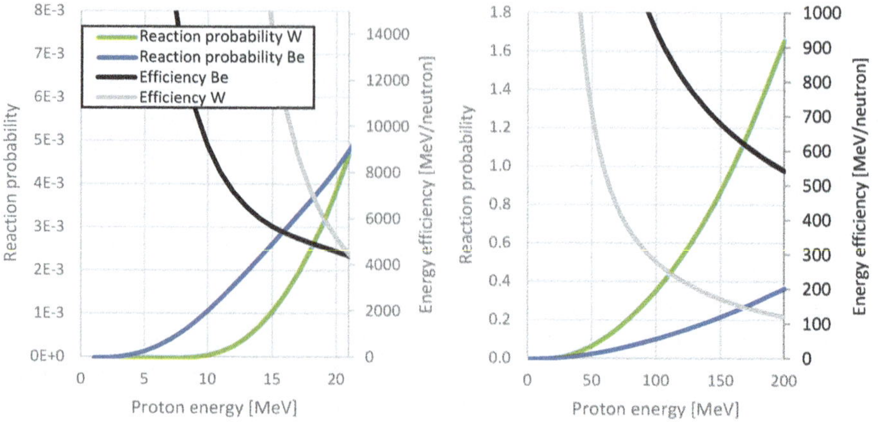

Fig. 4.12 Evolution of reaction probability W and energy efficiency H of Be and W reactions with protons. Up to 21 MeV the Be reaction is more effective. The calculation considered neutron production cross-sections from TENDL-2017

the one in tungsten by a factor 4.7, easing thermo-mechanical issues. We can see the energy efficiency at 15 MeV still ranges in the 6 GeV/neutron range, similar to the small scale D-T electronic neutron generator type. The reaction produces slightly lower neutron energy and has a different technical basis able to handle more beam power. Significantly better efficiency requires >40 MeV and a heavy metal target.

For the application, the delivered flux density is of higher importance than the source strength. The flux density derives from the source strength and the minimum working distance. With increasing beam energy we can roughly estimate a constant source diameter due to the increasing efficiency requiring less current at higher beam energy (Fig. 4.12), a growth of target length according to the beam range, and a growth of shielding thickness required for forcing the outside γ-radiation level to a given constant value. In conclusion, the minimum working distance increases with beam energy, reducing the ratio of maximum flux density to source strength. Nevertheless, the increase in efficiency enables higher source strength and typically results in higher flux density and more space for additional end-stations.

The γ flux grows according to the prompt γ-production cross-section resulting in an intensity I_γ. We assume a constant neutron source strength. Taking the γ intensity of 10 MeV protons onto beryllium as reference intensity with a requirement of 0.1 m lead shielding (Mauerhofer et al. 2017) (yes that is oversimplifying) we can calculate the required thickness d for a lead shield with a constant mass absorption coefficient of 0.1 cm^2/g and a lead density of 11.3 g/cm^3 according to

$$d[\mathrm{m}] = 0.1\ \mathrm{m} + \ln(I_\gamma / I_{\gamma 0}(10\ \mathrm{MeV} - \mathrm{Be})/113 \qquad (4.4)$$

From 10 MeV towards 200 MeV we can see an increase of the required shielding thickness from 0.1 to 0.27 m (Fig. 4.13) and a growth of source thickness by energy-

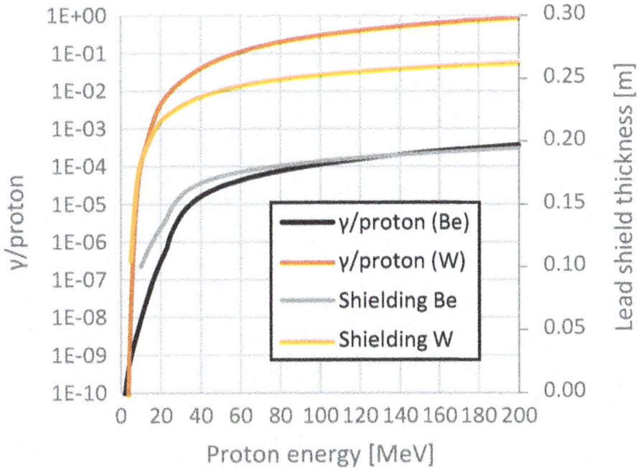

Fig. 4.13 Calculation results of the required lead shield thickness due to proton-induced γ-emission

loss calculations from 0.7 mm (Mauerhofer et al. 2017) to 26 mm (W) and 172 mm (Be) using SRIM (Ziegler et al. 2008), respectively. With this the minimum working distance grows from 0.107 to 0.3 m (W) resulting in a reduction of (accessible) flux density due to source size and shielding by a factor 8 from 10 MeV towards 200 MeV proton energy. In the same range our efficiency improves form 9 GeV/neutron (Be at 10 MeV) to 120 MeV/neutron (W at 200 MeV) by about 2 orders of magnitude leaving us a flux density gain of a factor 9.4 when comparing 10 to 200 MeV sources.

The calculations presented in this section should be understood as rough estimations. Different reaction in particular with deuterons, shielding materials, and radiation levels strongly influence the outcome. The used theoretical reaction cross-section probably significantly differ from actual cross-sections, but experimental data are so far not available for the full range of values considered here. Nevertheless, we can conclude a higher beam energy improves energy efficiency and with this source strength significantly. For the practically more relevant flux density, the growth of the accelerator neutron source with beam energy eats up a part of the gain of higher beam energies. The increasing costs and radioactive waste issues of increasing beam energy add up to the evaluation, opening up application scenarios for all types of sources.

4.3 Photons

Photons definitely represent the secondary particle type we have best control over in our current state of technology. Solid theories for the electron transitions in the atomic shell connected to line emission and photon absorption, models of Bremsstrahlung, and nuclear effects exist and are implemented in common codes such as GEANT4. In contrast to massive particles, kinematics does not strictly define the photon energy a specific ion beam produces, but we have more physical mechanisms and with this freedom of instrument design.

Why produce light in accelerators and not use a lamp?! Accelerators are the only devices capable of providing high-energy particles with very narrow distributions and extreme power density. The particle energy is decoupled from the atomic structure enabling reaching energies not accessible with thermal (incandescent lamp) or electronic band-gap (LED, plasma) mechanisms. The reasons and technology of this were discussed in Chap. 2. In short: Our technical flexibility and control of the charged particle beam together with the solid understanding of the photon production processes leads to highest photon energies, narrow photon energy distributions, and high intensity light sources. These three perks represent the main aspects of all accelerator based photon sources.

In applications of these photons, the energy requirements are typically given by certain processes or required range, therefore physicist condensed the power density and spectral width to the definition of the quantity *brilliance*:

$$\text{Brilliance}_R(x) = \frac{\text{Photons in a narrow wavelength band}}{\text{time} * \text{angular spreads} * \text{area}} \tag{4.5}$$

Equation (4.5) typically comes in the units of photons/(mm^{2}mrad^{2}second) for a bandwidth of 0.1% around the central photon energy. A high brilliance source is an intense collimated single wavelength source, comparable to a LASER, while a low brilliance source would be a divergent weak source of significant spectral width, comparable to an incandescent lamp. Equation (4.5) enables a quantitative comparison between technologically different setups. In the photon source community, it became the number one quantity of importance similar to the beam energy in the fundamental particle physics community or the triple product in the nuclear fusion community. Besides this also radiation pulse-length have an importance for time-resolved analytical processes.

The free-electron laser definitely offers the best light properties and output flexibility. Why the hell use any other light source type at all? The other light sources suffice for most applications and the FEL's specific and absolute costs exceed the other light sources by orders of magnitude. A medical imaging application would hardly profit from improved x-ray sources in its current form, since other technological aspects limit the spatial and contrast resolution. The economic and practical aspects of x-ray tubes outweigh their inferior light source properties. A statement like this has to be taken with care, since every technological advancement requires investment before it pays off. Nevertheless, the more performant and expensive free electron source types find application practically only in science.

4.3.1 X-ray and Bound Electron Sources

By far the most common accelerator based light sources in application are of the bound electron type. “Bound” depicts electrons, which are part of an atom and consequently rest in solid or liquid matter (also gases but they have practically too low density). The interaction with electrons bound to matter involves the energies of these binding states, in contrast to free electrons discussed in the following sections.

The one example everyone, except for the most lucky, know are the typical medical x-ray tubes. These tubes consist of an electron source, usually a hot emitter, a target of mostly W, Mo, or Cu due to their good heat conductivity and a DC voltage source accelerating the electrons from the source to the target with a few 10 keV and mA currents. Figure 4.14 depicts this layout using a thermionic electron source. The compact design consumes only electrical power, enabling enclosing it in a sealed case under vacuum. No additional vacuum or beam optics are required and the device is ready within a few seconds, forming the basis for their practical success. Electron beams offer practical advantages over ion beams by avoiding sputtering, a consumed resource (e.g. H_2 gas) in the source, and by having at least two orders of magnitude lower stopping power. Therefore, all devices operate on the technological principle depicted here.

Figure 4.15 shows exemplary x-ray emission spectra of such a tube for two different projectile energies. The spectrum combines the continuous Bremsstrahlung emission with the tungsten K-lines as given in Fig. 3.4. Each target element shows its own characteristic lines and the targets have to be selected according to the analytical requirements of x-ray energy. Figure 4.15 shows the four so-called $K\alpha_1$, $K\alpha_2$ and $K\beta_1$, $K\beta_2$ emission lines of W. The characteristic lines have a fixed photon energy, only their production cross-section increases with projectile energy. The continuous Bremsstrahlung aspect increases in spectral width and intensity with the projectile

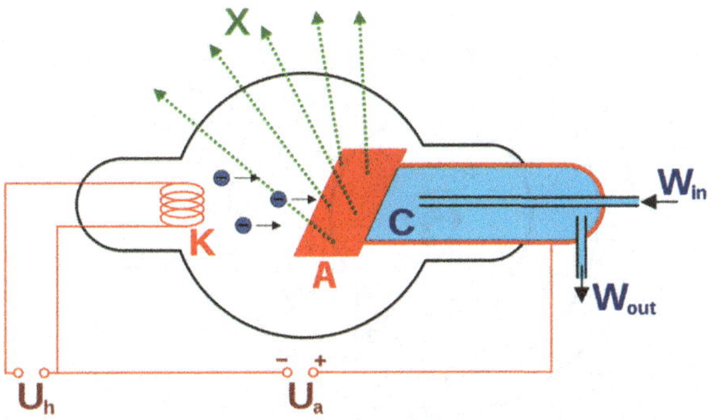

Fig. 4.14 Simplest design of a thermionic emitter (K) and impact anode target (A) x-ray tube with water-cooling (C). A sealed housing eliminates the need for vacuum equipment while still allowing x-rays to leave the vessel. The voltage U_a defines the electron impact energy. Reproduced from Wikimedia, public domain

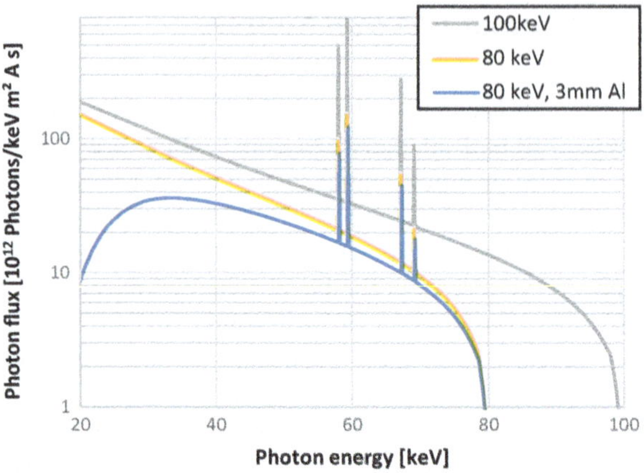

Fig. 4.15 X-ray tube spectrum with tungsten target using 100 or 80 keV electrons without or with 3 mm Al absorption filter. The spectrum comprises an overlap of Bremsstrahlung with characteristic emission. The two main K-α (≈60 keV) and –β (≈70 keV) line pairs dominate the photon spectrum only with an absorber. The maximum energy depends on Bremsstrahlung and is consequently projectile energy dependent. The characteristic lines change in intensity due to the energy dependent emission cross-section. Filtering reduces in particular the low energy part, resulting in a higher average photon energy (="harder" spectrum). Calculated using SpekCalc 1.1 (spekcalc.weebly.com) (Poludniowski and Evans 2007)

energy. The total photon flux increases with projectile energy, but the optimal energy depends on the optimal photon energy spectrum for a given application.

Figure 4.16 explains the physical mechanism behind the characteristic emission. The projectiles hit an electron bound to the atomic nucleus in the anode target. The release of this electron (ionisation) opens a vacant position which wants to be

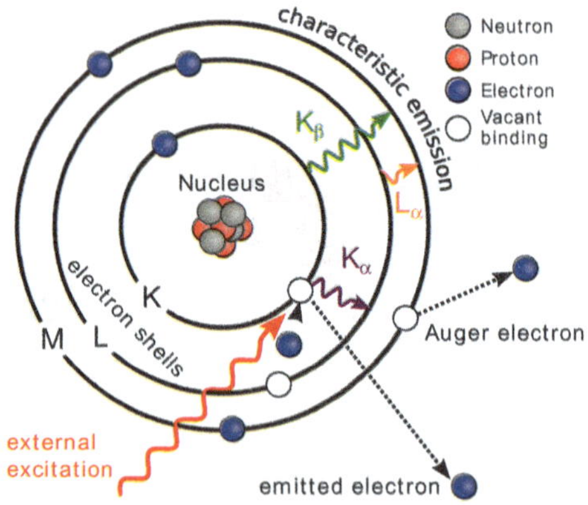

Fig. 4.16 Emission modes of bound electrons. An external excitation such as an electron beam kicks the system from the equilibrium state to an excited state which decays via emission of secondary particles. Picture adapted from original work of Cepheiden from Wikimedia commons CC-BY-SA3

filled due to the minimisation of energy. Electrons from other binding states of the atom with lower binding energy can transfer to this vacant position. If the released electron originated from the innermost shell of the atom this line is called K line. The next shell, the L shell, features less binding energy resulting in lower x-ray energy compared to the K shell. The origin of the transferred electron adds an α to the name (=Kα) for the nearest shell, a β for the 2nd nearest shell and so on. Due to the stronger binding of the inner electrons with increasing proton count Z of the nucleus, the Kα x-ray energy E_X scales roughly according to the Moseley scaling given in (4.6)

$$E_X \approx \frac{3}{4} * 13.6\,\text{eV}(Z-1)^2 \tag{4.6}$$

Bremsstrahlung adds a weaker but continuous part to the spectrum extending from the peak electron energy down to zero or, due to absorption in the exit path, a few keV, respectively. The term "brems" originates from the German word for deceleration and "strahlung" equals radiation. The origin of Bremsstrahlung lies in the deceleration of the projectiles in the target as discussed in Chap. 3. Consequently, Bremsstrahlung strongly scales with particle velocity, a fact favouring electrons over ions due to their lower mass. For x-ray sources the resulting Bremsstrahlung spectrum and intensity matters. Kramer's law describes the continuous target-material dependent photon flux spectrum J per projectile current according to (4.7)

$$J(\lambda, E_E) = \frac{K}{\lambda^2} * \frac{Z_{\text{Target}}}{\lambda^2}\left(\frac{\lambda E_E}{hc} - 1\right) \tag{4.7}$$

With Kramer's material constant K, the atomic number of the target Z_{Target}, the photon wavelength λ, the electron kinetic energy E_E given by the acceleration potential, Planck's constant h and the speed of light c.

Many applications, in particular in medicine, consider Bremsstrahlung as a spectral impurity, since a large spectral width reduces result quality by folding the contrast aspects of different photon energies to a single result. For example in medicine, wide spectra blur the contrast due to the energy dependent photon absorption length, while increasing radiation dose to the patient. Since the reasons of the spectral width lie in fundamental physics the only option is a spectral filtering. The easiest option for spectral filtering is putting a metal plate on the x-ray source. The metal absorbs photons similar to a high pass filter by exploiting the spectral "hardening" induced by the naturally stronger absorption of lower energy photons in matter (see Fig. 3.5). Figure 4.15 compares a filtered with an unfiltered spectrum, clearly showing the reduction of mostly the low energy flux. The filter also reduces the intensity of the desired characteristic emission by a smaller factor. Consequently, a filtered x-ray tube has to provide a higher primary photon flux for the same output flux. The resulting spectra still cover a large wavelength range. If sharper spectra are required, band-pass filters based on the reflection of x-rays on the crystallographic plane spacing allow for obtaining single emission line spectra. The crystals small width band-pass reduces

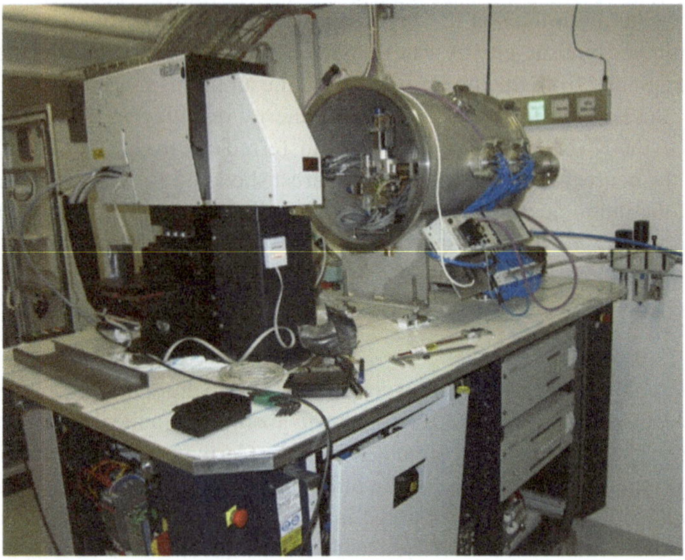

Fig. 4.17 The GALAXI x-ray source (left) represents the technical limit of bound x-ray sources. A liquid metal curtain acts as target and efficiently removes the heat to a cooler below the table. The 70 kV, 250 W source produces >10^9 photons/s. Copyright Forschungszentrum Jülich GmbH

the total photon flux significantly, making it only interesting for special scientific applications.

The standard medical x-ray tube used to acquire for example Fig. 6.3 uses about 30 keV with about 3 mA electron current. About 1 s of its photon flux suffices for producing the picture. Special scientific applications require higher fluxes. As with any target, the power load limits the x-ray tube output. Besides inclined geometries as shown in Fig. 4.9 or Fig. 4.14, liquid metals enable extending this technological limitation. The GALAXI device depicted in Fig. 4.17 represents an example for this advanced realisation of a bound electron source with a liquid Gallium-Indium jet target. The device was developed for material analysis using x-ray scattering. The low vapour pressure of Gallium together with the weak vacuum requirements of electron and photon beams make the use of a window between source and target unnecessary. Nevertheless, GALAXI uses a vacuum system for reducing photon scattering in air, improving beam brightness. GALAXI provides a total flux of 10^9 photons/s.

4.3.2 Synchrotron Sources

The technological limits of bound electron sources bring up the question of how to further increase brightness, enabling for example imaging shorter processes or for achieving intense photon beams even for monochromatic x-rays. Electrons with their strong Bremsstrahlung are the right Ansatz, but as we learned, the target results in technological limits. Subtracting the target leaves us only with a beam of free electrons, but how to accelerate/decelerate them sufficiently strongly without beam-matter interaction? Think about driving a car: You feel forces (which equals acceleration via $F = m * a$) when accelerating, braking, and/or cornering. Accelerating forces in accelerators are too weak with their current limits in the order of MeV/m as discussed in Sect. 2.2.2. Braking equals the bound electron sources, with their energy gradients by stopping in the order of 10 MeV/mm (for a few 10 keV electrons in metals). Consequently, the cornering option remains open, but we will need it in the quality of a race-track not a city cruise.

Technical setups for cornering were discussed in Sects. 2.2.2 and 2.3.2 in the context of beam optics. Think of the particle momentum as a 3 dimensional component vector. Changing the beam particle direction, using for example a dipole magnet, physically equals a forward deceleration and a perpendicular acceleration of the beam particles. Figure 4.18 illustrates the physics behind free-electron light emission by beam deflection. For example, a 90° deflection within 1 m curvature length decelerates a 1 GeV beam by 1 MeV/mm in the forward direction and accelerates it by the same amount in the perpendicular direction. The magnetic field strength and electron beam energy define the curvature radius and with this the magnitude of the acceleration. The curvature radius decreases with increasing magnetic field strength and increases with increasing beam energy, see Sect. 2.3.2, resulting in increased photon energy. These knobs allow varying the light wavelength and photon energy dynamically, respectively, theoretically in an unlimited but practically in a limited range. Free electron light sources lack a target for the light production, completely changing the technological basis and limits inherent to bound electron sources. Instead of the target heat load limitation, the magnitude of deflectional acceleration limits the output.

Equation 2.11 discussed the photon power emitted by beam deflection as a source of losses to the beam energy inducing an upper limit for the beam energy of circular

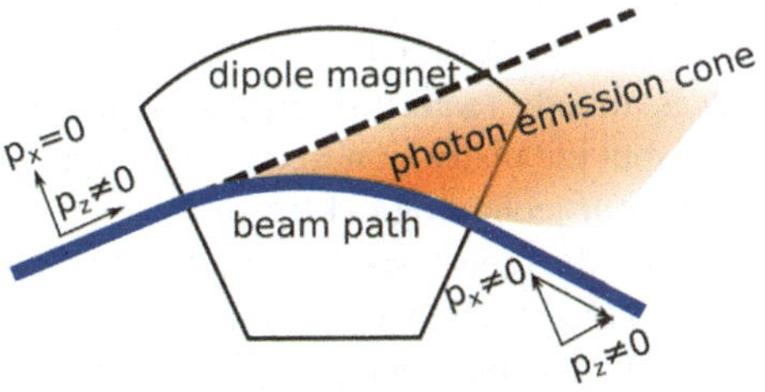

Fig. 4.18 The concept of producing synchrotron light based on charged particle acceleration in magnetic fields. A dipole magnet reduces the particle velocity in the forward direction z by transferring the forward momentum vector component to the perpendicular direction x; it accelerates and decelerates the beam. This produces radiation in the tangential forward direction at every point of the curvature forming a cone with a wavelength distribution defined by the magnetic field strength, particle type, and beam energy. The more kinetic energy the charged particles bear and the higher the photon energy, the narrower the emission cone

electron accelerators. At this energy, the synchrotron radiation emission intensity reaches the limits of input power of the accelerator, exactly the situation we are looking for now. This energy limit lies in the GeV range, slightly above our 250 MeV limit, but the strong beam energy scaling of emitted power with beam energy (2.11) demands this energy range. Varying beam energy results in varying photon flux and energy, requiring compensating the Bremsstrahlung losses with acceleration cavities to maintain electron energy and photon source properties. For GeV range and the requirement of constant beam energy we end up with the synchrotron accelerator type. This special class of Bremsstrahlung connected to free electrons in synchrotrons is usually just termed synchrotron light/radiation, while Bremsstrahlung usually depicts only the emission from bound electron targets.

The current technical limits of beam optics result in synchrotrons of a few 100 m circumference. The synchrotron being an AC accelerator, compared to the DC accelerators used for bound electron sources, delivers a pulsed beam current. The photon emission follows this AC electron beam pattern, making the synchrotron light a pulsed light with pulse length of a few 100 ps (GHz frequencies) and lower. The constant beam energy defines the synchrotron as a storage ring. For loading this storage ring with an electron beam typically a smaller synchrotron is connected to the storage ring which itself is fed with a LinAC and/or a DC accelerator. Once the storage ring is filled with electrons, the synchrotron radiation damping (see Sect. 2.3.1) and the accelerator confinement allow for a continuous operation with the same electrons for several hours. Light sources based on ions would require beam energies proportionally higher as the particle mass, see the *E/m* scaling in (2.11), resulting in significantly increased device costs. For this reason free electron light sources always use electron beams.

The cornering force induced by the magnetic deflection results in photon emission into a tangential cone. The opening angle of this cone shrinks inversely proportional to the relativistic γ-factor (the factor between relativistic and rest mass), resulting in an increase of brilliance with beam energy. Since every point along the curved part of the beam path represents an origin of one of these cones, a dipole magnet emits a photon fan with a width increasing with length and decreasing with curvature radius. Apertures allow selecting parts of this radiation fan for (several) specific end-stations. The emitted photon pulses already feature significantly higher brilliance compared to bound electron sources, but the restriction to the beam optical elements discussed in Sect. 2.3.2 so far limited our technological freedom.

Continuing the thinking of the race-track we could include in principle an infinite amount of curves in a closed race-track, the beam path can be more complex than an American oval race-track. The so-called wiggler (Clarke 2004) follows this thinking by applying the copy and paste method to the dipole concept. It adds several dipole magnets with changing polarisation in a row, generating a beam path in the shape of multiple S-curves as depicted in Fig. 4.19. The beam wiggles around the undeflected path, emitting radiation at the apex of each wiggling. The apex represents the point of maximum acceleration, since the second derivative of the position yields the acceleration. In case of a *sine* wave wiggling its second derivative is a—*sine*, the absolute value of maximum deflection and acceleration lie at the same point. This

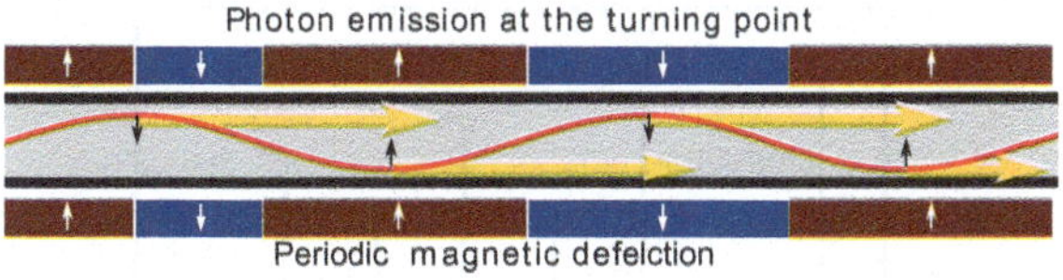

Fig. 4.19 Working principle of undulators and wigglers with the S-shaped electron beam path (red line), black arrows indicating the point and direction of maximum acceleration, and yellow arrows the emission of light. The beam actually moves into and out of the paper plane, but for illustration reasons the beam path was rotated by 90°

radiation adds up linearly in the sense of a sum of multiple dipole magnets resulting in a larger total emission of the wiggler compared to the dipole.

The undulator uses the same geometry as the wiggler with, at first sight, only a smaller period length and lower magnetic field than the wiggler. This shorter and weaker magnets result in a smaller maximum deflection (Sect. 2.3.2), the apex points of emission come closer together. In the GeV range, the electron velocity is very close to the speed of light c resulting in a relativistic length contraction given by the γ value with $\gamma \approx E_{Beam}/m_e c^2$ (remember: $m_e c^2 = 511$ keV). In the GeV range $\gamma \approx 10^4$, therefore a 10 mm period length becomes 1 nm as seen in the centre-of-mass system of the electron, an x-ray wavelength. Consequently, the emitted photons in an undulator have a spatial overlap and a fixed phase relation for a certain wavelength, therefore fulfilling the prerequisites for light interference. Constructive interference increases the photon brightness and brilliance for a certain wavelength and its multiples compared to a wiggler, while destructive interference weakens the parts of the photon spectrum not fulfilling the propagation conditions. This spectral selection increases the peak brightness for certain wavelength in the form of a quadratic scaling of brightness with the number of periods, but it also strongly restricts the instrument tuning by coupling the field strength, period length, and photon wavelength. Furthermore, the length contraction dictates small technical dimensions of the undulator parts for emitting x-rays and requires GeV electron beam energies in order to have a length contraction large enough to allow the parts to be made and aligned. These dimensional restrictions of the interference condition result in a technological limit of undulators for generating high photon energies/short wavelength. Wigglers and dipoles offer practical advantages for achieving higher and broader photon energies at a given beam energy. The undulator physics will be discussed in more detail in Sect. 4.3.3 where its idea is extend further for the free-electron Laser.

Figure 4.20 compares the emission spectra of these three so-called insertion devices. Wiggler and Dipole/bending magnets generate broad spectra in a wide energy range with a clear upper energy limit not shown in Fig. 4.20. The radiation spectrum of the technically simple and anyways present dipole magnet can be derived from theoretical models to a high accuracy making it ideal for calibration and less brilliance demanding applications. Still its brilliance exceeds the values of bound electron sources by orders of magnitude. The wiggler spectrum is basically

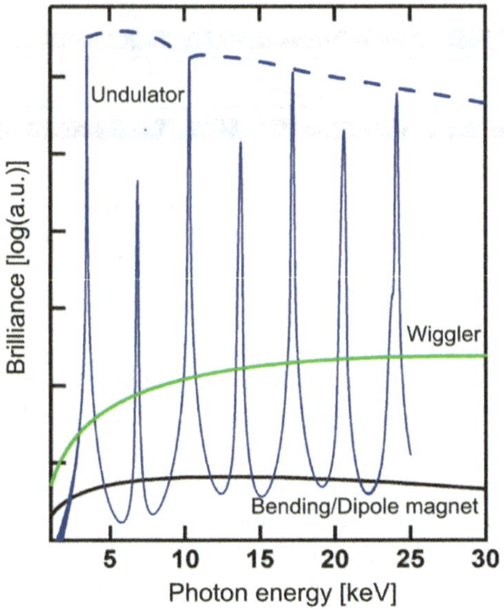

Fig. 4.20 Comparison of the photon spectra delivered by dipole magnets, Wigglers as the improved variant, and the sharp periodic structure of undulators. All spectral distributions have maximum photon energies not shown in the figure. The three insertion devices differ in peak brilliance (log-scale) by orders of magnitude with the undulator providing the highest brilliance. Translated and reprinted from (Möller and Falta 2010) with permission by Springer

identical to the dipole, but with increased brilliance. The undulator generates a resonance pattern of even multiples due to its interference physics with a peak width in the order of a few 10^{-3} of the peak energy. Its brilliance exceeds the values of the other two types by orders of magnitude. While the undulator generates significantly higher brilliance, it also has disadvantages for applications due to its limited wavelength flexibility and maximum wavelength. Each of the three options has its individual strength, consequently synchrotron facilities make use of all three options in specifically optimized end-stations.

The brilliance of synchrotron light receives increased importance when using monochromators, further reducing the spectral width down to the MeV scale. While monochromators are also connected to bound electron sources, their low brightness demands long integration times and mm-sized spot sizes for reasonable detectors signal intensity. With synchrotron light, these limits disappear, opening up applications with highly resolved imaging or time-resolved in-situ experiments using single wavelength photons. Compared to bound electron sources, which have a peak energy defined by the beam impact energy, free electron light sources have a technological disadvantage for producing photons above about 100 keV due to the required high beam energies and magnetic field strength in the superconductor range.

The example of the device BESSY-II with its vast amount of end-stations as depicted in Fig. 4.21 demonstrates the synchrotron device layout and its connection

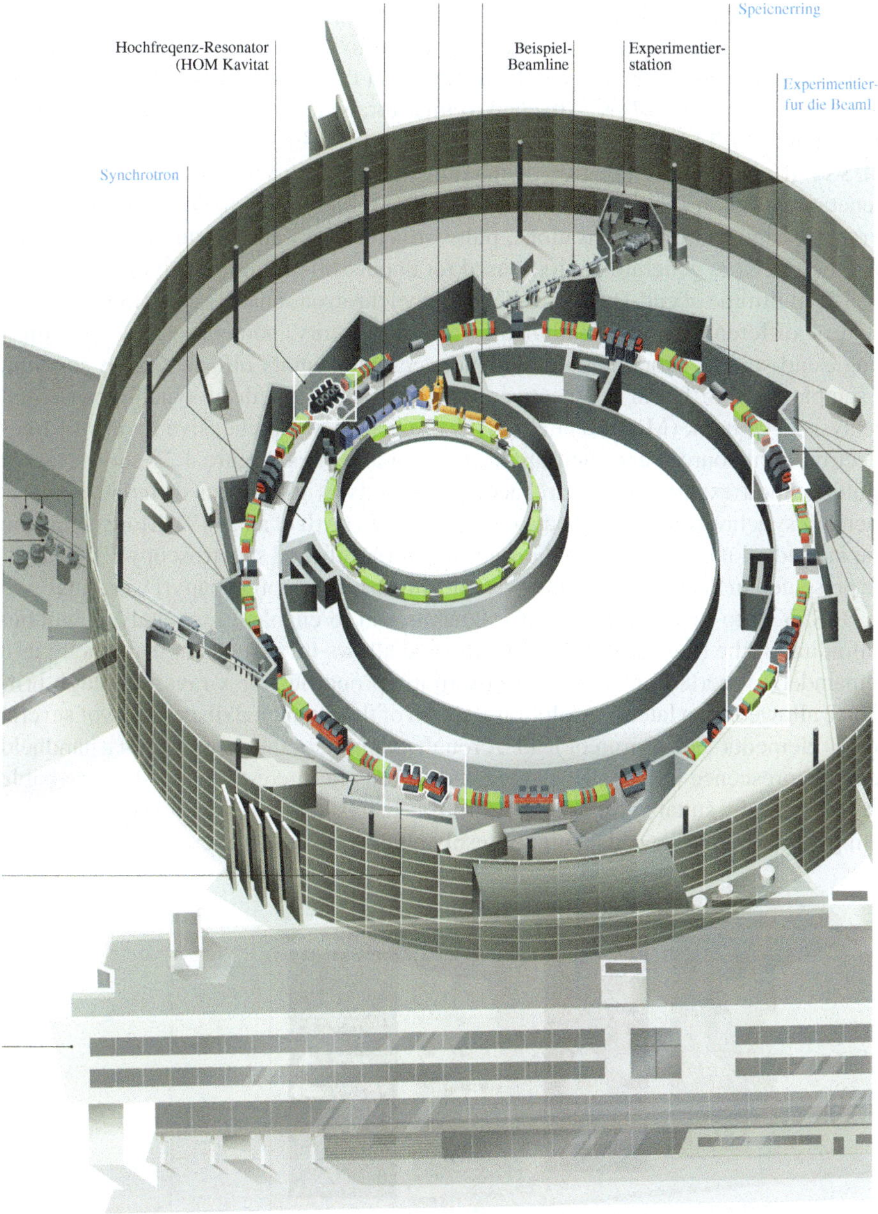

Fig. 4.21 BESSY II synchrotron lab layout. In the centre an accelerator and the storage ring provide the 1.7 GeV electron beam. On 45 tangential ports different experiments/end-station exploit the photon beam generated by Bremsstrahlung sources in the storage ring. The size and concept of synchrotrons allows serving more end-stations than any other concept. Credit HZB/Ela Strickert

with numerous applications. An inner 1.7 GeV electron synchrotron feeds the 240 m long outer storage ring with electron pulses up to an average beam current up to 300 mA. The pulse structure can be adapted via the choice of bunch length in the source region to 2 or 17 ps. Attached to this storage ring, multiple dipole magnets, one wiggler (7 T), and 12 undulators generate the light for about 45 end-stations. BESSY-II uses the high degree of instrumental flexibility of insertion devices for generating photon energies from 0.4 MeV (THz radiation) up to 90 keV (x-ray) with a range of intensities, polarisations, pulse structures, spectral width and so on. The insertion devices generate the light used for about 180 different analytical methods.

The multitude of end-stations attached to synchrotrons enables combining a multitude of different methods within the same laboratory, sometimes even within a single end-station. More methods add more complementary information by showing a sample from different perspectives, an additional strength of synchrotron based analysis. The book (Mobilio et al. 2015) provides a comprehensive review of free-electron light sources and their applications in analytics. Figure 4.22 shows such a multi-method example of an archaeological analysis using four different complementary synchrotron light-based analysis methods (Arlt et al. 2019). More details on the individual methods (except Fourier transform infrared spectroscopy FT-IR) will be presented in Sect. 7.1.2. In this example, historical Egyptian papyri were analysed at BESSY II to reveal hidden text fragments and to enable a text analysis without unfolding of the fragile artefacts. The method allows for visualising nanometre thin remainders of certain inks by their absorption properties for x-rays. The high brilliance allowed for a lateral resolutions <1 μm of the elemental distribution of several heavy elements such as iron or lead. A rough mapping of the papyri using a handheld x-ray fluorescence (XRF) device revealed regions of interest without letters visible by photography suitable for later synchrotron analysis. The synchrotron radiography using 19 keV photons adds a highly resolved qualitative map, revealing a letter and

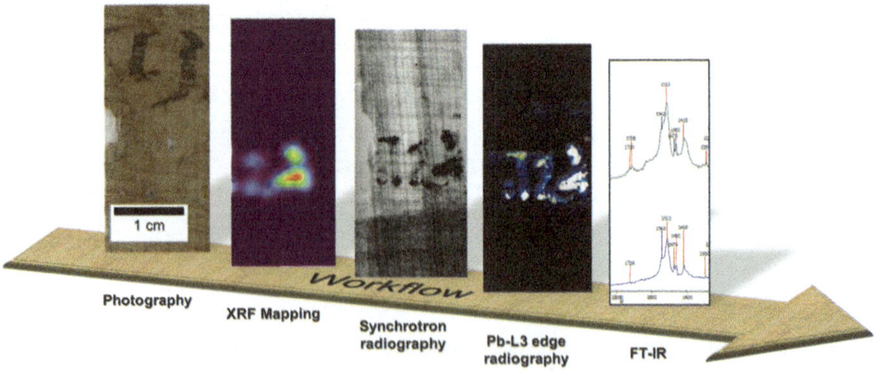

Fig. 4.22 A multi-method analysis of an ancient papyrus with a supposed empty spot as shown by photography. Combining several synchrotron-light based material analysis methods (Sect. 7.1.2) revealed signs which once stood in this now empty place. Furthermore, the methods could reveal the type of ink used. Credit HZB

structural features of the papyrus. The Pb-L3 absorption edge ($\approx$13 keV) radiography probes the sample with a monochromatic photon-beam, resulting in photon absorption mostly by lead atoms present in some ancient inks. This elemental sensitivity results in an ink-contrast map insensitive to dust particles and other impurities. Switching the photon energy to other elemental absorption edges yields additional information on the ink composition. The combination with the chemical analysis by infrared spectroscopy reveals the ink type chemistry.

Several technical improvements in synchrotron light sources promise significant performance increases of this light source type. The tight coupling of the photons to the charged particle beam shifts the technological limits of light source brilliance (4.5) to providing high energy and current beams with low emittance. Further reductions of emittance by new electron source types and improved insertion devices applied in so-called generation 4 synchrotron light sources exhausts the technological limit of increasing photon brilliance by reducing beam emittance by reaching the diffraction limit of the emitted photons at $\lambda/4\pi$. Improved electron source performance and beam optical setups result in higher photon energies and brightness over the years. A dual beam mode, the so-called transverse resonance island bucket mode (Goslawskiy et al. 2017), enables running two electron beams in the same storage ring enabling simultaneous generation of two different photon beams, for example short and long pulses or two different light polarisations. Two photon beams in the same experiment allow for shorter workflows and provide improved correlation of the results of different photon properties.

The usage time specific costs of synchrotron analysis stations exceed the costs of bound electron sources by at least one order of magnitude with typical values in the order of 300 €/h. For this money the synchrotron delivers higher beam quality and brilliance opening up unique methods not possible with bound electron sources. The high degree of organisation in the accelerator light community indicated by common programs such as Calipso-plus or LEAPS resulted in a very comprehensive set of specialised devices spread across numerous countries with a high impact in science, industrial and medical development. Mass-production or patient level applications have not been established to the author's current knowledge, but first activities towards e.g. semiconductor lithography or material analysis exist.

4.3.3 Free-Electron Laser

The free-electron laser (FEL) adds a new physical process to the light generation mechanism of the synchrotron light source (SLS), the Laser effect. Although first FELs and SLS were both employed in the 1970s, only the last decade's technological and theoretical advances allowed constructing FELs with improved performance in the VIS and X-ray wavelength compared to SLS (Fig. 4.23). Just 2017 a new device named XFEL, currently the most brilliant artificial light source, came into operation. Besides XFEL currently more than 20 smaller FELs are in operation worldwide. These FELs provide an incredible wavelength range for a single technology, spanning

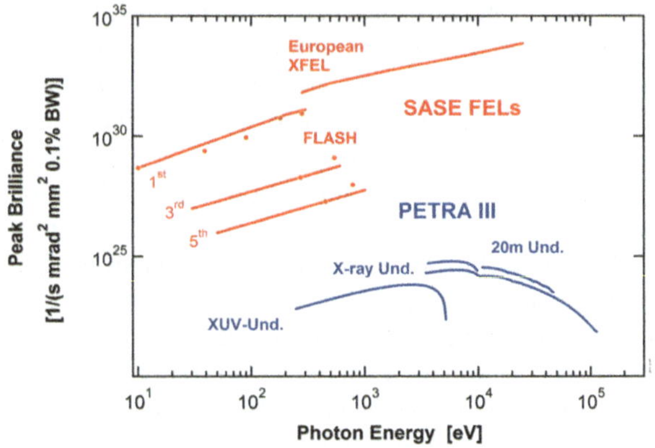

Fig. 4.23 Comparison of the brilliance for different photon energies of FEL and SLS devices. The best FELs reach 9 orders of magnitude higher brilliance than the best SLS. Reprinted from (Bari et al. 2017)

from the gigahertz region (some mm wavelength) up to x-rays (<0.05 nm) over all the devices. This flexibility combined with its high brilliance and short pulse duration makes the FEL technology (Schmüser et al. 2008) attractive for many scientific analytics.

XFEL as the most advanced technical realisation accelerates electrons to 17.5 GeV in a 2.1 km long linear AC accelerator. In the following 1.3 km of the device, several magnets and optical systems are operated to squeeze the light out of these electrons and deliver it to experiments. Throughout this volume a vacuum of $<10^{-7}$ mbar is maintained to keep the electron beam unaffected by residual gas. The electron acceleration is induced by 101 superconducting liquid helium cooled niobium resonator cavities. These cavities are superconducting to reduce electrical losses of the driving wave, which, in turn, increases the resonator quality factor and allows for higher duty factor, leading to a improved beam quality and brilliance. This chain of relations is typical for AC accelerators and is one of the technical requirements for XFELs performance/brilliance. Up to 27,000 electron bunches per second can be delivered to the following Laser part. The Laser part consists of 5 undulator magnet modules for different wavelength ranges, each consisting of several magnet cells. Within the cells, the undulator magnet gap and thus the wavelength of the emitted light can be adjusted with motors in a certain range. With this equipment XFEL can generate photons from 0.05 to 4.7 nm with <100 fs pulse length and a brilliance up to $5 * 10^{33}$ Photons/s/mm^2/mrad^2/0.1% bandwidth.

In contrast to the synchrotron, presented in the last section, the FEL produces photons with a *light amplification through stimulated emission* (LASER) process. The active medium of the Laser is the electron beam itself, so the accelerator can be considered as the equivalent of the optical pump in an optical Laser, hence the naming of *free-electron* Laser. The light production in a SLS is also based on free electrons, but it lacks the amplifying aspect and the coherence of a Laser. The physics behind this additional aspect of FELs is a coupling (bi-directional energy transfer) between the electron and the light wave by an overlap of both in an undulator (Fig. 4.24), which are also applied in SLS as discussed above.

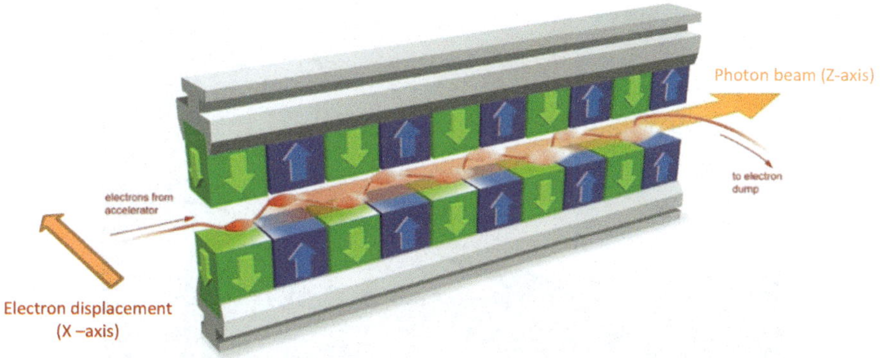

Fig. 4.24 The working principle of a high-gain FEL. Green and blue mark the periodic dipole magnets forming the undulator. The acceleration of the electrons on the sinusoidal track leads to the emission of photons at the apex of each period. The small track width allows for a coupling of the electron and photon waves inducing an amplifying feedback effect. Credit European X-Ray Free-Electron Laser Facility GmbH

The magnetic structure of the undulator accelerates the electrons periodically so they spontaneously emit photons. The acceleration is induced by a set of magnetic dipoles of periodically changing polarity with period λ_u. So far there is no difference to the devices applied in SLS, Sect. 4.3.2. The technical differences are in fact small, the main point being a higher total length, but the physical impact of these changes is enormous. It can be compared to a bridge picking up energy from the wind passing it. The more air is flowing over the bridge and the better it is coupling to resonance frequencies of the bridge structure, the more energy will be stored in the bridge in the form of kinetic and potential energy. The same way every electron bunch invests with every undulator period a small amount of energy into the FEL light. The Tacoma Narrows Bridge, see Fig. 4.25, is an unfortunate example of such an effective coupling (combined with low damping). First the bridge acquires more and more energy, but it remained intact. Only after reaching a critical stored energy level the breakdown happened. In this analogy a stable bridge (which avoids the resonant coupling with the wind) is the SLS undulator and the Tacoma Narrows Bridge is the FEL undulator. In the same way the wind couples energy into the bridge (before the final accident), the undulator brings the electrons in resonance and spatial overlap with the photons. This induces a coupling of photons and electrons increasing the energy stored in the photon field, which in turn increases the strength of this coupling. In addition to the quadratic scaling of undulator brightness with the amount of periods, this coupling adds a quadratic brightness scaling with the number of involved electrons once it reaches a critical level.

FELs are constructed either in a *low-gain* regime with weak coupling or a *high-gain* regime with strong coupling. Low-gain devices deliver gains of some percent for every passage of the electron beam through the light wave and hence require several passages of the electrons through the generated light to induce a significant

Fig. 4.25 Picture of the Tacoma Narrows Bridge from 1940. The wind is passing the bridge and exciting a natural resonance mode of the bridge with comparably high quality factor and low damping, respectively. The bridge finally collapsed when the driving force of the wind became too strong and too much energy was stored in the resonance

amplification. This multiple passes of electrons and light requires an electron storage ring and an optical mirror resonator to contain the light wave inside the FEL for further amplification. In contrast, the high-gain FEL (Fig. 4.24) does not require an optical resonator, as its amplification is sufficiently high for a single-pass system. This section focusses on the high gain FEL since it represents the latest technological breakthrough. In particular short wavelength (x-ray) and high light intensity, for which no suitable mirrors exist, require high-gain FELs. The scientific interest in these short wavelengths is one of the main reasons for the linear design and length of XFEL. The high-gain regime arises from an undisturbed feedback of the photon electric field on the electrons and a subsequent exponential intensification of the energy transfer, up to a certain limit. Naturally, this makes FELs high brightness devices, as we get increasingly more "bang" for each "buck".

The full mathematical treatment of FEL physics and the emission of radiation are out of the scope of this book. Moreover, some aspects of special relativity are required to understand the point of view of the electron beam. Nevertheless, for understanding of the technological aspects and physics of this accelerator application, the main findings and equations of the theoretical assessment will be discussed in the following part. The presented analytical equations are mostly based on approximations and idealisations, respectively. Consequently an actual device layout requires a sophisticated treatment by computer codes, but for understanding the FEL physics and scaling's, these analytical expressions are better suited.

Starting with the parameters describing the technical layout of an FEL we define the dimensionless undulator parameter K. K represents the amplitude factor of the oscillations of the electrons in the undulator, with higher displacements corresponding to higher values of K.

$$K = \frac{eB_0\lambda_u}{2\pi m_e c} \approx 0.934 B_0[\text{T}]\lambda_u[\text{cm}] \tag{4.8}$$

Here the undulator magnetic field strength B_0, the undulator magnetic field structure period λ_u, the electron mass m_e and the fundamental constants e (electron charge) and c (speed of light) are used. Furthermore, the energy of the electron beam is described by the relativistic Lorenz factor γ in multiples of the electron rest energy

$$\gamma = \frac{1}{\sqrt{1 - \overline{v}^2/c^2}} = \frac{E_{kin}}{m_e c^2} + 1 \tag{4.9}$$

with the average electron velocity $\overline{v}$ and their corresponding kinetic energy E_{kin}, respectively. It has to be noted, that $\overline{v}$ is naturally limited to c and because of the non-straight movement in the undulator field, the component in the forward direction z has to fulfil $v_z < \overline{v} < c$. With the help of these quantities we can describe the motion of the electrons in the forward direction z and in the perpendicular direction of the undulator displacement x (see Fig. 4.24) with respect to the travelled time t of the electrons in the laboratory system. As the electrons are typically at relativistic speeds, the time in the electrons inertial system is advancing slower than in the laboratory system.

$$x(t) = \frac{K\lambda_u}{2\pi\gamma}\sin\left(\frac{2\pi\overline{v_z}}{\lambda_u}t\right),\ z(t) = \overline{v_z}t - \frac{K^2\lambda_u}{16\pi\gamma^2}\sin\left(\frac{4\pi\overline{v_z}}{\lambda_u}t\right) \tag{4.10}$$

The equations demonstrate the sinusoidal movement of the electrons perpendicular to the central undulator axis with an undulator parameter K (equation 4.8) and electron energy dependent amplitude and frequency. The movement in axial/longitudinal direction is given by the average velocity and a small oscillation component. The accelerated charges/electrons spontaneously emit photons. This acceleration is given by the second derivatives of the positions given by (4.10) and therefore also sinusoidal. In order to induce the FEL amplification effect, the light wave (which has sinusoidal oscillations in its transversal electromagnetic components) has to have a frequency which is an even multiple of the x oscillation frequency of the electrons (4.10) with a fixed and aligned phase relation (coherence) as demonstrated in Fig. 4.26.

The resonance light wavelength λ_{photon} is given by the undulator period as seen from the relativistic electrons. The relative velocity makes the undulator appear a factor γ shorter to the electrons and due to the Doppler-effect, the wavelength of the light emitted by the electrons appears another factor γ shorter in the laboratory frame:

$$\lambda_{\text{Photon}} \approx \frac{1}{m}\frac{\lambda_u}{2\gamma^2}\left(1 + \frac{K^2}{2} + \gamma^2\theta^2\right) \tag{4.11}$$

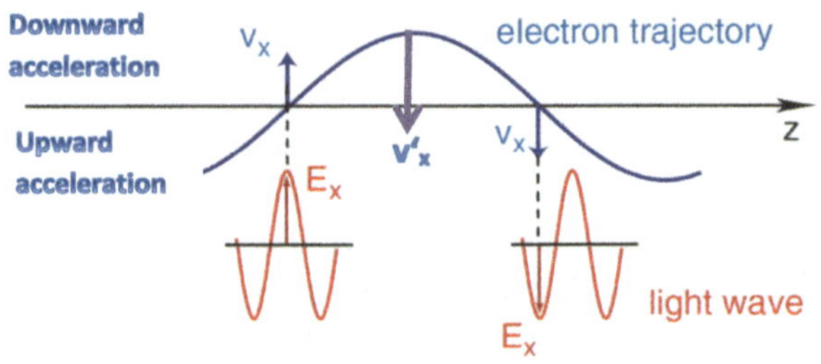

Fig. 4.26 Physical basis for the electron-photon coupling inducing the light amplification. The maximum in acceleration of a single electron in x-direction has to coincide with the photon wave electric field vector pointing in the same direction

with the angle towards the forward direction θ and the harmonic number m. The higher harmonics $m > 1$ are generated by the longitudinal oscillation (4.10). In accordance to (4.10), the importance of higher harmonics reduces with γ^2. Compared to dipoles and wigglers, a low light bandwidth is achieved naturally as the coupling is only effective for a certain resonance wavelength. This bandwidth of the individual harmonics is slightly degraded by the effects indicated in (4.11). The bandwidth exhibits angular (θ) and electron energy (γ) dependencies, connecting it to beam divergence, beam energy width, and sinusoidal track width ($\sim K/\gamma$) defined by the beam emittance and beam optical system (Sect. 2.3).

The coherence of light and electron wave can lead to the so-called microbunching. Microbunching is specific to the high-gain FELs, since the dipole magnets required to realise the multi-pass structure of low-gain FELs stir this tiny structure. The photon electric field concentrates the beam electrons to longitudinal bunches of one photon wavelength extend at the phase position of maximum energy transfer of the electrons to the photons (as indicated in Fig. 4.26) by slightly adjusting the electron trajectories. Accordingly, the more electrons are bunched together, the higher the FEL amplification and the higher the photon field amplitude, the stronger the bunch compression. In other words: A self-amplifying exponential growth sets in, piling up photons on the spots of highest intensity and leading to extremely short and coherent light pulses. This exponential amplification of the light power $P(z) = P_0 e^{z/L_{\text{gain}}}$ is defined by a characteristic length L_{gain}

$$L_{\text{gain}} = \frac{1}{\sqrt{3}} \left(\frac{2\gamma^3 m_e \lambda_u}{\mu_0 \hat{K}^2 e^2 \pi n_e} \right)^{\frac{1}{3}} \tag{4.12}$$

The microbunches increase their density n_e over the exponential amplification range of the FEL until counteracting effects become dominant and the bunches start oscillating between positive and negative energy transfer to the photons, a saturation is reached. In the transversal direction mainly the beam emittance defines the electron density limit. In longitudinal direction the space charge forces (given by beam current I_0) defocus the beam, limiting its density. As these processes are time dependent their magnitude in the laboratory system scales with the time dilatation of the beam electrons given by the relativistic factor γ (equation 4.9).

$$P_{\text{limit}} \approx \frac{\lambda_u I_0 \gamma m_e c^2}{6.93 * \pi e L_{\text{gain}}} \tag{4.13}$$

As the FEL is an amplifier it requires a base power to amplify, the so-called seed. The seed can be coupled in from an external light source of the same wavelength or from the spontaneous emission in the undulator (*self-amplified spontaneous emission*, SASE). The FEL amplification factor is independent of the seed intensity, but its saturation power limit is fixed (4.13), hence the device length required to reach saturation is seed intensity dependent (Fig. 4.27). An external seed, e.g. an optical Laser, can provide high intensity seeds, but it strongly reduces the wavelength flexibility of the FEL. SASE on the other hand provides a weak seed with a power P_I in the first harmonic ($m = 1$) given by the Larmor formula, similar to SLS undulators

$$P_1 = \frac{e^2\gamma^2K^2\pi}{3\varepsilon_0\lambda_u^2\left(1 + \frac{K^2}{2}\right)^2} \tag{4.14}$$

Nonetheless, little wavelength restrictions are implied and no external light sources are required with SASE.

For applications in science, the FEL offers unique features. The very short light pulse length down to fs allows for extreme time resolution enabling observing highly dynamic processes such as molecular reactions. At the same time the high brilliance and coherence of FEL light delivers high sensitivity for the analysis of small objects

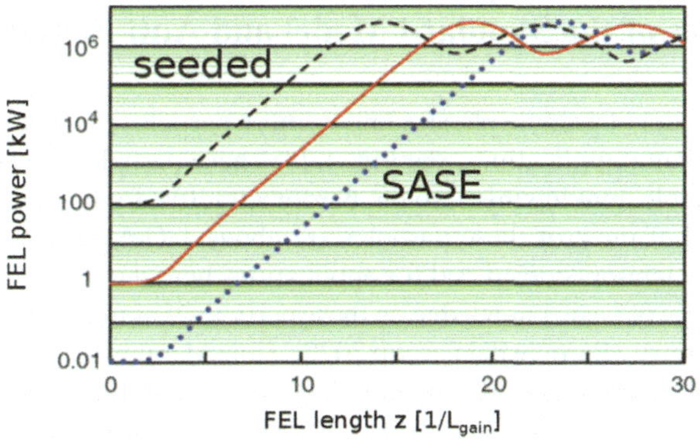

Fig. 4.27 Evolution of the radiation power with respect to the length of the FEL for 3 different seeds. The dotted line represents self-seeding, the full and dashed lines represent two different external seeds strength of 1 and 100 kW. After an initial settling phase the exponential amplification sets in and reaches saturation after 20, 15 or 10 gain length, depending on the initial seed strength. Reproduced from (Schmüser, 2008) with permission by Springer

such as individual molecules. The high brilliance and short pulses lead to instantaneous beam powers of up to several GW on sub-cm spot sizes. This power density destroys any investigated object, but on the fs (10^{-15} s) time-scale, significant changes or movements of particles cannot be expected, even on the atomic scale. The large range of possible light wavelength and the tunability within a single device make the FEL versatile and provide an additional degree of freedom comparable to SLS technologies. On the other hand, FELs are the largest, most expensive, and energetically most inefficient light sources.

Technologically, the FEL has several unique advantages and disadvantages compared to other light sources. Increasing the beam energy and γ, respectively, leads to shorter wavelength, (4.11), but also to longer devices, (4.12). As in many other accelerator applications, finally the performance is limited by the beam emittance. Special electron sources with low emittance and fast beam acceleration, counteracting the space charge forces, were developed for XFEL, but the emittance is still a limiting performance factor. A missed launch can mess up the whole race. Technically, the FEL can compensate higher emittance by higher kinetic energy (γ), but this in turn leads to larger and less efficient devices. The problem is less relevant for longer wavelength, but in this range other light sources compete to the FEL. Compared to energy efficiencies of some 10% achieved with diode lasers the FEL reaches only sub-‰ values. An efficient energy transfer to the light wave can only be achieved near the resonance energy γ_R, therefore especially in high gain devices most of the electron kinetic energy remains unused. Consequently, similar to the SLS only a few pilot applications in industry and medicine exist, but for science and development in any field the FEL provides a complementary addition for questions exceeding the possibilities of SLS.

4.4 Particles of the Standard Model and Anti-matter

The standard model of particle physics knows more particles than the ones discussed so far for accelerator applications, see Fig. 4.28. These particles are very short lived or hardly detectable, making them, so far, hardly attractive for practical applications. The rest masses of most of these particles are above 1 GeV and therefore beyond the range of accelerators considered for applications here, but they offer quite a few peculiarities. A lot of basic knowledge on these particles was made freely available by the particle scientists (Particle data group 2020). Consider this section as an overview on what is on the "market", in terms of particles, and where future developments could profit from an extended particle crowd. For more detailed readings regarding the standard model and fundamental particle physics the reader is referred for example to (Griffiths 2008; Larkoski 2019).

Let us begin with neutrinos, which are somehow a by-product of the β-decay and other nuclear reactions to allow nature maintain its lepton count. As such, neutrinos are produced in large quantities in the sun and other cosmic objects, but also in man-made nuclear reactors. Some 10^{10} neutrinos pass each cm^2 of our bodies every

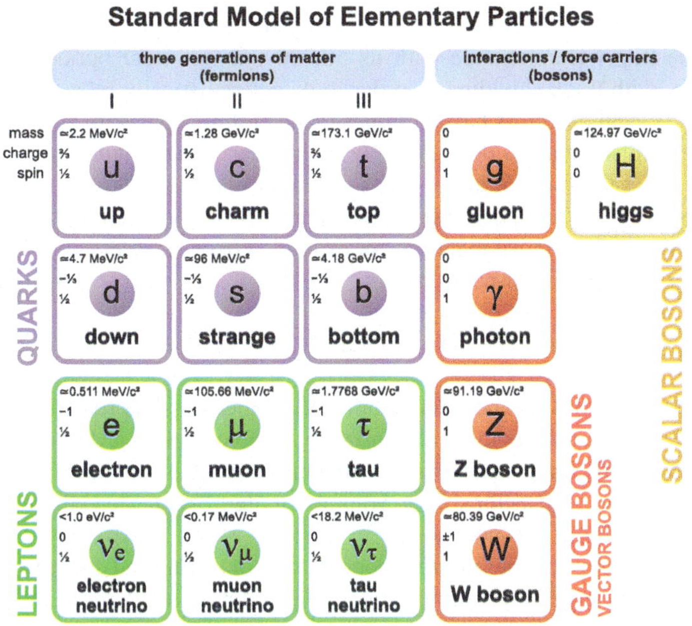

Fig. 4.28 Fundamental particles of the standard model. The particles constituting matter have three levels of generations with increasing rest mass of which only the first generations exists in everyday life. Protons and neutrons consist of up and down quarks, but quarks also form other massive particles. The bosons mediate the fundamental forces. On top of these particles all particles also have an anti-matter equivalent. Reproduced from Wikipedia public domain

second. These neutrinos originate mostly from the sun where electron neutrinos are produced to 90% by the fusion of two protons. Neutrinos are extremely light massive particles, yet until now their actual mass is so small that it could only be determined in the form of an upper limit given in Fig. 4.28, and therefore practically have velocities close to the speed of light.

In technical applications, neutrinos originate from the β decay with β^+ producing electron neutrinos and β^-electron anti-neutrinos similar to the decay of the neutron give in (4.5). Due to the absence of any binding mechanism of neutrinos to matter (at least none is known) or any electro-magnetic interaction (charge = 0) they cannot be focussed/densified or produced directly, but only by decay reactions requiring them for lepton count conservation.

$$n \rightarrow p + e^- + \overline{\nu_e} \tag{4.15}$$

Neutrinos exhibit extremely low reaction cross-sections with normal matter (Formaggio and Zeller 2013) and hence leave us unaffected. On the one hand, this fact is an advantage for us surviving the neutrino flux from the sun by providing neutrinos with virtually unlimited range through matter, exceeding even the size of earth already below 100 keV. On the other hand, it also complicates making use of this type of particles. Consequently, detecting neutrinos requires massive detectors, such as the Super-Kamiokande with its 50,000 tons of water buried 1 km below the surface, with thick shielding in order to detect statistically relevant amounts of neutrinos. Typical detectors have volumes of several m^3 up to km^3 in special scientific experiments. Neutrino detectors rely on nuclear reactions mainly with protons and neutrons in the form of an inverse β-decay or elastic scattering with electrons. In the following, the produced fast charged particles are detected in a usual charged particle detector. The active material, for example Ga or H, reacts with the neutrino forming for example Ge or neutrons:

$$\nu_e +^{71}\text{Ga} + 233.2\text{keV} \rightarrow^{71}\text{Ge} + e^- \tag{4.16}$$

Very recent scientific experiments indicate another possible reaction involving a coherent elastic scattering of the neutrino with a whole nucleus (Akimov 2017). This increases the interaction cross-sections with matter by up to two orders of magnitude, allowing reducing the detector sizes to volumes of liters or some ten kilogram [14.6 kg CsI in (Akimov 2017)], respectively.

First neutrino applications propose neutrinos for detecting and quantifying nuclear reactor activities, e.g. for preventing nuclear proliferation (Porter et al. 2010), via this unavoidable and highly penetrating by-product, or for communication through earth via a pulsed accelerator based neutrino source (Stancil et al. 2012). Improved technology potentially opens up new, distinct applications of neutrinos with their unique properties compared to standard particles. In these applications, accelerators potentially play an important role for producing neutrinos (indirectly).

In the group of anti-matter, the positron is definitely the application highlight. The positron is the anti-particle of the electron and hence has the same properties except for a positive charge. The positron is the only anti-matter particle observed (and applied) in our usual nature as it is produced in the β^+ decay. Its direct production in accelerators relies on electron beams or secondary reactions of photons producing electron-positron pairs. In ion accelerators producing proton rich nuclides decaying by β^+ (e.g. via ^{18}O(p, n)^{18}F) are the most suitable path for positron production. Despite these difficulties many analytical techniques rely on positrons due to their unique electron–positron anti-matter annihilation reaction with the characteristic emission of 511 keV photons (rest mass of electron and positron), see also Sect. 6.1.2.

For every lepton and quark denoted in Fig. 4.28 an anti-matter equivalent exists with inverse charge and quantum numbers, but identical mass, spin, and lifetime. The anti-matter particles react with their normal matter equivalent in a so-called annihilation, converting their complete energy and mass into photons. All anti-matter particles finally annihilate with the corresponding normal matter particle in our, by definition, normal matter dominated universe. Momentum and energy conversion force the release of two photons each with the energy of the rest mass of the particles plus half the remaining kinetic energy. We already came to known the positron (e$^+$), the anti-matter equivalent of the electron (e$^-$), in Sect. 3.1 where we learned about its production by ≥1022 keV photons. The positron annihilation with the electrons present in matter is not instantaneous when the positron enters matter, because only with sufficiently low distance and velocity between both reactions become probable. One could say it is not enough for an electron and a positron to stand in the same room, but they have to touch each other in order to start interacting. The positron, like all anti-matter, is not instable in itself, but only reacts in the presence of normal matter. Therefore anti-matter is an excellent probe for the density of the corresponding normal matter particle. A lower density of the normal matter results in a longer lifetime of anti-matter than a higher density, since a collision of both is less probable at lower density. The characteristic emission of two correlated photons from the annihilation enables a clear identification and localisation of the annihilation reaction. A few analytical applications exist including the material analysis technique positron annihilation lifetime spectroscopy (PAS) or the medical imaging via positron emission tomography (PET, Sects. 6.1).

The production of positrons requires accelerators. Most commonly, the decay of β^+ decaying isotopes (most prominently ^{18}F and ^{22}Na) produces positrons. The proton irradiation of stable isotopes produces these sources. Instead of this option, GeV electron beams directed onto targets to induce a pair production (in this case from electrons not photons) from which the positrons can be extracted. This source type yields typically below one e^+ per incident e^-. Other anti-matter particles such as anti-protons are produced only in accelerators in a similar fashion, but the electron features by far the lowest rest mass and therefore beam energy requirements. The generation of charged anti-matter particle beams from isotope sources delivers only low brightness. Pair-production sources deliver significantly higher anti-matter beam brightness.

This entire book discusses only the first of the three generations of matter, see Fig. 4.28. These higher generation particles decay quickly to at least one of the corresponding particles of the first generation and neutrinos. The muon (μ^-) represents the only exception in this line. Muons originating from the interaction of cosmic radiation with earth's atmosphere appear on the ground level with fluxes of some $100/m^2s$. The muon rest mass of 106 MeV/c^2 lies in the range of energies relevant for accelerator applications and its lifetime of 2.2 μs is the longest among the higher generations. Its 207 times higher mass compared to the electron gives muons some specific features compared to an electron, in particular when binding in an atom

$$E_n = \frac{m_e e^4 Z^2}{8\varepsilon_0^2 n^2} \tag{4.17}$$

The calculation of the electron binding energy for the proton yields the so-called Rydberg energy of 13.6 eV. By substitution of the electron mass for the muon mass, we can see that the binding energy is proportional to the mass of the negative charge carrier. The muons mass leads to a binding energy in a hydrogen atom of about 2.8 keV. The stronger binding shields the nuclear charge in a smaller radius, allowing for reduced nuclear reaction barriers. Dreams of muon catalysed nuclear fusion of hydrogen isotopes quickly die when considering the price tag of a single accelerator produced muon. The higher binding would also change the x-ray emission scheme of atoms depicted Fig. 3.4 allowing for new analytical tools. This simple estimation should not be overrated, but just give an impression of the aspects of muons for nuclear applications.

Currently applications of muons exploit the natural (cosmic) muons for imaging purposes or accelerator produced muons for material analysis methods such as the muon spin spectroscopy. The so-called muon transmission imaging or also muon tomography works similar to medical x-ray analysis but by exploiting the attenuation of muons. The flux and energy of cosmic muons reduces when passing matter. Due to their higher mass muons experience a significantly lower stopping power than electrons [enter mass in (3.7)] leading to high range. This range allows for imaging of large objects or geological formations. Triangulations convert the 2D to 3D images. A recent and highly visible work used this technology for investigation the Great Pyramid of Giza, see Fig. 4.29. An acquisition time of 2 month was required due to the low flux of muons, but finally a new chamber in the pyramid was discovered.

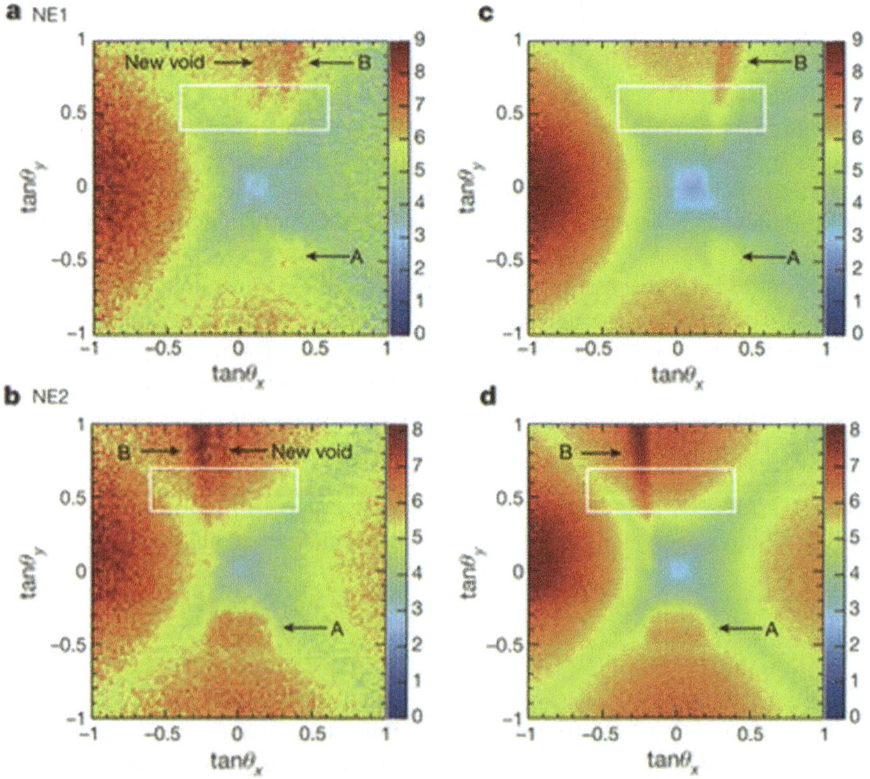

Fig. 4.29 Great Pyramid of Giza muon images acquired using two imaging plates (**a, b**) placed in the Queens chamber looking towards the pyramid top and a simulation of the known structure (**c, d**) as seen from the imaging plate positions. Spot A indicates the King's Chamber, spot B the grand gallery. A new void appeared next to the grand gallery. Voids absorb less muons, leading to a brighter spot on the film (Morishima et al. 2017). Reprinted with permission by Springer

References

D. Akimov, Observation of coherent elastic neutrino-nucleus scattering. Science **357**(6356), 1123–1126 (2017). https://doi.org/10.1126/science.aao0990

T. Arlt, H.-E. Mahnke, T. Siopi, E. Menei, C. Aibéo, R.-R. Pausewein, et al., Absorption edge sensitive radiography and tomography of Egyptian Papyri. J. Cult. Heritage **39**(S), 13–20 (2019). https://doi.org/10.1016/j.culher.2019.04.007

S. Bari, M. Beye, L. Bocklage, R. Gehrke, D. Horke, W. Laasch, et al., *PHOTON SCIENCE 2016—Highlights and Annual Report* (Deutsches Elektronen-Synchrotron DESY, Hamburg, 2017)

R. Behrisch, W. Eckstein, *Sputtering by Particle Bombardment* (Springer, Berlin Heidelberg, 2007)

J.A. Clarke, *The Science and Technology of Undulators and Wigglers* (Oxford University Press, 2004)

A. Eksaeva, E. Marenkova, D. Borodin, A. Kreter, M. Reinhart, A. Kirschner, et al., ERO modelling of tungsten erosion in the linear plasma device PSI-2. Nuclear Mater. Energy **12**(S), 253-260 (2017). https://doi.org/10.1016/j.nme.2017.03.014

J. Formaggio, G. Zeller, *From eV to EeV: Neutrino Cross Sections Across Energy Scales* (2013). https://arxiv.org/abs/1305.7513.

H. Gnaser, *Low-Energy Ion Irradiation of Solid Surfaces* (Springer, Berlin Heidelberg, 1999)

P. Goslawskiy, F. Kramerz, M. Ruprecht, A. Jankowiak, M. Ries, G. Wüstefeld, Status of transverse resonance island buckets as bunch separation scheme, in *Proceedings of IPAC2017*. Copenhagen, Denmark (2017)

D. Griffiths, *Introduction to Elementary Particles* (Wiley-VCH, 2008)

B.L. Herman, S.C. Fultz, Measurements of the giant dipole resonance with monoenergetic photons. Rev. Mod. Phys. **47**(3), 713–761 (1975)

IAEA, *IAEA TECDOC 1153—Use of Accelerator Based Neutron Sources* (IAEA, 2000)

A. Koning, et al., *TENDL-2015: TALYS-based Evaluated Nuclear Data Library* (Von, 2015). https://tendl.web.psi.ch/tendl_2015/tendl2015.html

P. Kudejova, G. Meierhofer, K. Zeitelhack, J. Jolie, R. Schulze, A. Türler, T. Materna, The new PGAA and PGAI facility at the research reactor FRM II in Garching near Munich. J. Radioanal. Nucl. Chem. **278**, 691–695 (2008). https://doi.org/10.1007/s10967-008-1506-9

A.J. Larkoski, *Elementary Particle Physics: An Intuitive Introduction* (Cambridge University Press, 2019). ISBN: 978-1108496988

E. Mauerhofer, U. Rücker, T. Cronert, P. Zakalek, J. Baggemann, P. Doege, et al., Conceptual Design Report NOVA ERA (Neutrons Obtained Via Accelerator for Education and Research Activities) A Jülich High Brilliance Neutron Source project (Forschungszentrum Jülich GmbH, Jülich, 2017)

S. Mobilio, F. Boscherini, C. Meneghini, *Synchrotron Radiation: Basics, Methods and Applications* (Springer, Berlin, 2015)

T. Möller, J. Falta, *Forschung mit Synchrotronstrahlung* (Springer Vieweg, 2010)

K. Morishima, et al., Discovery of a big void in Khufu's Pyramid by observation of cosmic-ray muons. Nature **552**(S) (2017). https://www.nature.com/articles/nature24647

Particle data group. (2020). Von. https://pdg.lbl.gov/abgerufen

G.G. Poludniowski, P.M. Evans, Calculation of x-ray spectra emerging from an x-ray tube. Med. Phys. **34**(6), 2164–2174 (2007)

A. Porter et al., *Reactor Neutrino Detection for Non-Proliferation With the NUCIFER Experiment*, (2010). https://doi.org/10.1109/TNS.2009.2035119

P. Schmüser, M. Dohlus, J. Rossbach, *Ultraviolet and Soft X-Ray Free-Electron Lasers: Introduction to Physical Principles, Experimental Results, Technological Challenges* (Springer, Berlin, 2008)

H. Seiler, Secondary electron emission in the scanning electron microscope. J. Appl. Phys. **54**, R1, S (1983). https://doi.org/10.1063/1.332840

D.D. Stancil, P. Adamson, M. Alania, L. Aliaga, et al., *Demonstration of Communication using Neutrinos* (2012). arXiv:1203.2847

J.F. Ziegler, J.P. Biersack, M.D. Ziegler, *SRIM—The Stopping and Range of Ions in Matter* (Chester, 2008)

Chapter 5
Technical Applications

Abstract All the physics and mathematics discussed above should yield a practical gain. This chapter discussed industrial and medical production applications of accelerators. The selection and layout of production routes for different radioactive source types are discussed together with the economic aspects and optimisation routes for production facilities. Besides the production of radioactive isotopes also direct applications of charged particle beams for surface near modification and machining exist. These applications extend from tribology to micro-electronics with connected technologies, limitations, and future options.

With this section, the second half of this book is about to start. In the first half we learned about the details of technological aspects and the physical basis of accelerator applications. From now on the focus will shift to exploiting this knowledge by thinking about practical applications. By considering the accelerator as a tool upon which we create methods allows focussing our thinking. Considering the accelerator as a separate tool means focussing the view onto a few main accelerator quantities. Most applications focus on beam energy and beam current, some also depend on beam emittance, beam energy distribution or others. None of these quantities is fixed for a certain accelerator or restricted by fundamental physics, but is subject to technical optimisation. The application/method can couple back to the accelerator and define its characteristics, as this book demonstrates.

The major difference between science and practical applications is the requirement of efficient and economical behaviour in the latter. This includes, among others, energy efficiency, continuous usage of the expensive equipment, and operating the equipment with maximum effectivity. Fear of damaging equipment or risking to invest in pushing a method towards the next level hinders technological advances.

Accelerators found some very unique applications in fundamental particle physics, but besides this most accelerator applications are in direct competition to other techniques. For the production of isotopes, fission reactors provide alternative means, in medicine classical surgery and drugs, and in analytics electron and ion based methods compete to optical and chemical methods. Advantages provide reasons for using accelerators instead of the competition in many fields, see e.g. (EuCard-2 et al. 2017). Often new methods develop by finding a niche, just like an

S. Möller, *Accelerator Technology*, Particle Acceleration and Detection,
https://doi.org/10.1007/978-3-030-62308-1_5

evolutionary process. Growing beyond the boundaries of this niche by clever optimisation and usability represents the basis for technological advancement which this book wants to promote by providing the overview of the accelerator applications and physics.

A practical problem dedicated people often face is the difficulty to implement a new idea. It lacks in funding, technical experience, or qualified workers. Sometimes it can be enough to formulate a precise and quantitative plan and present it to the management, but sometimes unfortunately even a good and profitable idea will not be realized. See for example the establishing of $LiCoO_2$ (Li–Ion) batteries from the basic discovery in 1980 (Mizushima et al. 1980) until the first Li–ion based pure electric car released around 2010. If the available competences prevent the realisation turning to established institutions might help to obtain the required competences. These institutions can be accelerator related manufacturers, but also universities, research centres or other companies using the same equipment and maybe even facing the exact same problem. The important lesson to learn is to stay dedicated, since everything that's valuable and worth it also costs.

5.1 Generation of α-β-γ-n Sources and Activation

Activation by nuclear reactions produces radioactive isotopes in targets and other parts of accelerators working with energies above a few MeV. These isotopes are inconvenient due to radiation protection, but we can also make use of some of them. This use separates the terms activation, radioactive waste, and radioisotope sources.

$$\frac{\text{Decays}}{\text{time}} \equiv \text{Activity} = \frac{\mathrm{d}N}{\mathrm{d}t} = -\frac{N}{\tau} = -\frac{N\ln(2)}{T_{1/2}} \tag{5.1}$$

$$N(t) = N_0 * e^{-t/\tau} = N_0 * e^{-t*\ln(2)/T_{1/2}} \tag{5.2}$$

In a way, the radioactive isotopes work as mobile accelerators since they provide at least a subset of accelerator products and particle energies. Their source strength, mathematically the change of the amount of nuclides N per time t, depends on the quantity of nuclei N and their mean life-time τ or half-life $T_{1/2}$, respectively, as given by (5.1). The half-life states the time after which $N_0/2$ of the initial nuclides decayed. Short half-lifes result in higher intensity, while longer half-lifes result in lower intensity, but over a longer duration. Equation (5.2) calculates the amount of nuclides N present at any point t in time, requiring only the current amount of nuclides N_0 and their half-life $T_{1/2}$. Practically relevant half-lifes lie between about one minute to 100 years, which covers 1432 of the 3896 known isotopes. The emitted particle species and energy depend on only the isotope. The three decay types responsible for the three arms of the nuclear warning symbol (Fig. 5.1) are α, β, and γ. Neutron emitting sources, such as Am–Be neutron sources, exploit the ^{9}Be(α,n)^{12}C reaction

Fig. 5.1 The nuclear symbol is typically used as a warning for radioisotope sources and has to be attached to the outer container

for transforming an α-source into a neutron source. Few heavy isotopes such as Cf-252 directly emit neutrons through spontaneous fission.

In detail, radioactive decays are more complex than simply stating the decay mode. In fact, the nuclide charts usually only display the dominant decay mode. Besides this main mode, the decay always emits photons and particles at several different energies, due to the nuclear structure and excitation levels of mother and daughter nuclide. Different excited states in general follow different decay routes. Nuclear reactions produce isotopes not only in the ground state, but depending on energy and reaction type excited states are produced. The decay spectrum in particular of heavier isotopes contains several lines splitting the decay energy (the mass difference between the isotope and its daughter) among them with statistically distributed ratios. Looking over the whole isotopic table we find energy limits for each decay type. Typically, γ-rays reach energies of keV to 3 MeV with values up to about 10 MeV in the decay of isotopes far away from the valley of stability, e.g. ^{36}K with a γ-ray at 9218.8 keV. The number of α-emitters produced from stable isotopes is quite limited with mostly polonium isotopes. Here an α energy of 7283 keV (^{211m}Po) represents the upper limit with typical values of 5 ± 0.2 MeV. The highest β/electron energies lies at 12,087 keV (^{48}K), but typical values are rather below 1 MeV.

Producing isotopes for applications dictates two necessary conditions for selecting the production route: First, the isotope has to fulfil the application requirements regarding half-life, chemical aspects, and emitted radiation and second, it has to be possible to produce the isotope via suitable reactions. In particular, medical applications require additional separation from the host material and purification. The criteria may sound trivial, but of the over 3000 available isotopes (Fig. 1.1) only a handful actually came into applications. Besides the necessary, we also have sufficient conditions regarding production cost, in particular defined by the reaction cross-section, and the target materials, e.g. enriched isotopes. Accelerators offer different options compared to fission reactor, allowing access to the lower half of isotopes in Fig. 1.1, the proton rich isotopes. The chain of reactions to reach a final production will be called the pathway on this nuclide map. This pathway is not necessarily direct but can also involve intermediate decays.

In technical applications, we start our pathway from a certain stable nuclide. These nuclides sit in the so-called valley of stability, the black line going through the centre of the nuclide chart (Fig. 1.1) starting with H and ending with ^{209}Bi. In principle, also the two practically stable elements thorium and uranium could serve as starting point, but the radiation protection legislation makes this significantly more complicated. Many isotopes are only accessible via production by accelerators. Cyclotrons are the standard solution since they deliver high energy and currents in compact on-site facilities and the production has only low requirements on beam quality. The beam current defines the maximum achievable activity due to the competition of production and radioactive decay. In particular, for short-lived isotopes this requires powerful accelerators. LinACs potentially offer higher beam currents, but with modern cyclotrons delivering several mA practical and safety issues represent the stronger limits than cyclotron beam power. Most isotope production relies on proton projectiles, but deuterons offer an interesting alternative. These technical aspects and also the physics of nuclear reactions favour ions over electrons for isotope production.

The wide range of known and fundamental energies and decay emission probabilities perfectly fits to calibration applications. Energy and efficiency calibration of radiation detectors usually relies on isotope source, since uncertainties such as an acceleration voltage are non-existing with radioactive sources and the values are exactly identical no matter where the product comes from. Also in industrial and medical applications radioisotopes are found. Mobile applications profit from the energy stored in the isotopes. The substantial particle energies emitted from isotopic sources would otherwise require bulky equipment and high input power, making an MeV particle source fit even into the human body. For example, a radioactive ^{60}Co source with its 1332.5 keV photons allows for contact-free analysis of the thickness of a steel sheet up to several 10 mm via intensity absorption. Producing this photon energy with an accelerator would require a 1–2 m long device.

Higher beam energies require larger accelerators, but higher energies also induce more reactions (see for example Sect. 4.2.1), some of them detrimental for the application in the sense of production of unwanted isotopes via (p,xn) and/or secondary radiation such as Bremsstrahlung. Especially threshold reactions (negative Q-value) or cross-section resonances drastically change the ratio of different reaction types. When several reactions are possible, the sum of all reaction cross-sections resulting in the same product is the production cross-section. The production of neighbouring isotopes can solve itself via largely different half-lifes, but isotopic separation via physical and chemical means is often a challenge. The choice of an isotopically pure target may be the easier option for preventing the production of problematic impurities in the first place.

Radionuclide labs and equipment have the means for detection and separation of isotopes of different elements via chemical (e.g. electrolytic plating) and physical (e.g. evaporation) means. Separating different elements is usually successful with elemental suppression up to about 10^6. The preparation of radioactive sources results in two general options relevant for safety: Open and enclosed sources. An open source can be touched, potentially resulting in contamination of man and machine. The enclosure avoids this by putting a sealed box around the radioactive source, easing handling and accounting. In order to make use of the radiation the particles have to penetrate the enclosure. This restricts the use to highly penetrating particles, namely higher energy photons and neutrons. Enclosed sources have their use as calibration sources for example for detectors, but also in medical context for local irradiation without allowing the radioisotopes to enter the metabolic system of the body.

5.1.1 Pathways on the Nuclide Chart

After selection of an isotope suitable for the given application, the first question for the accelerator technology is if the isotope can be produced or reached via ion irradiation or not. Here we focus on ion irradiation, neutrons will be handled later. Fig. 5.2 depicts the possible two body reactions of protons and deuterons. The (p,n) reaction as fundamental production pathways of protons lets us move like an "inverse β^+ decay" one isotope to the top and one to the left. The deuteron equivalent moves only one isotope to the top, since the deuteron contains an extra neutron. In addition, we have the α-producing reaction leading us to an element with lower proton count. (p,γ) reactions are in principle possible, but the reaction cross-sections are usually orders of magnitude smaller compared to the reactions with heavy products. Heavier projectiles add more neutrons to the equation allowing us to extend the accessible isotope space to the right. The sum of neutrons and protons of ion and target isotope minus at least one reaction product yield the heaviest accessible product.

The pathways can be followed by treating the nuclide chart like a board game map with the reaction notation telling us how many steps to go in which direction. We can select from α, p, n, d, and t product moves, multiples of each, and sums of any of them. Pathways of a proton induced reaction lead us firstly one step upwards for adding one proton. From there on (p,xn) reactions lead us x steps further to the

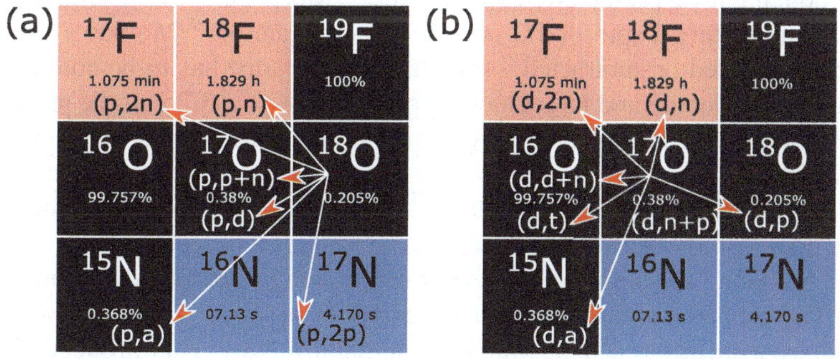

Fig. 5.2 Reaction pathways by protons **a** and deuterons, **b** starting from ^{18}O and ^{17}O, respectively. All isotopes to the right of the target remain inaccessible for protons. Heavier ions such as the deuteron contribute more neutrons to the system, allowing moving further to the right, but their higher stopping power will also decrease production efficiency. Reactions can be selected by *Q*-value related thresholds, but generally take place in parallel

left (p,p+α) leads us one proton downwards and two steps diagonally down-left for the α-particle. Deuterons extend the accessible isotopes by enabling the vertical path with the (d,n) reaction. Pathways oriented to more neutron rich isotopes (to the right) require heavier ions, requiring significantly more beam energy for enabling reactions.

Having understood available moves, the important question arises which of these moves are actually possible and how physics selects among the many possibilities. The Q-value in combination with the projectile energy $(Q + E)$ represent the selection criterion due to the energy conservation requiring $E \geq Q$. $Q > 0$ reactions are always allowed, but many of the reactions feature $Q < 0$ and these can only be induced with a beam energy $E > Q$. Among these possible reactions generally all take place in a given irradiation situation. The individual reaction cross-sections state yield the ratios of the different reactions. For thick targets, we have to consider the depth dependency of the projectile energy as discussed in Sect. 3.4. The conservation of spin influences the magnitude of the reactions.

Let us examine this aspect at the example of a β^- isotope and the connection of projectile and product via the spin conservation. Medical diagnostics make use of the excited isotope ^{99m}Tc with its 6 h half-life and the strong decay emission of a 140.5 keV photon. The production of ^{99m}Tc classically uses the decay of ^{99}Mo produced in fission reactors as a side product. The actually interesting isotopes is the second excited state of ^{99}Tc, the 142.7 keV (Spin 1/2-) state which then decays through the first excited state at 140.5 keV (7/2+) to the (9/2+) ground state. Spin differences >1 are considered forbidden transitions and the larger the mismatch the longer the half-life of the given transition. The mismatch of 3 results in a relatively long half of ^{99m}Tc catalysed by the intermediate 7/2 state. Without this state the half-life would be even longer. The ground state of ^{99}Mo has a spin of 1/2+, a good match to the ^{99m}Tc state (1/2−), explaining why ^{99}Mo actually decays preferably to the excited state. ^{99m}Tc forms a good example since it can be produced in both fission reactors and accelerators (Guérin, et al., 2010) with pathways depicted in Fig. 5.3. For reproducing it via protons we can apply the ^{100}Mo(p, 2n)^{99m}Tc reaction. Protons and neutrons feature a spin of 1/2 and the ground state of ^{100}Mo has a spin of 0. Spins can be added or subtracted, only the absolute value has to be conserved. In summary the projectile and target have a spin of $1/2 + 0 = 1/2$ and the products of $1/2 + 1/2 - 1/2 = 1/2$ favouring the production of the same excited state as with the decay of ^{99}Mo.

The last example implicitly assumed a ^{100}Mo target, but if we take a look at the isotopic composition of natural Mo in Fig. 5.3, only 9.63% of natural Mo consists of the ^{100}Mo isotope. This small content of the desired target reduces the production rate of ^{99m}Tc proportionally (3.20) and results in production of numerous other Tc isotopes with different half-lifes and emission modes. This product impurity potentially reduces the result quality of medical diagnostics applying ^{99m}Tc by mixing of signals. Enriched targets reduce this isotopic impurity contribution at the drawback of strongly increased target costs. Targets made from mono-isotopic elements (Bi, Ta, Na ...) avoid this issue, if available. Impurities exist in any material due to the nature of technical processing. Typically, the elements from the same column in the periodic table represent the strongest impurity, e.g. W in Mo and Mo in W. Purified target

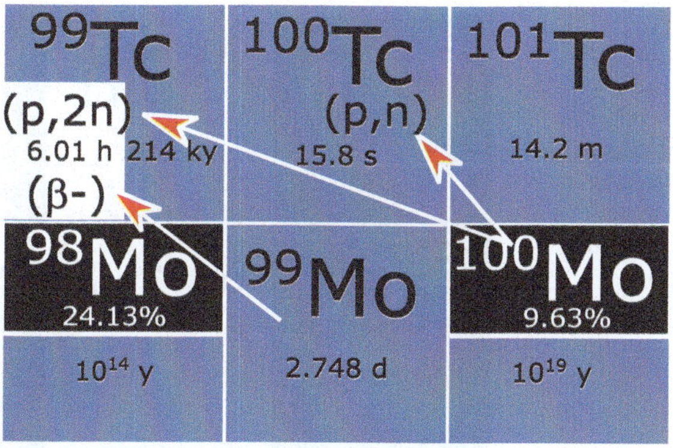

Fig. 5.3 nn^{99m}Tc a typical isotope used in medicine produced from the decay of ^{99}Mo produced in fission reactors can also be produced via proton or deuteron irradiation. The decay of ^{99}Mo(1/2+) results to 88.1% in ^{99m}Tc(1/2−) due to the conservation of nuclear angular momentum, ion reactions such as ^{100}Mo(p,2n)^{99}Tc have different pathways

elements help reducing this contribution to product quality, but they can never be fully avoided. In summary, the purity of the final product depends on the initial target purity, the isotopic purity and the folding of these with the production cross-sections and decays.

The isotope half-life provides an additional natural selection mechanism. Significantly shorter half-lifes enable an elimination of the undesired isotope by waiting a few half-lifes for it to decay. This requires producing an excess of the actual product, the more the closer the half-lifes. Significantly longer half-lifes contribute less to the total activity which may be tolerable for the application. A 10 times larger half-life results in a 10 times lower activity for the same amount of produced isotopes, see (5.1). The decays also produce daughter isotopes which might be of relevance as a product or impurity. Typically, the connections are short, since light projectiles only allow for one field steps on the nuclide chart, but in some elements (e.g. Fe or Mo) with instable elements in between small decay chains can develop.

Sometimes several choices of projectiles are possible for reaching the same isotope. Figure 5.4 demonstrates this at the example of accelerator production of ^{64}Cu. Let us discuss the four cases of interest to test our knowledge gained so far: The production from ^{64}Ni(p,n)^{64}Cu features a high cross-section of up to 800 mbar at 10 MeV (Sayed et al. 2015) and a low threshold of 2.5 MeV, actually the lowest among all reactions starting from ^{64}Ni. Its selection has the advantage of resulting in only little side reactions with short-lived products (^{61}Co), but its low isotopic abundance result in low reaction rates. The ^{65}Cu(p, p+n)^{64}Cu has a cross-section of only 400 mbarn at 23 MeV. Its threshold at 10 MeV exceed the one of ^{65}Cu(p, n)^{65}Zn resulting in a strong production of ^{65}Zn in this pathway, but the 500 times longer half-life of ^{65}Zn results in only little contribution to the total activity. The deuteron reaction ^{66}Zn(d,α)^{64}Cu has a positive Q-value but only a negligible cross-section of about 50 mbarn at 10 MeV. The many Zn isotopes complicate the situation

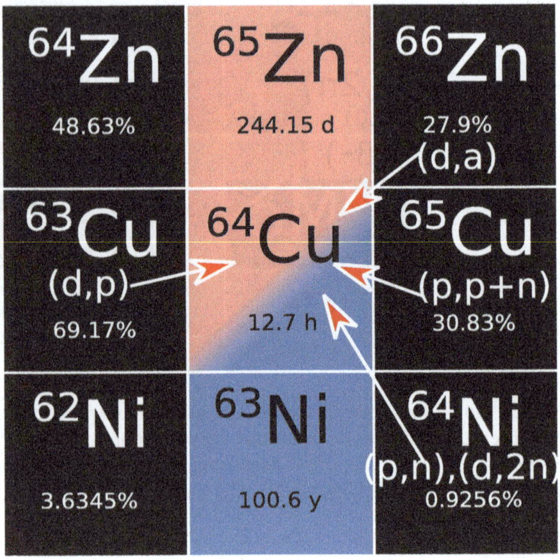

Fig. 5.4 The isotopes around ^{64}Cu allow for different pathways to be chosen. p and d projectiles allow for 4 different target isotopes The cross-sections and the generation of unwanted isotopes (also in view of target impurities) dictate the selection of the projectile

by producing a zoo of Ga isotopes with half-lifes comparable to ^{64}Cu. The reaction ^{63}Cu(d, p)^{64}Cu also features a positive Q-value. Its cross-section reaches a maximum of 300 mbarn only at 5.5 MeV opening up the production of Zn isotopes via reactions of ^{65}Cu in a natural Cu target. Anyways producing ^{64}Cu in a Cu target complicates the separation of the active from the stable isotopes, favouring a chemical contrast between target and product. Comparing the four options ^{64}Ni(p,n)^{64}Cu appears to be most attractive, in spite of the low isotopic abundance of the target.

Coming back to the example of Tantalum (Ta) used for beam optical apertures as discussed in Sect. 2.7 we can combine all our obtained knowledge. The technical requirements on apertures demand maximal heat dissipation capability (conductivity + melting point) at minimal material costs and (long-term) radioactive inventory. We would like to select an element for an aperture operated at a 15 MeV proton accelerator with kW beam power. The importance of the Coulomb barrier for nuclear reactions was mentioned before and in the MeV range we are still below the maximum barriers of the heaviest elements. Consequently, we start our search with heavier elements since these feature larger barrier potentials.

Au would be a good choice due to its good thermo-mechanical properties and its high barrier with only a single isotope and many stable isotopes around. Tungsten (W) as a significantly cheaper alternative with equally good thermo-mechanical properties has the disadvantage of several low threshold (p,n) reaction producing rather long-lived Re isotopes. The next idea is Ta, due to its acceptable price tag at reasonable thermo-mechanical properties. Table 3.1 states a barrier of 9.545 MeV for a proton beam on Ta, but in contrast the (p,n) reaction has a threshold of $Q = -0.976$ MeV, reaching 1 mbarn at 5.45 MeV. Figure 5.5 demonstrates the possible pathways. The given barrier potential corresponds to the cross-section maximum of about 100

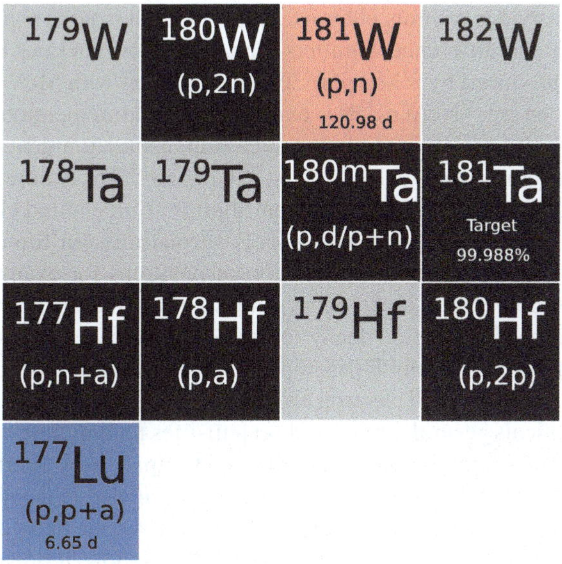

Fig. 5.5 Graph of Ta activation-decay scheme for E_0=15 MeV proton irradiation. The graph shows reaction pathways with $Q + E > 0$ below the respective product. Greyed isotopes and anything above or to the right of the pane is not reachable. Black isotopes are stable, red and blue decay via β^+ and β^-, respectively. Ta can be considered a low activation element, since only ^{181}W contributes to the long-term radioactive inventory with most path' ending on stable nuclides

mbarn. Ta features practically only the ^{181}Ta isotope, limiting the nuclear activation in quantity and diversity as given in the example of Fig. 5.5. The comparably low (p,n) cross-section produces ^{181}W, an unstable isotope with an inconvenient half-life of 121 days. ^{181}W decays by electron capture with Q = 186 keV back to ^{181}Ta, leading to low dose rates and shielding requirements. Try and compare the situation to other elements, for example Fe or Cu, and list their strength and weaknesses for specific applications.

5.1.2 *Examples of α, γ, β^-, β^+,n Sources from Accelerators*

Having discussed the principal path' accessible on the nuclide chart this section will discuss a few selected examples. In general, the application defines the requirement of chemical element and minimum half-life, e.g. due isotopic separation and preparation of the products or due to economic minimum service times of devices using a radioactive source. For applications, generally shorter half-lifes are desirable since this results in higher activity per produced isotope. The main question to be answered for selecting an isotope is the type and energy of radiation the source has to emit.

Practically, every decay emits photons. This makes all isotope sources γ-sources. The details make the difference. Emitters with rather low energies in the 100 keV range, e.g. ^{55}Fe produced by ^{55}Mn(p,n)^{55}Fe, and emitters with MeV scale energies, e.g. ^{22}Na produced by ^{22}Ne(p,n)^{22}Na, exist. For both aforementioned examples a single line dominates the emitted photon spectrum, a situation generally desirable. In particular heavier isotopes such as ^{160}Tb (^{160}Gd(p,n)^{160}Tb) feature several intense photon lines and strong Bremsstrahlung contribution from emitted electrons.

Medical diagnostics applies β^+ sources for Positron Emission Tomography (PET). Besides this also a few scientific application of positrons for example in positron lifetime spectroscopy exist. The β^+ decay follows either the Electron Capture (EC) and/or the positron (e^+) emission decay route. Equation (5.3) depicts the difference between both for the prominent PET isotope ^{18}F. With a certain probability, for ^{18}F = 3%, the nucleus absorbs a shell electron and decays to ^{18}O without positron emission. Radiopharmaceuticals generally require short half-lifes for diagnostic signal strength as well as for effective irradiation of tumours, see Chap. 6. Very short half-lifes lead to cost explosions for transportation which will be calculated in detail in Sect. 5.1.4. This aspect results in a technically induced lower limit for the half-life. With a half-life of 110 min ^{18}F perfectly fits into this application window. It releases 634 keV in the positron route (3-body decay = broad spectrum). The production usually works via the ^{18}O(p,n)^{18}F reaction with several 10 MeV protons from cyclotrons.

$$\begin{aligned} &^{18}\mathrm{F} + \mathrm{e}^- \rightarrow {}^{18}\mathrm{O} + \nu_e \quad (\mathrm{EC{:}}\ 3\%) \\ &^{18}\mathrm{F} \qquad\;\; \rightarrow {}^{18}\mathrm{O} + \mathrm{e}^+ + \nu_e \; (\mathrm{positron{:}}\ 97\%) \end{aligned} \tag{5.3}$$

The positron emission route of β^+ subtracts 1022 keV from the EC decay energy for producing the positron and not consuming the electron. Consequently, not all isotopes marked in red (β^+) on the nuclide chart can emit positrons. For example, the decay energy of ^{55}Fe of 231 keV is insufficient for positron emission, resulting in 100% EC decay. ^{64}Cu features both routes and also the β^- decay route (Fig. 5.6). This multitude of decays limits the radiation purity of ^{64}Cu as a positron emitter when compared to ^{18}F with its positron emission dominated decay. In the medical

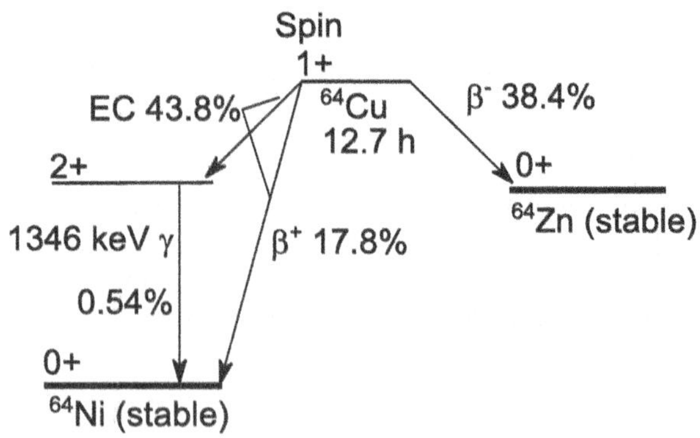

Fig. 5.6 Decay scheme of ^{64}Cu. Only 17.8% of the decays emit a positron. EC dominates the overall decay. The path over an intermediate 2+ state of ^{64}Ni takes place in only 0.54% of the decays

context, the mixture of radiations increases the patient dose for a given PET signal, but the chemical aspect of copper and its metabolism nevertheless add an interesting aspect against ^{18}F. Section 5.1.1 discussed its production options.

Among the accelerator produced isotopes β^- decay is rather rare due to the pathway options leading mostly towards the proton-rich half of the nuclide chart (e.g. (p,xn)). Only a few cases of neutron-rich stable isotopes in advantageous positions have β^- products on their proton rich side in the nuclide chart. The application of electrons has an application field in medicine, where low energy electrons enable a local tumour irradiation by body-internal sources. Section 5.1.1 discussed ^{99m}Tc as a β^- emitter already above as an example of a rather pure photon emitter.

Only a few accelerator produced α-decaying isotopes exist, since these generally lie on the upper edge of the nuclide chart, the fission domain. From the stable elements only bismuth has a path leading to the α-emitting polonium isotopes ^{206}Po to ^{210}Po via proton or deuteron reactions. Reactions of lead with helium ions potentially also lead to these polonium isotopes, but due to the high stopping power of Helium ions in matter the reaction are generally less attractive than the proton reactions. From Uranium and Thorium also pathways exist with (p,n) and (d,2n) reactions and subsequent β^- decay resulting in the relatively efficient α-sources ^{232}U and ^{238}Pu. These release several α's per produced isotope due to long decay chains through the valley of α-emitters down to polonium. The elements have the drawback of the complicated handling due to their classification as nuclear fuel. The use of α-emitters will be discussed further in the frame nuclear batteries in Sect. 8.2.

No direct neutron emitters such as the fission produced ^{252}Cf can be generated from accelerators. The only option is using an α-emitter such as ^{210}Po or ^{232}U mixed with beryllium for exploiting the $^{9}Be(\alpha,n)^{12}C$ reaction. So-called AmBe sources exploit the same reaction with beryllium using the α-emission of ^{241}Am. Due to the widespread availability of fission produced isotopes the accelerator route is so far unattractive.

Recently the nuclear data section of the International Atomic Energy Agency released a software enabling screening of arbitrary isotope production using with any light ion projectile (nds.iaea.org/mib). The software's name Medical Isotope Browser suggests a certain application, but it can also be used for assessing production of technical sources. The provided results extend from amounts up to radiation safety aspects, the input data can be of different type, and the projectile energy can be scanned over a range for finding optimized quantities. These features make this tool a versatile and convenient option for finding new interesting products and investigating and optimizing details of existing products. Since this software uses only semi-empirical cross-sections, the results require experimental verification.

5.1.3 Comparison to Production by Neutrons

A large part of the nuclides associated with nuclear technology originate from fission reactors, in particular the pressurized water uranium reactors (PWR). This reactor

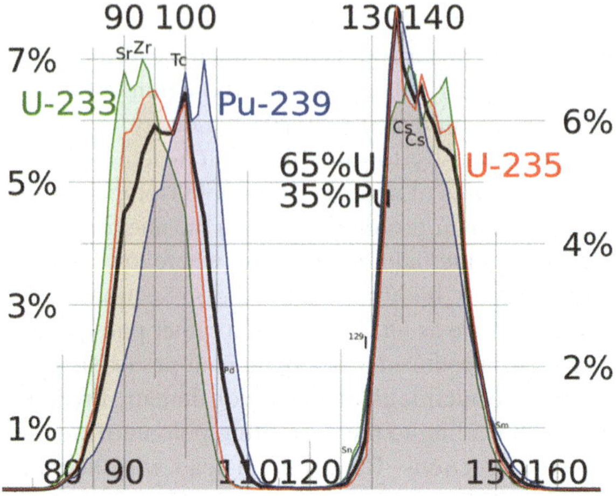

Fig. 5.7 The masses of nuclides and their part in the total produced isotopes produced by fission of the three nuclear fuels ^{233}U, ^{235}U, and ^{239}Pu (coloured lines) and the fission reactor mean (black line). The fission reaction breaks the fissile isotope into two parts of mostly 80–110 and 125–150 amu. Published by JWB at en.wikipedia under CC BY 3.0

type dominates the nuclear power generation and scientific applications. Although variants and vastly different sizes exist, the physics of the nuclear reaction in these reactors is the same since about 60 years. These reactors generate a typical set of nuclides, for example ^{137}Cs, ^{131}I, ^{90}Sr. For example ^{90}Sr could be produce by the 4 product reaction $^{235}U(n, 2n)^{144}Xe^{90}Sr$ with a $Q = 175.7$ MeV. Fig. 5.7 depicts the two regions of produced isotopes in this reactor type. The origin of this isotopic signature lies in the neutron energies and the fissile fuel Uranium.

The fission reactions require emitting several neutrons per reaction to be self-sustained, resulting in broad primary neutron energy spectra due to the n-body reaction nature (Sect. 3.3.2). The water moderator surrounding the fissile fuel further thermalizes this spectrum due to the increase of fission cross-sections towards lower neutron energies (similar to Fig. 5.8). This results in typical neutron spectra extracted from the reaction zone as depicted in Fig. 5.7. PWR's emit neutrons from thermal energies (MeV) to several MeV, but thermal neutron energies dominate the flux. Accelerator based neutron sources (see Sect. 4.2) also produce broad spectra, but with more weight on the fast neutron part of the spectrum. The additional conversion step from ions to neutrons of accelerator neutron sources in general results in a reduction of overall efficiency for isotope production compared to direct ion irradiation, favouring the direct application of ions. Currently this leaves the field of neutron irradiation to fission sources.

The emitted neutrons enable external production of additional isotopes not directly generated by the fission reaction as fission fragments. Since neutron reactions are not limited by the Coulomb barrier neutrons can interact with any target at any energy. The thermal and mostly also the epithermal neutrons cannot induce negative Q-value

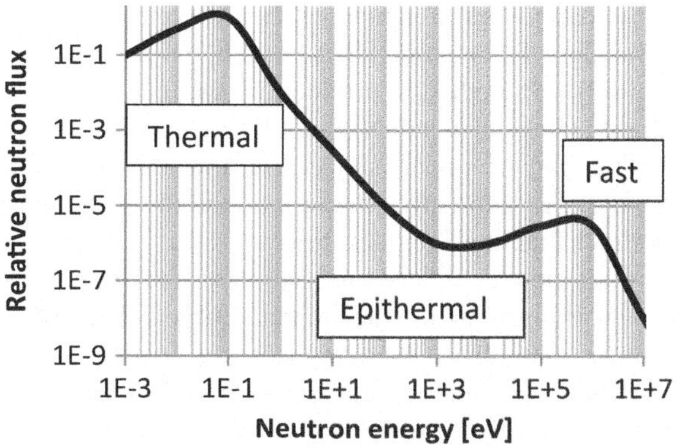

Fig. 5.8 Qualitative sketch of a uranium pressurized water fission reactor neutron spectrum. Most of the neutrons initially bearing up to ≈10 MeV are moderated to the thermal energy range in the order of a few 10 MeV

reactions. Besides fission reactions thermal neutrons can mostly induce only the (n,γ) reaction also called neutron capture. For some isotopes (n,p), (n,d), and (n,α) also feature positive Q-values, but even in these cases usually the (n,γ) cross-section dominates at thermal energies. On the nuclide chart the (n,γ) reaction represents a move straight to the right towards the β^- decaying half (Fig. 1.1). Fast neutrons can induce negative Q-value reactions. For stable isotopes this opens up the (n,2n) and (n,xn) reactions which for any stable isotope target have to have negative Q-values in the order of MeV (Fig. 8.1). Since the second emitted neutron is initially bound in the stable nucleus, it logically requires energy to remove it from there. The (n,2n) reaction moves us one step straight to the left on the isotope chart.

Ion irradiation offers a clear cut-off in the particle energy defined by the set acceleration voltage, resulting in well-defined reaction conditions. The broad neutron spectra make the production of isotopes from fission (and other) neutrons less controlled than those induced by mono-energetic ion beams, resulting in a larger variety of unintendedly produced isotopes. The long range of neutrons and their interaction with surroundings anyways result in broad neutron spectra even in the theoretical case of a mono-energetic neutron beam as for example provided by a D-T neutron generator (Sect. 4.2.1). Therefore, isotopes, e.g. the medically relevant ^{131}I, have to be acquired by chemical and physical separation from the fuel rods or neutron targets.

The accelerator equivalent of fission is spallation (Sects. 4.2 and 8.1). Spallation also produces a double peak of products, but in principle from an arbitrary starting nucleus instead of only uranium. The resulting product requires a chemical separation of the interesting isotopes from the target material, since numerous elements and isotopes are present in a mostly homogeneous mixture.

Several currently relevant isotopes cannot be produced by accelerators, for example ^{90}Y which lies outside the possible pathways of H and D on the nuclide chart. Some isotopes such as ^{131}I can be produced via both neutrons and accelerators using e.g. ^{130}Te(d,n)^{131}I, but the ground state spins of this ion reaction mismatch with 1 on the input side and 7/2 + 1/2 on the product side, potentially preferring a different reaction. The numerous other iodine isotopes accessible via (d,xn) from stable Tellurium isotopes would results in a rather low purity, while (n,γ) provides less access to detrimental isotopes in this case. In this case neutrons offer a qualitative advantage. The isotope ^{64}Cu has a half-life of 12.7 h, consequently the production profits from a high local production rate and fast chemical separation of the ^{64}Cu from the host material (Cu or Ni), yielding a potential advantage for accelerator production in an application near device. The lower costs of accelerators compared to fission devices will generally enable maintaining a denser grid of accelerator production facilities than neutron facilities.

With all the criticism of neutrons and fission, their strength and complementary working with ions has to be admitted. The difficulty of producing neutron rich isotopes, namely β^- decaying isotopes, using ions can hardly be expected to change due to the fundamental movement options on the nuclide chart as discussed above. Neutrons hardly produce proton rich (β^+ decay) isotopes, since this would require a strong high energy flux inducing (n,2n) reactions which would require completely different reactor types such as fast breeders or generation 4 concepts. In some cases, e.g. for positron emitters produced by ion irradiation, the pathways are clearly limited. In a few cases the neutron and proton induced isotopes compete regarding their costs and properties in the sense of a commercial competition, for example in the case of ^{64}Cu production which is possible using both neutrons and ions.

5.1.4 Optimization of Production and Cost Efficiency

Scientist are proud of good ideas and clever solutions, but to bring these to application it has to be possible to make a profit with it. In this section we will work towards a case study for cost calculation of producing isotopes with accelerators. The application/usage defines the value of a product. The (preparation) costs form from numerous aspects. For example we discussed already the 1€/eV rule (now rather 25c/eV) for accelerator beam energy as being the main contributor for investment cost of the accelerator itself. On top of this base cost levels due to vacuum components, radiation safety, maintenance, and personnel add up. A higher degree of integration and automation saves investment and running costs, but not every aspect scales identical with the production goals of a certain application. The price (which should not be confused with the value) of a product, a market property, derives from the balance of supply and demand and the competition for the product. These aspects belong to economics, we as the accelerator applicants focus on optimizing the technology, dictating the product costs.

First of all we should get an overview of the contributors to our product costs and decide on a common denominator to have them comparable. Here we chose the hour as common denominator, yielding the following equation for our product costs:

$$\text{Costs} = \frac{\frac{\text{Operational costs}}{\text{hour}}}{\frac{\text{Isotopes produced}}{\text{hour}}} * \frac{\text{isotopes/product}}{\text{handling losses}} * e^{-\frac{\text{lead time}}{\text{decay constant}}} \tag{5.4}$$

Equation (5.4) shows a possible cost calculation basis with several contributors. A similar term could be found for other accelerator products such as analysis services or patient irradiation. The last term might be the easiest one, since it describes the physical decay of the produced isotopes on the basis of the time required between the production of isotopes until their final use in the form of a product (lead time).

The operational costs summarise everything consumed by the production and the investment for all devices and personnel required for conducting the production, including the accelerator and post-processing facilities, if required. First of all it contains salary of personnel, power consumption, and maintenance. The investments required for buying the accelerator and preparing a laboratory (building, shielding), or a corresponding rent for the laboratory, have to be distributed over a reasonable lifetime. These properties depreciate over time by wear and technological advancement, reducing their value over time, often independent of usage (this part belongs to maintenance). Lastly it contains the costs for targets, namely the mother isotope, its isotopic enrichment, and price. The isotopic enrichment strongly scales the price, for example a factor 4 between 97 and 99% enrichment of ^{18}O in H_2O. From 97 to 99% the production rate will only increase by 2%, but the amount of impurities reduces by a factor 3. Only if the impurities significantly reduce the product quality or induce radiation safety problem such a price increase can pay off. Usually only minute quantities of radioactive isotopes are required, therefore the most important factor for isotopes is the usage ratio, the amount of mother isotopes lost or remaining unused during the production process.

The production rate of isotopes depends on the choice of beam energy, current, and target. The operational costs and in particular the investment part scales with these technical properties. The physics behind it was discussed in Chap. 3. To summarise: The beam energy E defines a reaction probability which monotonically increases with E (for thick targets). Multiplication of this number with the flux of incident projectiles and the enrichment of target isotopes yields the number of nuclear reactions per time. For example ^{64}Cu could be produced using either 3 MeV on natural Nickel due to ^{64}Ni being the only of the stable isotopes with a (p,n) reaction barrier below 3 MeV or a beam energy >10 MeV could be used on enriched ^{64}Ni, avoiding the production of un-wanted isotopes from the four other Nickel isotopes. Enriched ^{64}Ni represents an extra cost factor, but the efficiency factor H (see Sect. 3.4) is considerably lower at 3 MeV, see considerations in Sect. 3.4. How will these aspects weight against each other or do they even matter in the final costs or do other quantities dominate?

Lastly, the demand of isotopes per final product (e.g. an injection dose) and the losses of material upon conversion of the (raw material) isotopes to the product influence the costs. In the example of ^{18}FDG ($C_6H_{11}FO_5$) for PET, every product FDG molecule contains only one ^{18}F isotope. The final product, an injection dose of FDG in this example, contains many of these molecules in a quantity defined by the application requirements. Multiplying the costs per isotope derived from the first ratio in (5.4) with the corresponding amount of molecules in each product yields the cost per product. Not all product ends up in syringes, in this example, due to purification and handling, a part of the isotopes are lost/retained in the apparatus.

Let us now apply these theoretical considerations to a specific case study. Our example calculates cost for an accelerator facility with estimates for synthesis capacity and the impact of transportation distance for an ^{18}F FDG factory. A neat aspect of this example is the exponential increase of cost with transportation distance due to decay of the relatively short-lived ^{18}F. We derived the optimal beam energy regarding energy efficiency from depth dependent stopping calculations in Sect. 3.4 to 12 MeV for the $^{18}O(p,n)^{18}F$ reaction with a reaction probability of 0.33%. ^{18}O enriched water costs about 120€ per gram when ordering larger quantities with 97% enrichment from major isotope/chemicals suppliers. We assume the ^{18}O not reacting will be completely recycled into the process. The conversion and separation consumes about 20% of the produced isotopes. Trained personnel for nuclear environments and accelerator operation costs about 60€/h in the public sector in Germany in 2019. Operation and production control require at least 3 people, accounting to 180€/h. The cyclotron itself accounts for 12 M€ considering the 1€/eV rule. The shielded laboratory building including a chemical processing plant for FDG synthesis accounts to 8 M€. We assume a lifetime of this complex of 30 years over which we distribute these costs (depreciation model). This cyclotron type delivers up to several mA beam current with a down-time of 10% in particular for ion source maintenance. The electric power requirements of the base device including ion source, magnets, vacuum, and controls account to about 80 kW plus the beam power, which is generated with an efficiency of 1/3. Therefore, a 12 MeV at say realistic 400 μA proton beam requires about 95 kW electrical power. The total operational cost account to 453 €/h, with the device and building accounting for 55% of this. Per mA and hour the accelerator produces 6×10^{16} ^{18}F atoms at 12 MeV from which the FDG will be formed. Every application of ^{18}F FDG for PET requires up to 10 pMol = 6×10^{12} FDG molecules (= 600 MBq activity) at the patient location. Figure 5.9 compares different lead times as a sum of chemical processing, transport, and application. 2 h more for transport represent the difference in time required to transport the FDG

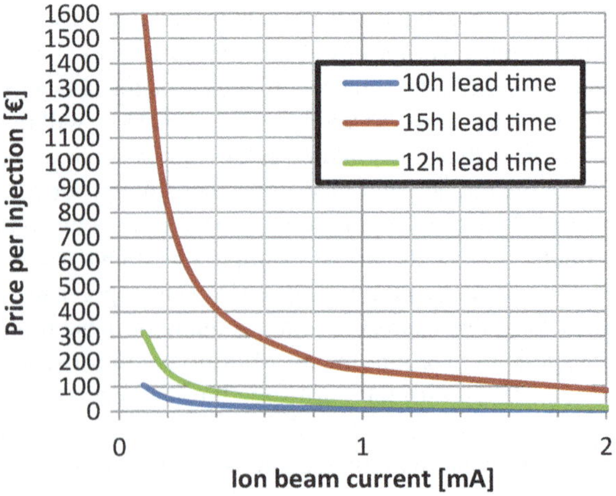

Fig. 5.9 Cost estimate curves of the specific cost of a single ^{18}F FDG dose for PET depending on accelerator beam current and the lead time from irradiation till injection. Higher beam currents and shorter lead times clearly reduce the specific costs. Already 2 h more triple the costs due to the 1.83 h half-life

from Forschungszentrum Jülich to Cologne or to Frankfurt. Consequently, in Frankfurt the same product would cost 3 times more than in Cologne, in Berlin (15 h) even 16 times more. These cost calculations justify a rather dense grid of ^{18}F sources with strong advantages for smaller and cheaper sources installed at the location of application.

Please note these calculations represent rough estimates based on experience and a few numbers from relevant manufacturers. Nevertheless, the case study highlights the importance of powerful accelerators, since a large part of the costs (e.g. personnel) scales not at all with the accelerator power. Beam current is the cheaper option compared to beam energy since the accelerator size will scale only weakly with beam current but linearly with beam energy (constant specific acceleration). Furthermore, the cost structure calculated here dictates the technological development goals to be pursued by revealing the major contributors to the product costs. In this case, due to the short half-life of ^{18}F, transport time is the major contributor. Consequently, smaller accelerators with higher current and lower beam energy inducing less radiation safety issues and building requirements would be a possible solution to improve the production site density in a country.

5.2 Rare Isotope and Radioactive Decay Tracers

From our daily life we know: A process is easiest to understand and the result becomes most credible if we see it happening live. Two reasons draw responsible: Firstly, the live observation actually contains the most complete set of information, we are sure nothing was missing or overseen. Secondly, we are certain that a straight path between start and end state exists. This may sound trivial, but imagine we left our pizza on a table with three dogs in the room. After we come back from the rest room, the pizza is gone, but there are at least five possible event chains connecting the same starting and end-point. The dogs may even select one among them to blame, implanting an expectation of what happened in us, but building up experience or claiming someone guilty from incomplete information remains unsatisfying. For these reasons, scientists developed the concept of an in-situ experiment. Simply spoken we are looking inside the situation, more accurately spoken the observed process remains (mostly) unaffected by our measurement and the time resolution of the measurement is faster than the characteristic progression time of the process. For watching dogs eating pizza, the human eye is (just) fast enough, but in technical and natural processes already placing a proper probe often constitutes a major challenge.

Imagine a bearing working for example in a combustion engine or an electric motor. We want to analyse the rate of abrasion of material from the bearing in a car engine while it runs through different scenarios, for example city traffic and autobahn. We cannot measure it directly by mass-loss or thickness changes since it is hidden in the engine. The contrasts of measuring the whole engine are too small since we compare μg of wear particles to 100 kg of engine weight. We could remove the material from the lubricant, but separation takes time and still we could not derive the

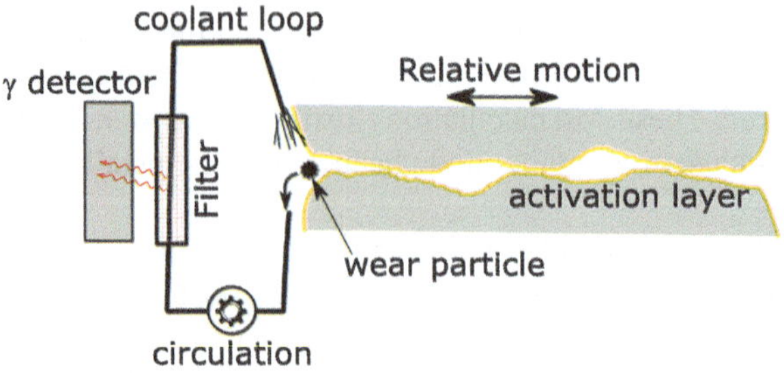

Fig. 5.10 Scheme of a radiotribological experiment. A system transports particles released by wear towards a photon detectors. The detected signal enables wear quantification

exact location the material originated from. For solving this ambiguity, the material needs to be marked and traced. Upon erosion of the marked material it can be traced like a child dropping breadcrumbs in the forest. Scientifically, the breadcrumbs are named *tracer*, with the advantage of nuclear breadcrumbs being inedible, ensuring a known amount reaches our measurement. In Sect. 5.1 we received the tool for marking materials with tracers on the atomic scale through nuclear reactions.

In Chap. 6 we will learn that medicine applies this technology quite intensively for diagnostics, but a few examples also exist in the technical context. This technical applications consider processes of material migration, for example corrosion, evaporation, or plasma sputtering. This section focusses on nuclear activation based tribology, also called radionuclide tribology (RNT), but many aspect share among the different fields. The science of wear and surfaces in relative motion defines tribology. Figure 5.10 depicts the basic setup for a RNT experiment. Wear as a factor of component lifetime and production costs affects all applications of moving parts. The friction between surfaces in contact removes material over time with a rate measured in thickness or weight per area and time depending most importantly on the contact pressure, lubricants, relative velocity, the materials, and the surface roughness. In particular, bearings with their contact surfaces being hidden in a larger part with lifetimes often in the order of 10,000 h complicate a precise and direct measurement of their erosion rates. An in-situ analysis of the erosion would enable accelerating the testing and parameter studies yielding information on static and dynamic processes. Parameter studies could for example include varying lubricant properties, the friction velocity, or start-stop scenarios.

Radioactive sources provide a solution, since modern detectors can quantify them down to quantities of single atoms. Radioactive materials can be (electro-) plated onto the affected surfaces, but this potentially changes the surface properties. Accelerators enable placing these probes directly into the affected surfaces by activation of the investigated materials without relevant change of the material and its wear properties. Depth dependent activation (Sect. 3.4) enables a depth localisation and quantification of the activation and focussing of charged particle beams (Sect. 2.3) allows for a lateral localisation down to μm scales. In contrast to neutron activation, ion activation usually results in less different products with shorter half-lifes easing detection and increasing measurement contrast. This so-called Thin-Layer Activation (TLA) achieves sensitivities down to a few 10 ng (Racolta 1995). The surface near activation of ions results in reduced radioactive inventory compared to a bulk activation, reducing overall costs. Activation reactions exist for practically all technically interesting materials using p, d, or ^{3}He ions in the order of 10 MeV (Racolta 1995).

The wear removes small amounts of material from the contact surfaces. In order to detect these they have to be transported away from the wear zone to a detector. Measuring the amount of activity still present in the part would yield a much lower signal to background ratio than measuring the removed part in a remote location starting with zero activity. Typically, a lubricant transports the eroded activity. A γ-detector placed in the lubricant system then allows quantification of the activity. The time-dependent signal increase of the detector corresponds to the amount of wear. Assuming a constant depth distribution of the activation/isotopes yields a proportional/linear increase of the signal with wear. The detection accuracy and the integration time required per wear parameter set depend on the activity in the sample. Higher activity increases the signal level, but safety and cost aspects together with possible material property changes due to the radiation damage demand as low activity as necessary.

In the case of a fluid lubricant transporting the wear, the fluid integrates the released activity. The nuclides released by wear add up to the existing inventory in the oil, their release rate represent the slope of the actual measured signal which is proportional to the total activity in the oil. The actual sought quantity is a derivative of the experimentally measured quantity. Measuring derivatives is always a bad situation, since deriving the measured values exaggerates the statistical noise, reducing the signal to noise ratio of the sought value. In order to keep the activity close to the detector and allow for a credible calibration of the detection efficiency, the floating wear debris has to be reproducibly located close to the detector. Additionally, the closer the activity to the detector the better the geometrical detection efficiency and the better the detection limits. Filters or magnets can aid in fixing the radiation to a defined point in the lubricant system, but their efficiency strongly depends on the wear particle properties, for example the particle size to filter porosity dimension. Loss of lubricant, remote filters, or sedimentation potentially reduce the detection efficiency. Clogging of filter can also increase the geometric efficiency, faking an increase of wear rate over time.

Since the application defines the materials of interest, also the possible tracer isotopes are given. Only the beam energy enables a slight selection freedom by enabling more $Q < 0$ reactions with increasing beam energy. Let us consider the example of automotive engines by an analysis of specific activity and its differential evolution in the circulating lubricant oil. In this case typically steels will be eroded. The activation as calculated by depth dependent reactions, see Sect. 3.4 in (3.20), varies with stopping power and reaction cross-section. In (Rayaprolu et al. 2016) it was shown how higher beam energies result in more uniform depth profiles of activation in tungsten due to the flattening of cross-section and stopping. For iron, as the main constituent of steel, the activation by a few MeV protons yields predominantly ^{57}Co and ^{58}Co, both with strong γ-lines of 122 and 811 keV, respectively, and half-lifes of a few month. These isotopes originate from (p,n) reaction with the rarest iron isotopes since these feature the lowest reaction thresholds (Q-value < 0). Depending on the alloying elements present in the steel additional isotopes occur via corresponding reactions. To name only a few: Chromium leads to ^{54}Mn, vanadium to ^{51}Cr, and tungsten to ^{184}Re, enabling to separate different steels by the different activation products connected to the composition. An energy resolved detection of the photons enables quantification of selective processes preferentially removing certain elements.

Let us consider the method in the frame of a case study of radiotribology of an iron material. This could be pure iron, a steel, or any other material with a correspondingly lower iron percentage. Other constituents will not be considered due to the independent nature for nuclear analysis. For simplicity, the calculation assumes pure iron which naturally contains 2.119% ^{57}Fe and 0.282% ^{58}Fe. The activation using cyclotrons in the 10–30 MeV range yields activation thicknesses in the order of 1 mm. Figure 5.11 shows us the amount of ^{57}Co (Half-life = 271.8 d) and ^{58}Co (Half-life = 70.9d) isotopes produced and their distribution throughout the sample depth for a 12 MeV proton irradiation. The production strongly varies with depth. The calculation allows for a direct connection of depth and eroded isotopes. A constant distribution and with this a linear connection of wear rate and counting rate increase

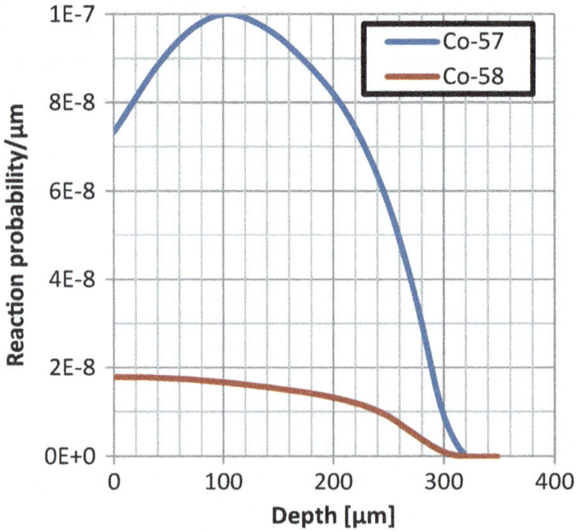

Fig. 5.11 Depth dependent reaction- probability of the $^{57}Fe(p,n)^{57}Co$ and $^{58}Fe(p,n)^{58}Co$ reactions in pure iron irradiated with 12 MeV protons. The calculation considers the isotopic ratio in natural Fe. The irradiation produces isotopes down to 0.3 mm with a total range of 0.35 mm. Variations in cross-section and stopping power vary the isotope production density. Stopping from SRIM2013 and cross-sections from TENDL-2015

can only be assumed for ^{58}Co down to about 50 μm where increasing cross-section and stopping power cancel each other. A varying depth distribution would require using Fig. 5.11 as a depth scale for the later erosion. Considering Fig. 5.11 the selection of 12 MeV appears actually quite optimal. A lower beam energy would reduce the total amount of activity, but the plateau of ^{58}Co eases the result interpretation and 12 MeV lies close to the cross-section peak of ^{58}Fe(p,n)^{58}Co resulting in a maximum sensitivity.

For the sake of simplicity of a constant production rate, we focus on ^{58}Co, although ^{57}Co offers 5 times higher signal levels. The photons emitted by ^{58}Co results in a certain detector efficiency of say 10% (only 1/10 of the decays result in a signal). The efficiency generally decreases with photon energy, also favouring the lower energy photons of ^{57}Co. Furthermore, we assume a geometrical efficiency of 50% equal to all wear particles come to rest on one side of the detector. ^{58}Co emits a 811 keV photon with 99.5% probability per decay. If we irradiate the material with 1 mC of protons=6.25×10^{15} protons this produces 1.1×10^{8} ^{58}Co/μm with the reaction probability of 1.8×10^{-8}. In total, we would receive $1.1 \times 10^{8} \times 0.5 \times 0.1 \times 0.995 = 5.5 \times 10^{6}$ counts/μm of wear. The assumptions represent a rather optimistic case, but the numbers would allow resolving a loss of 0.1 nm with 550 counts/second resulting in a statistical uncertainty of 4.2% [see Sect. 2.5 in (2.43)] for a wear rate of 0.1 nm/s.

Instead of activating the sample, we can also build the sample itself from traceable metastable and rare stable isotopes. This could be ^{13}C enriched molecules consumed by a plant or deuterated water participating in a chemical process instead of hydrogen. Naturally, only 1.1% of all carbon nuclei are the ^{13}C isotope and only 1.5×10^{-4} of all hydrogen nuclei are the deuterium isotope. Rare stable isotopes have the advantage of being relatively cheap compared to accelerator produced isotopes (e.g. 99% enriched D_2O has a price tag of a few 100 € per litre), but since they occur naturally a certain background exists reducing the possible contrast compared to radioactive tracers. Being stable, these isotopes do not emit radiation and hence their detection requires active methods. The chemistry of isotopes differs only insignificantly, therefore the actual process to be investigated remains unaffected by the isotopic changes, as it should be. At the same time, this prohibits a detection contrast using chemical analysis methods. The isotopes have to be followed by methods with a nuclear or an atomic mass contrast. Chapter 7 will discuss several options accelerators offer for tracer analysis, but also numerous conventional methods exist such as mass spectrometry or chromatography.

5.3 Material Modification

Humanity is addicted to forms. We shape nature to make it more beautiful or functional and we want it to stay that way. Our need for more, better, and cheaper requires testing and extending the technological limits. Unfortunately, approaching technological limits counteracts the wish list by melting, creep, fatigue, bending, dimensions, and so on. For this reason, central development challenges for technical products include bringing the material in shape and developing materials, which stay in shape in spite of the harshest treatment and environment. With the ongoing climate change another factors adds itself to the top of the wish list uninvited. Sustainability and its equivalent resource efficiency require making things smaller, increasingly specialised, and longer lasting.

Optimized shapes enable reducing resource usage without decreasing component performance. Technical limitations of mechanical shaping, casting, and joining techniques limit the freedom of shape. New technologies such as 3D printing extend the shaping options. 3D printing start filling application niches by offering advantages in certain applications. Ongoing progress reduces costs and extends the application range. Accelerators with their possibility for extreme power load densities contribute to these advancements. The potential for reducing waste by avoiding to mill down a material and instead printing it directly in shape sounds appealing, but as always in technology the details are important. Technological disadvantages of 3D printing include the increased energy usage for melting a material instead of milling it and the post processing of 3D printed objects due to the different surface and bulk properties. In the end, the application decides on the balance of benefits and disadvantages.

The same methods continue to be relevant on the micro-scale. Bringing the material into shape means sputtering, heating, radiation damage and implantation on this scale. Radiation induced changes also change a materials interaction with chemicals, further extending the processing capabilities. The development of microchips to a large extent relies on increasing flexibility and decreasing structure sizes on the microscale. These manufacturing technologies rely on accelerators in many different aspects. The impact of microchips and their performance on our society and industry emphasizes the importance of micro-machining technology.

One of the aspects of the semiconductor technology as a basis for microchips is the tailoring and synthesis of specific material properties by accelerators. Without these processing steps, silicon is just a badly conducting metalloid.

5.3.1 Doping by Implantation and Activation

Doping changes the properties of silicon from being a useless metalloid to being a semi-conductor. A band-gap between conduction and valence band draws responsible for the low electrical conductivity of pure silicon. Electrical conduction requires electrons in the conduction band, but the band-gap energy (1.12 eV for silicon) prevents electron from leaving the valence band. Only at higher temperatures the high energy tail of the thermal energy distribution allows a few electrons to pass the band-gap. Doping the silicon introduces new intermediate energy levels into the band gap. The exact position in the band gap and whether these levels contain extra electrons or free electron-states (holes) depends on the dopant element. This intermediate level shortens the energetic distances, significantly increasing the electrical conduction via these electrons or holes. Doping with elements providing excess electrons produces a negative (n-type) conduction, while excess holes result in positive (p-type) conduction.

So far so boring. The difference to a normal conductor lies in the possibility to control the amount of these excess electrons and holes via the dopant concentration. With small dopant concentrations in the order of 10 ppm applying a voltage allows for a complete depletion of these small amounts of charges. Without its extra charges, the depleted layer goes back to the originally low conductivity. The directionality/polarity of this depletion depends on whether it is a p- or n-type semiconductor. Combining a p-type and an n-type semiconductor results in a transfer of charges, since the n-type can provide electrons to fill the holes in the p-type semiconductor. This transfer results in a depletion layer. Applying a voltage over this p–n junction results in either a stronger depletion of the charges and a further reduction of conductivity by applying a positive voltage to the n-type side (reverse biasing) or in a refilling of the free charges and an increased conduction by applying a negative voltage to the n-side. This unidirectional isolator feature is responsible for the unique possibilities for electronics offered by semiconductors.

In Sect. 2.5, we discussed several different types of radiation detectors, in particular silicon and germanium based. Production of a semiconductor radiation detector starts with the respective single crystals. The detection exploits exactly this depletion zone effect of a p–n junction operated in reverse biasing. Charged particles again separate the combined electron-hole pairs, resulting in a current of the otherwise isolating depletion zone. For radiation detection, usually very thick depletion zones are required. The depletion zone represents the effective detection volume, requiring thick zones for effective particle energy absorption.

The small dopant quantities would not even be considered as a part of the composition in most other materials, but with accelerators they can be easily controlled when measured as an ion current. For a well-controlled doping the original material has to be extremely pure and crystalline. Practically single-crystals represent the ideal basis. Nowadays silicon single crystals disks, so-called wafers, can be made with several 100 mm diameter and sub-mm thickness for only a few 10 €. These single crystals are then doped with known amounts of for example phosphorous (n-type) or boron (p-type) in so-called implanters. Figure 5.12 shows such a silicon wafer preparation and implantation facility. Typically acceleration voltages of 100 keV DC implant the dopants about 100 nm deep into the silicon, see Fig. 5.13. Post implantation annealing removes the defects induced by displacement collisions (Sect. 7.4) of the ions entering the solid and increases dopant distribution homogeneity. The strength of ion doping lies in the accurate control of ion dose, species, and range defined by acceleration voltage, ion optics, and sample current.

A spatially homogeneous doping is the challenging part of doping. Different dopant concentrations will result in different conductivity, increasing for example the electrical losses in regions of lower dopant concentration. The depth homogeneity suffers from straggling as depicted in Fig. 5.13. A peak forms in depth using a single

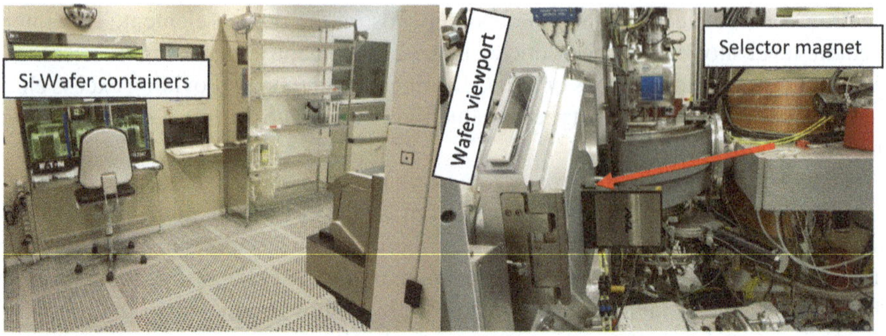

Fig. 5.12 A 70 kV implanter with attached clean room for 300 mm wafer doping (left). In the clean room the operator attaches wafer boxes to the automatic manipulator of the implanter (right). Its ion source produces a multitude of species from gases and vapours. A magnet selects the ion species and directs the beam (arrow) onto the wafer. Copyright Forschungszentrum Jülich/Sören Möller

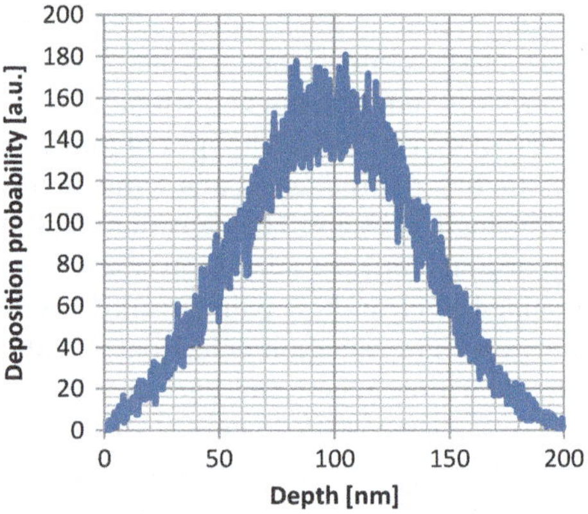

Fig. 5.13 SRIM 2013 calculation of 77,000 phosphorous ions implanted with 70 keV into silicon. The implantation shows a maximum of deposition probability at 97.5 nm depth which extends down to 200 nm due to straggling. According to SRIM every projectile induces 960 displacements into the silicon lattice using a displacement threshold of 15 eV

beam energy, but ideally the dopant should reach a constant concentration throughout a relevant depth. The typically Gaussian beam profiles (see Fig. 2.21) limit the lateral implantation homogeneity. Beam optical shaping can only partially compensate for this. Scanning the ion beam using deflection beam optical elements over the wafer provides a fast option for redistribution of the implantation dose, but for large wafers, required for industrial scales, the deflection and with this also the impact angles become significant. Variations of the impact angle reduce the implantation depth through the geometrical path effect. A translational or rotational movement of the target itself avoids this, but this movement is generally slower and technically more challenging in the vacuum environment of an implanter.

The range of doping represents an economic limitation of ion implantation. The high stopping power of heavy ions such as phosphorous with a few 100 keV allows only doping the near surface of a wafer. Doping the whole silicon ingot (from which the individual wafers are cut out) in one step would significantly reduce the required effort and improve the dopant homogeneity at least within a single ingot. Unfortunately, the beam energy required to implant boron or even phosphorous several ten millimetres deep into silicon are not feasible. The economic need for large wafers is in conflict with this idea, at least for ion beams.

Doping by nuclear reactions/activation represents a feasible alternative due to the low amounts of dopants required. Currently only fission reactors are used for this. Here the ^{31}Si produced via ^{30}Si(n,γ)^{31}Si decays to stable ^{31}P. This methods represents a prime standard for doping and several patents were issued in the 1970s and 80 s, but the reality of the limited availability and the costs of the required large thermal neutron sources limits the economic relevance.

Doping by light ion beam irradiation reduces the required beam energy for reaching larger depth. A patent (Patentnr. US09192452, 1998) was filed for phosphorous doping using the ^{30}Si(d,p)^{31}Si reaction with a subsequent decay of ^{31}Si to the stable ^{31}P similar to the neutron reaction. Unfortunately, the required isotope ^{30}Si is the rarest among the silicon isotopes, leading to poor reaction probability in natural silicon and correspondingly high ion dose requirements. For this doping process, standard cyclotrons would be sufficient for activation of a few mm of silicon, but the economic gain of treating large volumes vanishes with these physical limitations. The typical problems of inhomogeneous depth profiles of ion interactions also remain as discussed in the case of iron activation in Sect. 5.2. Doping via implanters on the wafer level requires less investment due to the lower beam energy and the 100% implantation efficiency compared to the sub-% efficiency of nuclear reactions.

Accelerators also provide options for technical developments and quality control in the doping and semiconductor context. Ion beams, in particular Rutherford-Backscattering spectrometry (RBS) enable analysis on the ppm level of dopant concentrations with depth resolutions down to nano-metres (see Sect. 7.1.5). Deficiencies in the ion stopping models and also the spatial aspects of homogeneity can be measured for doping analysis and optimisation after implantation (ex-situ). These methods allow for the quantification of the implanted element and additionally for the remaining crystallinity via ion channelling.

5.3.2 Welding, Cutting, and Additive Manufacturing

Welding and cutting as classical manufacturing technologies and the still new field of additive manufacturing all require introducing a high power density into the material to remove or melt material. At several occasions in this book the extreme power densities connected to accelerated beams impacting targets was mentioned. Consequently, charged particle beams can be considered to take over manufacturing duties. For the introduction of power/heat into a material no difference exists between ions and electrons, except for a higher range of electrons. For manufacturing it makes no difference whether the beam deposits the power within 1, 10, or 100 μm, making this aspect irrelevant. Technically, the easier handling of electron beams and electron sources favour electrons and this is also what is applied in the accelerator based manufacturing field.

Considering additive manufacturing, also called 3D printing, Laser based methods are currently more common. In contrast to Lasers, electron beams require the manufacturing to be in vacuum due to the energy loss of electrons in atmosphere and the interaction with the targets potentially releases significant radiation dose rates via Bremsstrahlung. These aspects make the Laser based technology generally the cheaper option. Applying electron beams has to offer quality or processing speed advantages to balance this. Lasers have the fundamental disadvantage of high and variable reflection coefficients, in particular for metals. Charged particle beam technologies feature 100% energy absorption, independent of incident angle or surface conditions. Typically, the manufacturing methods apply electron energies of some 10 keV equal to a range in the order of 10 μm with beam powers up to 100 kW in a high vacuum chamber. Focussing the electron beams results in local power densities easily above 10 GW/m^2. Beam optics enables scanning speeds in the order from μm/s to km/s, depending on the setup details. Varying the length of exposure t of a certain position enables fine-tuning the heat penetration depth. Equation (5.5) states this penetration depth depends on the square-root of the thermal diffusivity κ of the material and the exposure duration t. The beam range defines a minimum heat penetration depth, but according to (5.5) only heating pulses shorter than 10 μs actually allow keeping the heat within this depth (for a typical steel κ). Using beam scanning of pulsed beams enables setting of the exposure time as discussed in Sect. 2.2.1.

$$\text{Thermal diffusion length} = \sqrt{2\kappa t} \tag{5.5}$$

Firing such an intense electron beam onto a material easily melts the material. Electron Beam Additive Manufacturing (EBAM) or also Electron Beam Melting (EBM) represent the accelerator variants of additive manufacturing based on melting powders. The reproducible and surface morphology independent power absorption of 100% together with the control of heat penetration optimized regarding powders and manufactured part provides control over the material micro-structure. This allows for improving the mechanical properties and the surface quality. The high beam power density enables a high productivity since the processing of a certain powder volume equals a certain amount of energy invested for melting it.

The capability for melting also provides an option for high accuracy welding with very small welding width and depth. This localisation of the affected zone enables joining materials with incompatible thermal expansion coefficients, which would crack when using larger welds. Vacuum parts are one of the application fields, but also tiny components such as μm thick membranes or foils profit from the spatial resolution of the electron beam.

The localized energy deposition enables cleaning, heating, annealing, or melting the surface. The choice of temperature via adjusting the beam power density selects between the different functions within one device. Moving the beam over the surface enables texturing the target surface in an arbitrary pattern. This can increase for example the effective surface area for functional purposes such as joining with another material or it can harden the surface via the rapid cooling inherent to a thin heat affected surface layer.

5.3.3 Surface Modifications and Nano-Machining

The last section discussed the preparation of materials on the macro-scale, in this section the focus shifts to the micro-scale. The short length scale implies modifications of the surface or at least surface near modifications down to a few μm, in contrast to the bulk related methods discussed above. The importance of the surface for technical applications extends beyond thin film technology. Complete technical setups, so-called Micro-Electro-Mechanical-Systems (MEMS), imprinted on the surface enable extremely compact and resource efficient sensor and actuators as depicted in Fig. 5.14 (Huang 2019). These systems combine electronics with mechanical systems. The inertial measurement unit installed in smartphones represent a widespread application of MEMS. The wide range of sensors fitted into modern smart-phones ranging from acceleration over orientation up to atmospheric pressure requires small and highly integrated systems and contribute important features to the device impossible with equivalent conventional scale devices.

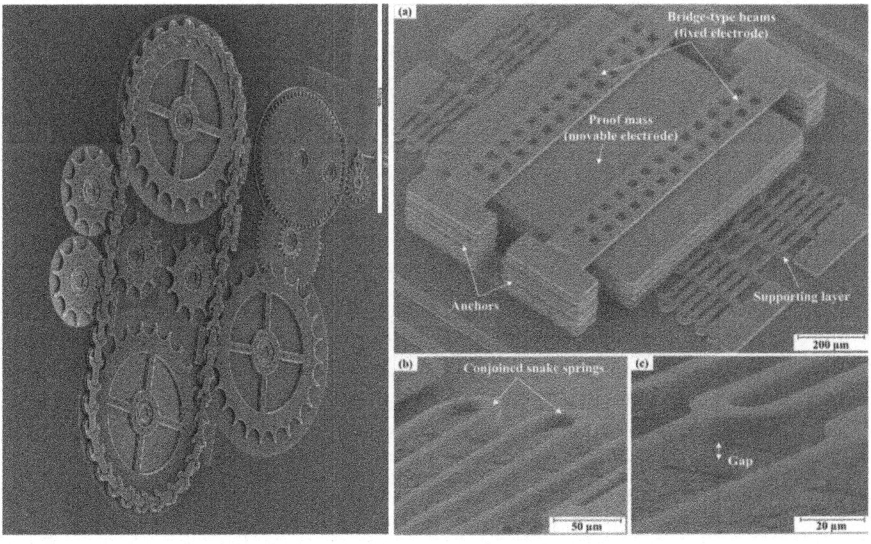

Fig. 5.14 Left: 1 mm sized MEMS system chain drive-train example made from silicon. Courtesy of Sandia National Laboratories. Right: Inertial switch prototype. Reprinted from (Huang 2019) with permission by Springer

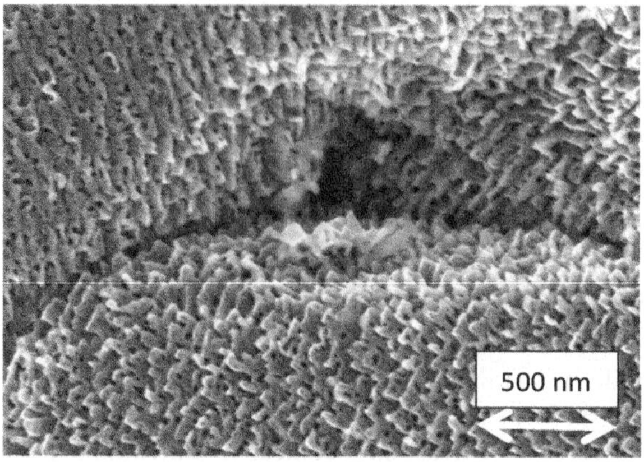

Fig. 5.15 Formation of so-called fuzz induced by implantation of He into W. The structure develops naturally due to the formation of sub-surface gas bubbles and their subsequent bursting in *W* and many other materials. The fuzz nano-structure forms an open porosity with extremely high surface area, low heat conductivity, and a porosity of up to 98%

Furthermore, the surface represents the first contact point for the interaction of a material with any external process. The absorption of light depends on the surface structure and composition, condensed to the surface emissivity ε. Figure 5.15 shows an example of a self-arranged surface structuring induced by helium ion implantation increasing for example ε from about 0.1 in the regular surface state to ≈ 1. Chemical interactions, for example reactions in a catalyser, scale with the available microscopic surface area, which can be orders of magnitude larger than the macroscopic surface area, given a smart microstructure surface morphology. Surface hardening and structuring reduce mechanical friction and increase wear resistance resulting in improved component lifetime and reduced costs by requiring only a modification of the surface instead of more expensive wear resistant bulk materials.

Several accelerator driven technologies exist for these surface modifications. An important aspect for the manufacturing of complex structures is aspect ratio of the manufacture structure, e.g. a hole or trench. Figure 5.16 depicts four ideal and real examples. A beam inducing sputtering will start drilling a rectangular shaped trench. As the sputtering reaches higher depth, the angular emission distribution of the sputtered particles (cosine like emission, Sect. 4.1) results in a certain amount of prompt re-deposition at the trench side-walls. Consequently, a non-rectangular shape develops. The manufacturing speeds typically scales with the achieved beam current. In particular, for small structures the beam emittance and current density limit this value as discussed in Sect. 2.3.

The established technology for the manufacturing of nano-structures in semiconductors is lithography with photons in the order of 10 eV, depending on the structure size. Lithography combines the physical marking/modification of a photoresist by beam bombardment with subsequent chemical etching steps removing the marked or the non-marked part of the photoresist, respectively, depending on the selected

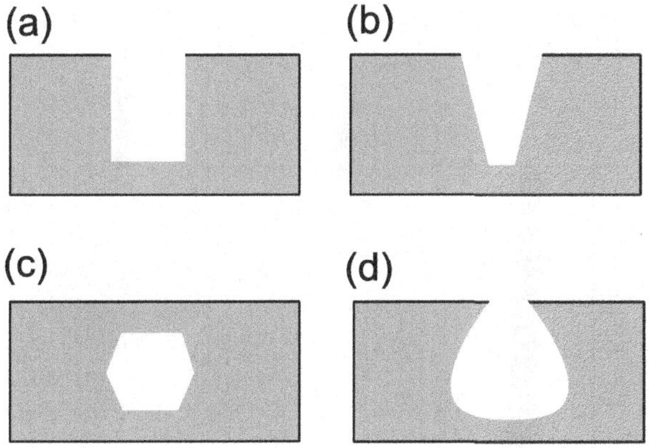

Fig. 5.16 Four options for micro-machining aspect ratio and the problem of actual component separation and distance. **a** The ideal rectangular trench. **b** A real drilling hole produced by keV FIB with a characteristic form given by trench-side interactions. **c** An ideal 3D machining modification leaving material above the part removed at a designated depth. **d** A real 3D machining feature with an entrance channel for the beam

working mode and resist. The part of the surface still covered by photoresist will not be removed by the subsequent etching step with a different etching agent due to the protective function of the photoresist. Typically, first a mask is produced using a CAD model and a small beam inscribing the nano-structure onto photoresist covered metal (e.g. Cr) layer deposited onto a transparent material (e.g. quartz glass). The masks imprinted with the foreseen structure enable a quick irradiation of wafers in the production stage in the order of minutes using a broad photon beam, which illuminates the wafer where the metal was removed in the mask. The photon wavelength limits the minimum structure size possible to manufacture using this method.

Ion- and electron-beams offer the potential for strongly reduced wavelength, see the de-Broglie wavelength (7.1). This would further decrease the possible structure sizes enabling manufacturing more complex and energy efficient microchips. All types of charged particles are applicable offering specific advantages in terms of range and trench aspect ratio as shown in Fig. 5.17. Accelerators could also serve as photon sources (Sect. 4.3) producing shorter light wavelength than currently available, extending the limits of the classical photolithography. So far, accelerators have only little relevance in the industrial mass fabrication, since photo-lithography provides significantly higher production speed due to its large irradiation area enabled by the masking technology. This compares to the relatively low speed of accelerator beam scanning for the interesting small structures and the limits of beam current delivered to nano-metre sized spots (nA range). Masks are difficult to apply for charge particle beams due to the high penetration depth of electrons and the sputtering induced by ion impact. Moreover, the energy invested per incident particle clearly favours 10 eV range photons over several 10 keV range ions or electrons. The speed of particle beam lithography technology currently lacks about 7 orders

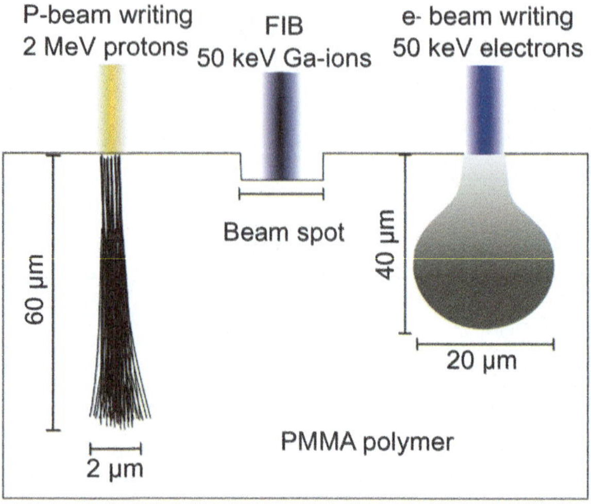

Fig. 5.17 The lithographic aspects of different beam types. Protons with their low straggling allow for a damage/developing profile located at the track end (Bragg peak), deep in the material, with only little lateral extent and a high trench aspect ratio. 10 keV range heavy focussed ion beams (FIB) sputter only at the surface, effectively drilling a hole instead of modifying the resist. Electrons create a plume of damage due to their large straggling effect resulting in comparable depth and lateral extent of the developing effect

of magnitude in manufacturing speed (measured in area/time) behind photon lithography. Bridging this large gap via technical improvements appears challenging and would require orders of magnitude improvements in all involved fields, namely beam optics, particle sources, resists, and instrument layout. However, accelerators can also serve as photon sources (Sect. 4.3), possibly producing shorter light wavelengths, thereby extending the limits of the classical photo-lithography.

The discussed charged particle lithography still relies on the same technological approach as photolithography, but in principle the whole technology could also look different. Removing target material directly without a photoresist via ion induced sputtering (Sect. 4.1) represents an alternative route. Analysis methods make use of this so-called focussed ion beam (FIB) method, see Sects. 7.1.3 and 7.1.4. FIB, for local 3D analysis. While heavy ions are established, technical solutions for light ions were only developed only in the last 10 years, e.g. the Zeiss Orion NanoFab device. Furthermore, the ion-matter interaction changes materials via induced heat and radiation damage (Sect. 7.4). Any of these contrasts can be exploited by chemical and physical removal methods.

Following this idea MeV protons, preferably applied due to their high mass over charge and the highest beam currents generated in the accelerator sources (Sect. 2.4), have the potential of becoming a next generation tool for nano-scale 3D machining. The main advantages over electrons are the smaller amount of large angle scattering events in matter, leading to high aspect ratios of cuts, and the higher stopping power. In contrast to keV ions which allow only surface-near 2D machining, MeV level

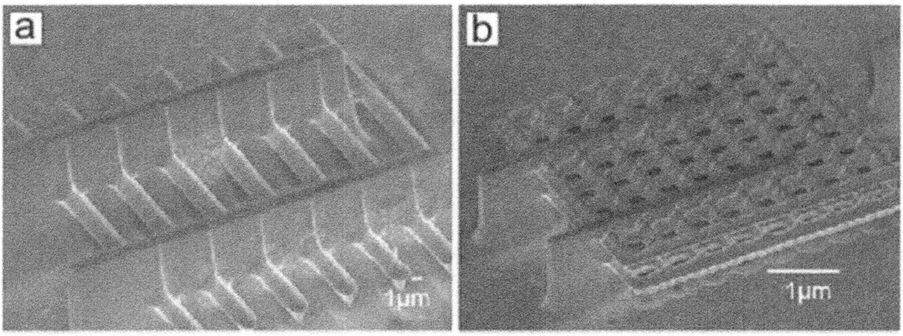

Fig. 5.18 SEM pictures of 2 MeV proton beam nano-machined 3D structures in silicon. **a** Free-standing silicon wires at 6.95 μm depth. **b** Two levels of photonic crystal slabs at 2.4 and 7 μm depth. Reproduced from (Dang, et al. 2013) with permission by Springer

ions offer increased penetration depth, allowing for 3D machining by shifting the Bragg Peak to different depth via a modification of the ion impact energy (Fig. 5.17). These are actually the same advantages relevant for proton therapy of cancer, see Sect. 6.2.1, although on a smaller depth scale. The combination of both allows for producing complex and 3 dimensional structures as shown in Fig. 5.18 by variation of the beam energy and modification of, in this case, porosity of silicon.

The limit of MeV ion-beam machining so far lies in the achievable lateral resolution and the processing speed. While focussing of keV ions and electrons to sub-nm beam spots became state-of-the-art, MeV proton beams required for relevant 3D machining depth reach only the order of some 10 nm spot sizes with < nA currents. The main obstacles related to the beam spot size lie in the beam emittance, the size and type of focussing optics, and the constructive challenges of compact arrangements of ion optical elements (see Sect. 2.3.2). The 3D nano-machining potential of this technology is driving the development, potentially creating a new market if the technological limits can be significantly extended.

Besides these designed surface modifications nature also develops spontaneous patterning, nano-structuring, and ripple formation by ion-beam and plasma impact through natural self-organised processes (Bradley and Harper 1988; Bernas 2010). These have the advantage of requiring neither scanning nor masks. Plasmas provide an equal flux of ions and electrons to surfaces, but since electrons hardly influence the material structure plasma practically represent a technical way for providing low energy ion beams, reducing instrument costs. Ion impact-energies in plasmas range from eV to <1 keV, with typical values in the 10–100 eV range. Surface barriers require at least a few 10 eV or the impacting ions will be mostly reflected, reducing the effective ion flux. Ion beams originate from plasma sources, hence only above 1 keV the formation of an ion beam makes sense. If the process just requires injection of ions into the material not the interaction with energetic particles, the plasma offers the more efficient option due to less projectile energy compared to ion beams. Ion beams offer deeper implantation range and more structural damage due to the higher beam energy, surpassing for example the newly formed surface structure for continued effect.

Spontaneous structures can feature a variety of morphologies depending on the target material, beam flux, beam energy, impact angle, and temperature. Figure 5.15 depicts a neat example of so-called tungsten fuzz. This morphology develops through implantation of He into W (and several other metals) in a surface temperature window of 810 K to about 2000 K in W. Models suggest, the He atoms are mobile enough in the W lattice to form sub-surface bubbles which deform the metal locally after reaching a certain pressure with increasing ion fluence. The bubbles can even burst under the gas pressure induced by the forced He influx. This process continues resulting in an accumulation and pile-up of these deformations, the fuzz grows with ion fluence. The fine irregular structure results in a strong absorption of light making the fuzz appear black in spite of its purely metallic nature.

References

H. Bernas, *Material Science with Ion Beams* (Springer, 2010)

R. Bradley, J. Harper, Ripple topography induced by ion bombardment. J. Vac. Sci. Technol. A **6**, 2390–2395 (1988)

Z. Dang, A. Banas, S. Azimi, J. Song, M. Breese, Y. Yao, S.P. Turaga, G. Recio-Sánchez, A. Bettiol, J.V. Kan, Silicon and porous silicon mid-infrared photonic crystals. Appl. Phys. A **112**, 517–523 (2013). https://doi.org/10.1007/s00339-013-7782-4

EuCard-2, A. Faus-Golfe, R. Edgecock, N. Hall, *Applications of Particle Accelerators in Europe.* bit.ly/2vVewNs. Retrieved on 14 June 2018 (2017)

B. Guérin, S. Tremblay, S. Rodrigue, J.A. Rousseau, V. Dumulon-Perreault, R. Lecomte, E. van Lier, A. Zyuzin, E.J. Lier, Cyclotron production of 99mTc—An approach to the medical isotope crisis. J. Nucl. Med. **51**(4), S. 13N–16N (2010)

Q. Huang, *Micro Electro Mechanical Systems* (Springer, Singapore, 2019)

K. Mizushima, P.C. Jones, P.J. Wiseman, J.B. Goodenough, LixCoO_2 (0<x<l): a new cathode material for batteries of high energy density. Mater. Res. Bull. **15**, 783–789 (1980)

P. Racolta, Nuclear methods for tribology. Appl. Radiat. Isot. **46**(6–7), 663–672 (1995)

R. Rayaprolu, S. Möller, C. Linsmeier, S. Spellerberg, Simulation of neutron irradiation damage in tungsten using higher energy protons. Nucl. Mater. Energy **9**, 29–35 (2016)

A. Sayed, A. Elbinawi, M. Al-Abyad, U. Seddik, I.I. Bashter, Evaluated activation cross-sections and intercomparison of the production parameters for the medically relevant radioisotopes 64Cu and 86Y. Pramana **84**(4), 569–579 (2015)

Chapter 6
Nuclear Medicine

Abstract Nuclear medicine applies radiation for diagnostic and treatment purposes. X-ray based diagnostics are the working horse of numerous medical fields spanning from dentists to oncology. The contrast mechanisms of this method will be discussed from a physical and technological perspective. Developing this technology from 2 to 3D (tomography) only slightly changes its limits, but widely extends its costs and benefit. Here the external X-ray sources can be replaced by internal radionuclides, which follow metabolic processes, generating a completely new type of contrast. In medical treatment, also external and internal radiation sources are applied. Radiation therapy focusses on oncology applications. A main technological limit for the success of this therapy form lies in the correct application of radiation doses only to tumour tissue. Combining metabolic and diagnostic targeting methods with radiation therapy enables treating a larger range of cases.

Live Long and Prosper (Star Trek or the Promise of Science).

Nuclear medicine, the second major topic of this book. The reader has to accept this is not a medical, but a physics book. As usual the reader is referred to more specialised books for digging deeper into the medical aspects and more detailed information can be found in general books on the topic, e.g. (Zimmermann 2006; Cherry et al. 2012). Quite a few books also focus on specific applications such as orthopaedics (Elgazzar 2004), paediatrics (Treves), or specific treatments such as thyroid cancer (Bell and Grünwald 1999). This chapter will give a brief overview of the available methods and concepts from a physical and technical point of view. It will deliver an understanding of the underlying technologies of accelerators in medicine and present physical aspects promising success on the medical front and a little basic biology required for understanding the beam-matter interaction with biological targets.

Earlier we discussed radiation protection and its detrimental health and legislation aspects. In short: Radiation protection legislation does not apply for medical treatments. The medical personnel requires dose monitoring, but the patient is excluded. The positive aspect of applying radiation will have more advantages than disadvantages. The idea behind this appears logical and necessary: The treatment improves the health more than it damages it. If you have ever seen an episode of *House, M.D.*

S. Möller, *Accelerator Technology*, Particle Acceleration and Detection,
https://doi.org/10.1007/978-3-030-62308-1_6

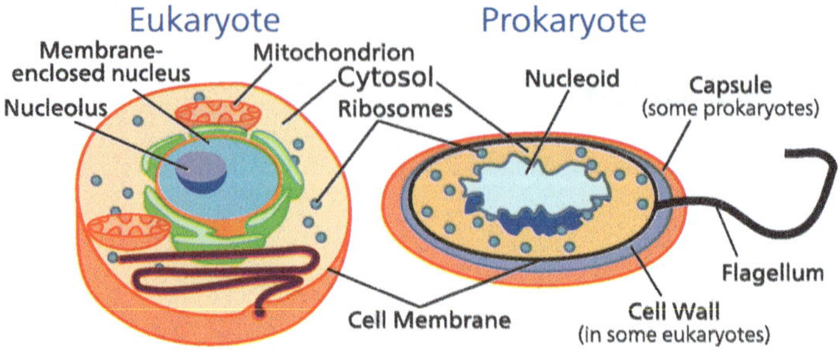

Fig. 6.1 Basic structures of cells. Eukaryote is our human "standard" cell type while prokaryote cells are simpler such as bacteria. Cells represent coupled matter complexes with chemical and physical means of communication

you will recognize they always start with CT, PET, angiograms and many other diagnostic methods before the actual play against the mysterious disease starts. Most of these terms will be discussed here as they relate to accelerators and nuclear technology. Nevertheless, the doses received by the patient can be substantial. An average X-ray computed tomography (CT) induces about 10 mSv of dose (see Fig. 2.52), 5 times more than the natural yearly dose. Positron emission tomography, where a radioactive substance is injected into the blood, yields similar doses. Sometimes these radioactive substances even appear in contamination monitors at airports or radiation protection zones. Due to the few applications to a single patient over his/her life, the numbers compare to natural life-time doses for rare usage. Upon higher doses induced by intense therapy or repeated diagnostics things change and the risk of developing secondary cancer (cancer as a result of the medical application) increases with dose. Some organs are more prone, such as the bone marrow developing leukaemia (Curtis et al. 1992), but also children suffer from increased risks for example for developing thyroid cancer (Iglesias et al. 2017).

In order to understand the risks, but more importantly the benefits of nuclear medicine we have to shortly discuss the working of multicellular organisms such as the human. Our basic building block is the cell, a microscopic sized biological entity. Figure 6.1 shows its basic structure. Two main types exist, but they have many aspects in common. Most importantly, all cells feature the DNA (Deoxyribonucleic acid) or the similar RNA (Ribonucleic acid), the software of the cell. Ribosomes convert this to the cells craftsmen, the proteins. The DNA with its tremendous amount of information allows for producing a zoo of proteins for different tasks, forming its proteome, the economy of the cell. Our eukaryotic cells have more functions and a more complex setup than prokaryotic cells, enabling operating a complex organism such as the human body with its numerous specialised cells (skin cell, liver cell, muscle cell …). All our different cell types have the eukaryotic structure and origin, but became specific cell types via cellular differentiation. The cell membrane protects its insides from the outside, but also mediates its bio-chemical interaction with the surroundings. Blood vessels transport oxygen and other required substances such as energy (e.g. sugar) and hormones close to the cell. The diffusion processes of these supplies through cellular agglomerates bridges the last few 10 μm from blood vessel to the cell and its neighbours. According to its function, the cell's membrane allows entrance of specific supplies via corresponding receptors (docking ports).

Living means consuming and circulating substances inwards and outwards through the cell membrane, summarised under the term metabolism. Metabolism represents an important aspect for nuclear medicine, since it defines one way we can interact with cells. An interesting aspect: Besides the physical half-life of a radioactive substance another residence time exists, the biological half-life. Biological half-lifes range from below minutes (breathing) over days (water exchange) to about 50 years (bone constituents and heavy metals). The household of natural ^{40}K is a good example for the metabolic contribution. The human body acquires large amounts of ^{40}K via food, e.g. upon eating a banana, see Fig. 2.52. In spite of this and the long half-life of ^{40}K of 1.28 giga-years, the ^{40}K activity of our body throughout our live remains relatively constant at 6 kBq. Our body constantly circulates the potassium with an optimal rate of about 4.7 g/day corresponding to a half-life of 58 days therefore, as all isotopes of a certain element behave chemically nearly identical, the amount of ^{40}K remains constant in our body. The additional ^{40}K uptake leaves our body by excretion.

No matter which type of cell and which kind of specialisation it was adapted for, beam-matter interaction works universally. Shooting projectiles through cells will always induce elastic scattering reaction, energy-loss and, with sufficient energy, even nuclear reactions. Cell size and especially the elemental composition affect the beam-matter interaction. For the beam cell boundaries and functional building blocks have no relevance, it will continuously slow down and statistically fly through multiple cells depending on the elements (and sometimes isotopes) it encounters. From the cells view, it makes a strong difference if beam-matter interactions displace or ionise atoms in the DNA or some random water molecule in the cytosol. Just like in the human body, see Sect. 2.7.1, not every part of the cell is equally sensitive to radiation damage. Since radiation occurs naturally, cells in themselves, but also the complex organism have developed means of repair and removal of defective cells and their components.

The inner metabolism opens up a path to change the cells elemental and isotopic composition. Figure 6.2 shows a cut-out of the elemental composition of DNA. We see mainly H, C, N, O and a few P atoms in the backbone. The cell itself contains much more elements. Depending on its function, we see for example iron in red blood cells, iodine in thyroid cells, or zinc in immune cells. Figure 7.30 shows an ion-beam compositional analysis of a plant cell with numerous detected elements. Even more elements can be found in the mineralised part of the skeleton (hydroxy-apatite), blood (e.g. plasma and erythrocytes), and other non-reproducing parts of the organism. Although H, C, N, O by far make up the largest part of biological matter, a large part of the periodic table can be found in the human body, with many of the elements having physiological meaning. Both aspects, the metabolism and the beam-matter interaction, form the basis of any diagnostic and therapy concept.

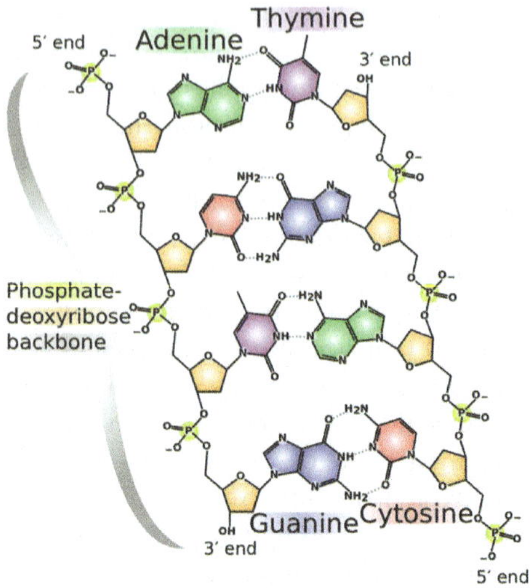

Fig. 6.2 Molecular composition of a DNA macro-molecule. The four building blocks (G-C-A-T) encode the DNA information. The chemical notation omits the C atoms at the unlabelled nodes. Published by Madeleine Price Ball at Wikipedia under CC BY-SA 3.0

Interestingly, the different types of radiation have established their specific applications. Imaging/diagnostics, discussed in Sect. 6.1, relies almost completely on photons (X-ray, γ-ray) due to their high range and quantity attenuation in matter. Electrons, ions, and neutrons find their application in destructive treatments of cancer discussed in Sect. 6.2 due to their low range and decay in energy depositing sufficient amounts of energy in a localised volume. The technical reasons should not be underestimated. X-ray tubes easily produce high intensities of photons below 1 MeV. This is still the cheapest and most practical option among all radiation sources. The rise of proton therapy can be considered quite new. It came up about 40 years ago, while X-ray tubes are rather 120 years old, therefore the technological maturity of ion beam applications is lower than that of X-rays.

Practically this might be more important than it seems. Coming from the idea to a product requires more work in the medical context than in other fields discussed in this book: Clinical studies. The complex constructs are designed to avoid human risks in a multiple stage effort. Qualification of new methods and treatments starts with preclinical studies regarding toxicity via animal testing. Testing with animals has certain drawbacks not only due to the different physiology, but also the lack of information regarding mood, pain, fitness and so on. Clinical studies (studies with human patients) follow three distinct phases. Phase 1 dedicates to testing the side-effects and proving their relevance. This involves testing doses of the product, finding the critical dose when side-effects start to appear with effects an animal cannot communicate (headache, itching ...). Phase 2 wants to prove the positive effect of the new product. Again doses and possibly also formulation are checked. This phase requires test person having the illness (e.g. cancer) and a control group treated in apparently the same way but without the active component (blinded experiment). Phase 3 exploits the knowledge of all prior phases for a large scale double blinded study. The new product has to provide statistically sound evidence of the effectivity of the new product in comparison to existing options. Mathematically over 1000 test persons are required to avoid excessive uncertainties in the results. This scheme makes the development of new pharmaceuticals and therapies safe but also expensive and time consuming (about 10 years). Nuclear products offer advantages in this process because of the solid physical understanding of their behaviour and the easy diagnostics of extremely small quantities as we will see. On the other hand the complications in nuclear safety and the investments for new accelerators hinder the development.

6.1 Imaging Diagnostics and Their Information Properties

Medicine knows numerous diagnostic tools, since a sound analysis of the patients problem is the first step to a successful treatment. Common diagnostics include for example chemical analysis of blood composition, microscopic analysis of tissue samples, or imaging techniques. The medical specialties of radiology and nuclear medicine cover all imaging diagnostics. Accelerator technologies have a special meaning for radiology: Only accelerators allow producing photons of sufficient energy to pass through the human body. Consequently, practically all imaging technologies use photons in the keV to MeV range. These technologies are nearly non-destructive, meaning the body can tolerate the induced radiation within certain limits, see Sect. 2.7.1. The alternative of blind explorative surgery usually represents the more risky alternative. Radiology includes numerous other technologies such as ultrasound and magnetic resonance imaging, but here we will focus on accelerator based methods.

The first point to discuss about imaging is what it actually images, the physical contrast. Thinking of photography, we would directly state it catches the colours and the brightness similar as our eyes do. Somehow true, but physically speaking it's a lot more complex. Colour photography combines three images at different wavelength each with its own contrast in reflection and light emission of the objects in the field of view. What would we say does an electron microscopy image show? Definitely it does not show colors, because it's an electron image not a photon image. What does black, white, and grey state in these images? More on this later in Sect. 7.1. Accelerators open up possibilities for numerous contrasts. The two options of medical diagnostics are morphological/anatomic and functional information. Morphological diagnostics usually requires external (field of radiology) radiation source while functional diagnostics use internal sources (field of nuclear medicine).

The morphology describes in particular density, composition, and microstructure of the constituents of the body. Morphologically an organ might look the same whether it is dead or alive. In contrast, a bone looks functionally the same whether it is broken or not, but morphologically it will look quite differently. For diagnosing the organ, a functional contrast will be more informative. Functionally it makes a huge difference whether the organ is dead or alive, as a living organ will consume (e.g. sugar) and interact with its surrounding by its metabolism. An organ such as the brain will consume more sugar and other substances upon activity than on stand-by. Tumors typically consume more sugar than the related healthy tissue. Also infections/inflammatory diseases will change the substance budget. It could even be altered intentionally by drugs. A functional contrast consequently resembles the local functional "strength" in-situ. In between these black and white extremes a lot of shades of grey exist. Furthermore, in many cases contrasts will be induced in both functional and morphological images, but generally in different magnitude.

In order to detect/diagnose a medical problem from such an image sometimes minute differences need to be identified. The analyst needs to look at the right spot and identify the contrast connected to the disease or the changes from the normal, respectively. No matter how good the physician interprets the image, physical limits apply as to every measurement. Aspects of resolution, background, and statistical effects as discussed earlier, e.g. in Sect. 2.5 equally apply to medical analytics. The detectors convert the photons to a signal, but signals from noise and a multitude of objects of which many are uninteresting (the background) arrive simultaneously. This forms the signal-to-noise-to-background ratio which has to be $\gg 1$. In the morphological contrast the signal strength difference between separable objects is known as density/attenuation resolution. In other words, how much smaller (considering absorption) will the signal be if we double the density of a given object? Usually the signal will not reduce by a factor 2, but less! If the difference reaches the noise level of the detector, the two situations cannot be distinguished/resolved anymore.

The spatial resolution receives special importance in medical diagnostic. On the one hand, a bad spatial resolution reduces the contrast due to mixing of signals from affected and unaffected volumes. On the other hand a good resolution typically requires more measurement time, increasing the patient dose, or reduces the signal to noise ratio, since the same signal intensity distributes over more image pixels. The limits depend on the contrast type, but also on technical aspects, in particular the detectors. The higher the photon energy to be detected, the smaller the detection efficiency (see Fig. 2.43) and the larger the detectors become. Lower efficiency translates to lower signal-to-noise ratio and larger detectors correspond to worse spatial resolution, but low photon energies might absorb too strongly in the body, reducing the signal level and probing range. Whether or not a certain feature can be detected by a diagnostic method and which emitter and photon energy are optimal can often be calculated in advance in technical applications, but in medical diagnostics the situation is more diverse and harder practical limits apply due to radiation protection and patient geometry. However in particular low contrast situations are prone to unclear or even false diagnosis.

Let us consider a simple example of a small tumour to be found in the body. Even if the tumour cannot be spatially resolved it may become visible if it exhibits sufficient contrast. Imagine the situation of a detector with 10 mm spatial resolution. Inside this resolution bin, let us call it a pixel, an object with a diameter of 5 mm and a 10 times stronger signal response (contrast) than the surrounding matter is situated. Calculating the 2D projection areas yields 78.5 mm^2 of the resolution pixel versus 19.6 mm^2 of the object. Multiplying the ratio of 0.25 times 10 in signal yields a 2.5 times higher intensity of this pixel compared to the surrounding pixels. Of course we assumed a 2D situation or, in other words, an equivalent extent of both types of matter in the third direction (the information depth), a linear response, and so on, but the systematics of the situation is as demonstrated. Modern devices can discriminate 10^{-3} difference in absorption corresponding to an equivalent ratio of detector noise to dynamic range of the intensity signal, e.g. an intensity signal of 0–1024 digital levels with 1 level noise.

In the end, the physician inspects the image, looking for aspects different from the normal. The contrast and the resolution power of the imaging diagnostics govern the limits of what can be diagnosed. Diagnosing a bone fracture via X-ray imaging represents the ideal case, the bone has a strong contrast to the gap of the fracture, only reduced by the flesh background covering the broken bone as a whole. The modifications inflammation induces in tissue features less contrast; here the detection depends on the attenuation resolution and maybe additional diagnostic results to know what you are looking for. Small entities with sizes below millimetre, such as initial forms of tumours, and limited contrast bring all current imaging methods to their limits.

The combination of methods yields synergies by combining spatially highly resolving with contrast rich methods, solving the problems of the individual methods. The combination of the functionality of positron emission tomography (PET) in addition to the spatial resolution of CT X-ray tomography is a typical example for this approach. This systematic approach not only applies to medical diagnostics but will again be an important factor for the information properties of material analysis. For the individual methods several tricks were developed to mitigate resolution and contrast problems: Contrast agents temporarily change the density of a certain part, e.g. in angiography. Adapted device settings such as probing wavelength or exposure time generally improve the situation, following the idea of colour photography. New detectors and evaluation methods improve efficiency and reproducibility.

6.1.1 X-ray Absorption Imaging and Tomography

Figure 6.3 shows an example of the typical X-ray image most people in the industrialised countries known from themselves. This technology is now about 120 years old and could be seen as one of the major breakthroughs in modern medicine as it allowed for the first time to look into the body without opening it. The contrast of X-ray imaging originates from the different photon absorption coefficients in particular between soft tissue and bones. It features no depth information due to the physics of photon energy-loss, see Sect. 3.1, making it a 2D imaging method. Usually X-ray energies of the Kα bands of Al to Mo, Fig. 3.4, induced by electron impact are exploited for X-ray generation. Deep analysis requires higher energy X-rays up to about 200 keV. In this energy range, only low energy secondary electrons are emitted which induce the radiation dose connected to this method.

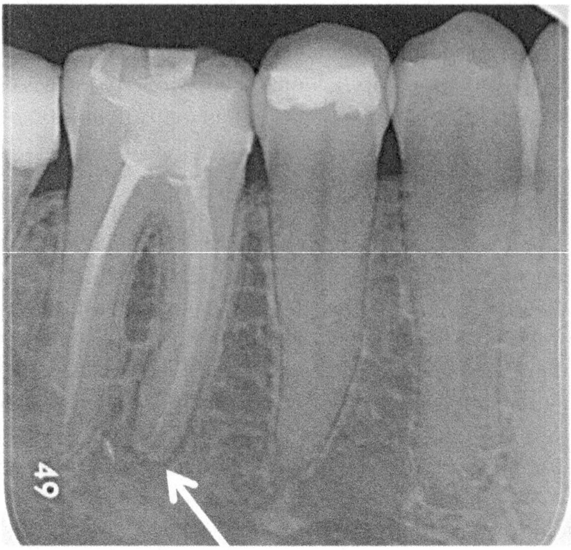

Fig. 6.3 X-ray image of the authors back teeth taken through the cheek with an in-mouth/intraoral film. The picture demonstrates the different contrasts provided by different materials. The material used for filling the root after the root canal surgery strongly absorbs the X-rays (white film) while tissue and dentine are semi-transparent (dark film). Overlapping parts, for example the four root canals of the left tooth, appear as ghosts above each other. A common problem of this projecting imaging lies in the missing depth information: The arrow indicates an inflammation at the root tip, but it cannot be said whether it affects the frontal or the backside root canal or both. The small leak in the filling responsible for the inflammation remains invisible due to limited spatial resolution

All variants of the X-ray imaging method exploit the contrast induced by the photon absorption in matter, see Sect. 3.1. For the calculation of this absorption in the human body we require information about X-ray energy, absorption coefficients and matter density in the human body, and the length of the path through the body and its constituents. Absorption coefficients are for example available in online databases such as NIST (Hubbell and Seltzer 1996) or in the report (White et al. 1989). Absorption contrast is generated by different coefficients μ for different kinds of tissue/matter weighted by the proportion of the depth this tissue type has compared to the total body depth. Figure 6.4 demonstrates how this forms the intensity image visible in diagnostic images such as Fig. 6.3.

Equation (3.3) calculates the intensity I after each depth slice of thickness d the X-ray pass through in relation to the initial intensity. Considering the situation mathematically, for several stacked slices we have to insert the final intensity I of the preceding slice as the initial intensity I_0 of the subsequent slice, resulting in a product over n slices (6.1). Exponent rules allow transferring the product into a sum of the exponents, but a sum of plain values equals a final value. Since the many different slices $\mathrm{d}_n\mu_n$ produce the same I as the single slice $\mathrm{d}\mu$, (6.1) demonstrates mathematically the equivalence of the absorption signal of an arbitrary depth profile

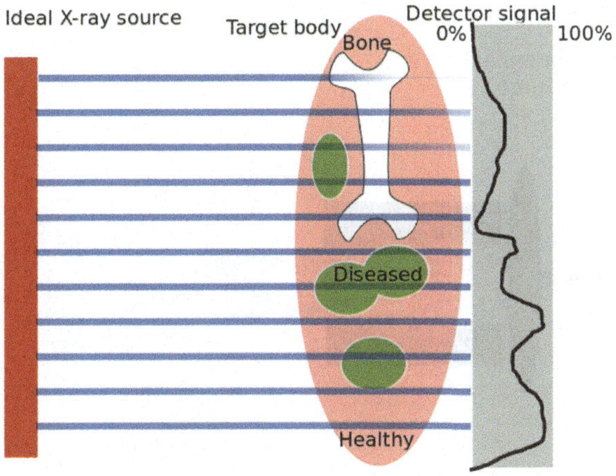

Fig. 6.4 Schematic of X-ray imaging. A collimated X-ray beam passes through the body seeing different absorption in bones (high), healthy (low), and diseased (here medium) tissue types. The remaining/transmitted X-rays induce the film/detector signal according to the initial intensity minus the line-integrated absorption. The old fashioned X-ray films darken due to X-ray impact, the exact opposite behaviour as with digital detectors where dark represents zero signal

to a flat depth profile. Therefore a depth information cannot be extracted from 2D photon based imaging.

$$I = I_0 * \prod_n \mathrm{e}^{-\mathrm{d}_n \mu_n} = I_0 * \mathrm{e}^{-\sum_n \mathrm{d}_n \mu_n} = I_0 * \mathrm{e}^{-\mathrm{d}\mu} \tag{6.1}$$

For this reason, the lateral resolution becomes important in order to allow identification of structures. The lateral resolution depends on extent and divergence of the radiation source, the same way as we know it from visible light casting shadows on the wall. The reasons are purely geometrical. Figure 6.5 depicts the typical X-ray tube situation and the challenge of forming a high intensity collimated beam from an isotropic emitter (it is not an X-ray laser). High intensity and small spot sizes are possible with electron sources (Sect. 2.4) providing a small X-ray source region, but the isotropic emission of Bremsstrahlung cannot be avoided. Modern devices reach resolutions down to μm in technical applications, but with the restrictions of the (moving) human body and the geometry about 0.1 mm are a practical limit. Photon detectors for the required energy are larger than visible light camera chips, hence featuring fewer pixels. With sub-mm resolution this limits the field-of-view due to the amount of pixels on the camera chip. Consequently, diagnostics still applies films featuring down to μm resolution on large areas at tolerable costs. In conclusion, the detection properties move in a triangle between large field-of-view, high spatial resolution, and low costs.

A technical trick for obtaining the missing depth information uses several images acquired from different angles. Via a mathematical algorithm, the approximate 3D

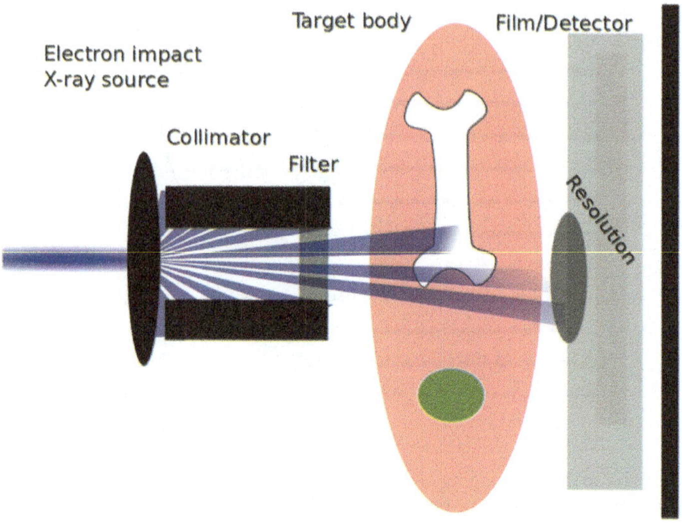

Fig. 6.5 In X-ray tubes electrons produce the X-rays as a divergent beam from a focussed electron beam. These X-rays pass apertures and filters for improving the beam quality, at the expense of intensity. The collimator cannot be infinitely narrow, therefore limiting the spatial imaging resolution due to mixing of information of the probed volume

situation behind the 2D images can be reconstructed, giving this method the name computer tomography (CT). The more 2D images the clearer the 3D reconstruction. Taking several images increases the received patient dose proportionally, leading to significantly higher doses in the order of 10 mSv compared to single 2D images. Dose and imaged volume are proportional to each other: The method is not more efficient in acquiring images, it just acquires more images resulting in more information. This dose level is not critical in a single application, but it is at the lower edge of the range where health effects appear. Nevertheless, significant increases of tumour occurrence were observed (González et al. 2009). Statistically the additional information pays off and the medical claim of a health advantage dominated method prevails. In particular, after several close applications or in combination with other exposures CT diagnostics induces relevant biological doses.

The requirement of digital processing requires applying digital detectors making the method also more costly compared to 2D imaging. The computing power requirements grow with spatial resolution due to finer grid requirements, reaching the level of 30 min for processing even with modern multi-core clusters. Modern scanners combine the X-ray source and the detector array on a single ring in 180° to each other. The X-ray source produces a line illumination, exposing all detectors at once as depicted in Fig. 6.4. This generates a 1D image slice of the body. Rotating this scanner ring around the patient adds another dimension, providing a 2D slice. Moving the patient axially through scanner ring adds the third dimension. Technically a continuous and simultaneous movement is advantageous, resulting in a spiral CT

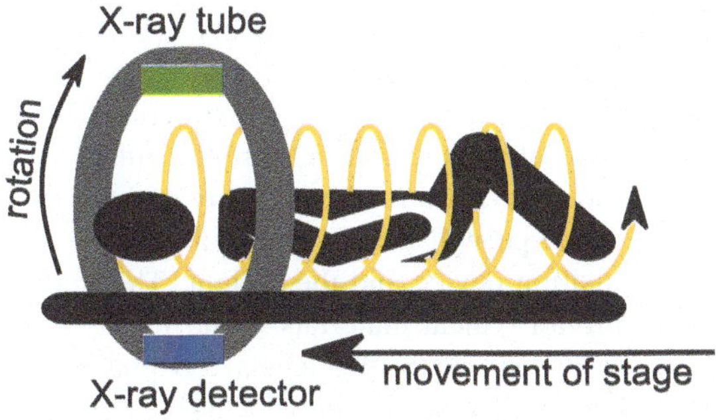

Fig. 6.6 Spiral operation mode of a CT with scanner ring. X-ray source and detector rotate around the patient while the patient moves through the scanner ring. The patient-stage movement velocity through the ring defines the slice (axial) resolution and the received dose

acquisition path as de-facto standard (Fig. 6.6). The spiral density yields the axial resolution with typical values of 1–5 mm, depending on diagnostic requirements.

Figure 6.7 shows a typical result of this measurement. The computer generated 3D image was re-sliced to a consecutive set of 2D cross sections. This result presentation is often used, since 3D models of a dense object imply the difficulty of defining transparency for enabling a view to the insides where medical imaging expects to see something. The interpretation of these medical X-rays images requires a physician trained in seeing deviations from the normal. This requires normalisation/calibration of the image for a reproducible interpretation of the results. The images have to look the same, no matter where and with which device they were made, otherwise the physician will have to learn analysing the image with every different X-ray device and patient diameter (absorption) again. The so-called Hounsfield scale uses a water

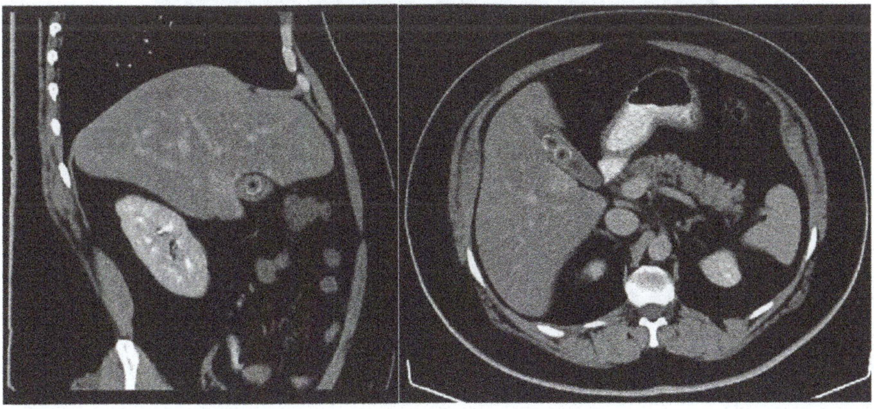

Fig. 6.7 Abdominal CT of a male patient with gallstones with 5 mm slice thickness. Side-view (left) and top-view (right). Iodine contrast agent was given to the patient. X-ray tube voltage of 130 kV. Total patient dose of 25 mSv

and an air reference body for this purpose. Air equals a Hounsfield value of -1000, water of 0, and bones up to 1500. Looking at (6.1) we require the original intensity I_0 and a scale for the product $d\mu$ in order to be able to separate d and μ in the actual measurement. The detector efficiency can be absorbed in I_0 as it only scales the intensity. Two known parameters are mathematically enough to solve (6.1). After the calibration each tissue type will have a given absorption value in the image, independent of the acquisition device and the patient. The scale has a range of –1000–3000 for describing all typical materials found in the body, equivalent to a typical 12 bit detector intensity scale (=4096 levels). Displaying only a sub-range of interest within this total dynamic range enables focussing the diagnostic contrast on certain features and tissue types.

Contrast agents allow for a temporary change of the specific absorption coefficients by dissolving a substance of higher absorption coefficient in them, e.g. heavy metal containing drugs or heavy gases. In particular for the investigation of blood vessels, a so-called Angiogram as depicted in Fig. 6.8, iodine containing molecules are injected for increasing the X-ray absorption, making the blood vessels distinguishable from the surrounding tissue. Extra contrast can be induced with a correct alignment of the X-ray energy to the (K-line) absorption edge of the contrast agent. For iodine, this increases the absorption coefficient by a factor 5 at 33.3 keV.

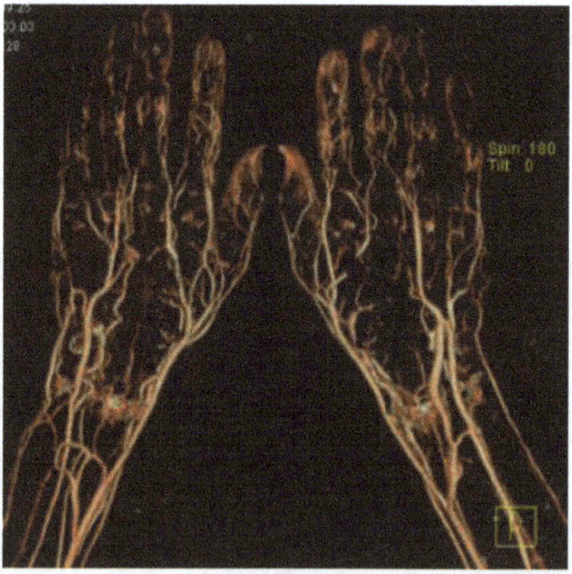

Fig. 6.8 CT angiography of the hands. A contrast agent injected into the blood distributes naturally in the circulatory system, revealing even the smallest blood vessels due to the absorption enhancing heavy metals in the agent. Due to the short acquisition times, X-ray imaging even allows a time resolved measurement of the distribution process in the circulatory system. The method allows for example localisation of blood clots or regions of abnormal blood vessels. Published under public domain by MBq at German Wikipedia

Calibrated 3D absorption information has additional value besides the tomographic imaging. Many methods (e.g. most other methods presented in this chapter) require information on charge density distribution for charged particle stopping power *S*. Photon absorption and charged particle stopping power equally scale linearly with matter density, since both only depend on the number of electron passed. Therefore, density information transfers between both. In the typical region of X-rays of 10–200 keV the photoelectric effect dominates the absorption coefficient, see Fig. 3.5. This effect scales with the mean nuclear charge *Z*, since this equals the number of electrons in an atom. The stopping power *S* scales with Z^2, see (3.6). Analysis at multiple wavelength improves the accuracy of *S*. Equation (6.2) describes the connection between absorption coefficient and ion stopping power with *Z*. This type of connection allows recalculating CT images to 3D stopping power data as input for ion therapy planning.

$$\mu \sim Z \sim \sqrt{S} \tag{6.2}$$

120 years of X-ray imaging require a new technological perspective. Quite recently the Phase-contrast X-ray imaging (PCI) recommended itself as this upcoming technology offering increased soft tissue contrast (Weitkamp et al. 2005). Instead of relying on X-ray absorption, this method images the phase shift of a quasi-coherent X-ray beam induced by the interaction with the imaged object. In principle, this method does not require a tissue dose, since absorption is not an integral part of the method. Differences in tissue refractive index induce this phase shift via their influence on the speed of light. The improved soft-tissue contrast originates from the high cross-section of these phase shifts, orders of magnitude higher than the absorption cross-section. So far, the technical challenges of extracting the phase information prevent a wide-spread application, but several solutions are being investigated for this potential future of X-ray imaging.

6.1.2 Emission-Computed-Tomography

The term emission depicts the source of the photons exploited in this class of diagnostics via emitting isotopes located in the body itself with detectors outside the body. Computed tomography tells us the methods acquire a 3D information, usually in the form of several 2D slice views of the body. The generation of the 3D and also the 2D slices requires computers since usually hundreds of detector positions are summarised, similar to the external CT discussed in Sect. 6.1.1. Having the photon sources (emitters) in the body enables a functional imaging by placing the isotopes into metabolically active molecules. Upon incorporation of these vectors into the body, e.g. intravenous or via ingestion, the metabolism transports them through the body. So far so uninteresting, but diseases are a problem to the body also because they change the metabolism or produce regions with different metabolism. These changes are specific to the disease and hence also the vector has to be designed to follow the specific metabolic difference of the sought disease (may it be an inflammation, cancer, or something else) in order to induce a contrast in the distribution of the emitting isotopes. The vector contributes the functional aspect, e.g. via a sugar, and the emitting isotopes witnesses the vector location. The optimal contrast will be achieved if the isotopes remain in the location of consumption of the vector, the so-called metabolic trapping. The design of the vectors in combination with the available emitting isotopes therefore represents a major interdisciplinary technological challenge of emission-computed tomography.

The photon detection typically employs large ring-shaped arrays of scintillators connected to photomultipliers, see Sect. 2.5 in order to achieve a high detection efficiency at the MeV scale photon energies without the high costs of semiconductors. The detector system has to deliver photon detection rates (intensity) together with a localisation of the photon origin, the position of the radioisotope, in order to generate a tomographic diagnostic image. The detectors cannot deliver a position of origin or even the impact angle information, the detector signal is just a yes or no and the radioisotopes emit isotropically in all directions. Geometrical efficiency will increase the signal in detectors close to the origin due to increasing detector solid angle, but absorption in the body and geometrical aspects blur the information, hence even a 2D array of detectors hardly delivers a useful image. Restricting the detector response to a line (1D) (Line of response, LOR) along which the original decay has to be situated enables generation of useful 2D images. 3D (tomographic) information requires imaging from several locations/angles around the patient, leading to the typical detector ring geometry. A mathematical tomographic reconstruction convolutes the different 2D images to a 3D image. The 3D reconstruction is mathematically demanding and introduces artefacts, depending on the amount of detectors and available 2D slices. External CT ensures this line of response via the collimated X-ray beam, but requires moving the photon source for the 2D image aspect. With internal sources only the detector has to be moved around the patient (cheaper option) or numerous detectors have to be positioned around the patient for a simultaneous acquisition of several images (faster option).

Physics allows for two options to filter out the line of response from the random emission of radioactive decay, depicted in Fig. 6.9. The Single-Photon-Emission-Computed-Tomography (SPECT) method uses collimators in front of the detectors which only allow photons from a certain direction to enter the detector (Rahmima and Zaidib 2008). Basically it works like a pinhole camera. Similar to the pinhole camera this masking is at the expense of losing the largest part (>99.9%) of the photons in the collimator (the reason for using cameras with lenses). Several standard isotopes from fission and accelerator origin exist in clinical use. Significant absorption of the emitted photons in the body reduces the detector signal and induces a depth dependent sensitivity. Therefore the emitted photon energies should be >100 keV. External CT enables a correction of the depth dependent absorption coefficient. The isotope ^{99m}Tc, generated from the decay of neutron-irradiated molybdenum, is a standard isotope with the strongest line at 140.5 keV. Some isotopes used for therapy also emit high-energy photons, in particular ^{131}I (see Sect. 6.2.2), enabling an imaging without additional tracer injection. The increasing density of isotope producing accelerators

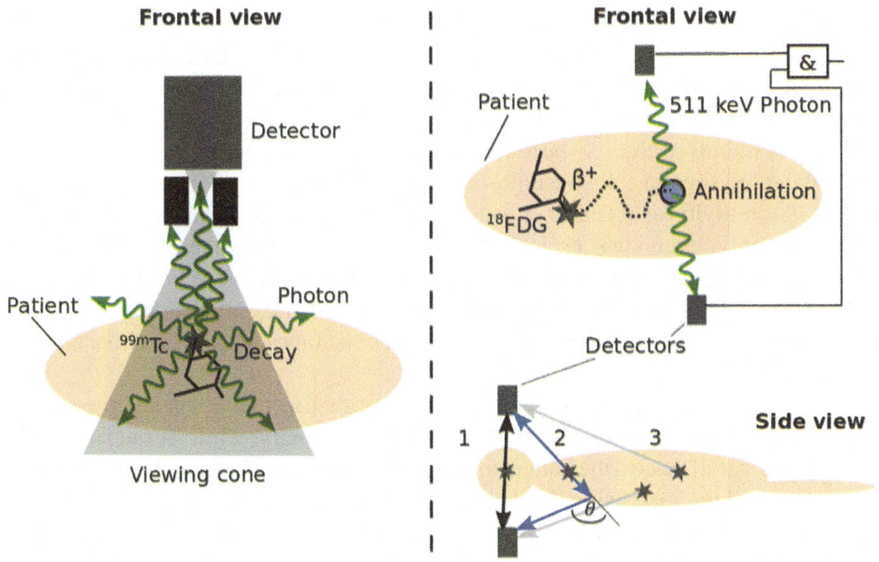

Fig. 6.9 Left: SPECT detection geometry with its collimator defined viewing cone. Right: PET requires coincident detection of two 511 keV photons from positron annihilation. Below: Case 1 shows the ideal situation for clear 2D imaging. Moving the detectors or the patient allows for recording several 2D slices. The coincidence in case 2 has a scattering event in between, but the system can only assume the event to originate from a non-scattered line. Energy resolved filtering allows mitigating case 2. In case 3 a random coincidence of two uncorrelated photons produces a false signal. With detection in two or more detector rings (3D mode) the probability of detecting both photons strongly increases, but also the detection probability of case 2 and 3 events increases

makes the use of shorter half-life isotopes economically viable, providing a wider range of isotopes for clinical diagnostics.

The second option exploits two correlated photons forming a line. Many decays emit several photons, but these are uncorrelated. A simple Dual photon emission computed tomography yields no advantage. The only process capable of emitting correlated photons is the annihilation of matter with anti-matter, in practice the annihilation of positrons with electrons. The β^+ decay produces positrons which annihilate with an electron producing two 511 keV photons due to the rest mass of the electron and positron. Fundamental physics dictates the conservation of momentum, hence both photons have to be emitted in opposite directions (180° apart) in the centre-of-mass system. Their correlation forms a line, fixed by detecting both photons simultaneously. The lab and the CMS are equal only for resting positrons, but due to the annihilation cross-section increasing towards lower energies deviations from the 180° are usually $<1°$.

Since only positrons allow for this photon correlation, the method is called Positron Emission Tomography (PET). Its information properties are similar to SPECT, but with the coincidence detection, the collimation is not required anymore. This increases the counting statistics at a given injected activity by orders of magnitude. Together with the metabolically interesting aspect of the standard PET molecule, Fluordesoxygluconase (FDG) enriched with ^{18}F, this draws responsible for the large success of PET. The body consumes FDG in the same way as regular glucose, but the fluorinated product cannot be further used like the regular products of glucose, therefore the ^{18}F becomes metabolically trapped and accumulates the faster, the more sugar a cell consumes. Besides FDG, numerous other ^{18}F molecules such as Fluorethyltyrosin (FET) or sodium fluoride exist with specific sensitivity for certain cancer types and tissues. ^{11}C, ^{15}O, and ^{13} N also feature interesting diagnostic properties and open up more vectors and metabolic functions to be diagnosed via PET (Miller et al. 2008). The significantly shorter half-lifes of these isotopes in the minute scale strongly limit the availability of these isotopes for practical reasons, see Sect. 5.1.4.

Emission tomography generally requires short half-life nuclides to acquire images in a time frame the patient can remain relatively stationary, while at the same time limiting the overall dose to the patient after the measurement. Technically, the detector count rates and the increasing dead-time at high count-rates limits the dose and the measurement time. Modern detectors require 10–100 ns after each event to become ready for the next photon, corresponding to limits in the order of some 10^6 counts/s (with about 10% dead-time). Arrays of several detectors and the limited detection probability in the order of 10% (the detector efficiency) increase the required activity. Correspondingly, radiation doses of some 100 MBq have proven to be a good compromise for radioisotope injection. These activities induce patient doses in the order of 5–30 mSv.

Figure 6.10 shows examples of three brain PETs for cancer diagnosis. The high contrast allows an easy identification of regions with increased FET uptake. The limited spatial resolution compared to CT results from several instrument and physical effects, all contributing similar uncertainties in the order of mm. The finite detector size limits the localisation of the LOR due to geometrical effects. The detection of scattered photons blurs the distribution of events, since these events do not lie

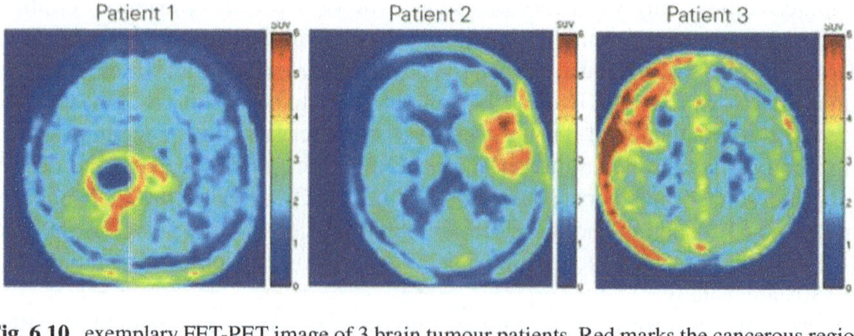

Fig. 6.10 exemplary FET-PET image of 3 brain tumour patients. Red marks the cancerous regions of high ^{18}F uptake with about 3 times higher uptake values compared to healthy tissue. In the right most image somehow the ^{18}F accumulated at the cranial bone without a relation to a tumour. Reproduced from Rickhey et al. (2008) with permission by Springer

on the assumed LOR. The detector energy resolution limits the minimum separable Compton scattering angle. The detector energy resolution allows for identification of these scattered photons due to their energy-loss according to the Compton effect responsible for the scattering, see (6.3).

$$\Delta E = E - \frac{E}{1 + \left(\frac{E}{m_e c^2}\right)(1 - \cos(\theta))} \tag{6.3}$$

For 511 keV photons this energy-loss accounts up to 340 keV, but the actual problem lies in small angle scattering events with keV energy losses which cannot be resolved by the detectors with their typical energy resolution in the order of 10 keV. Therefore, two photons detected in temporal coincidence are not necessarily correlated. The finite coincidence windows make these so-called false positives possible. Electronics and detector response have their processing time, resulting in a coincidence gate length. The more activity present, the higher the probability of two decays emitting photons within the coincidence gate time. Again, the system assumes a location of the decay within the regular 180° correlation given by two opposing photon detectors, while the true location can be anywhere. Catching all four photons (2 from the e^+ annihilation + two Compton photons) would resolve the false positive, but the full solid angle cannot be covered by detectors and not all detectors can be gated with all other detectors individually for resolving also coincidences other than 180° (factorial scaling of the required gates with the number of detectors).

The β^+ decay emits the positrons with energies of a few 100 keV, resulting in a positron travelled range of 1–2 mm before annihilation. The stopping power of tissue for electrons or positrons, respectively, in the MeV range is in the order of 2 MeV/cm. This range slightly disconnects the position of the decayed isotope from the origin of the radiation, reducing the spatial resolution. Physically, the positrons travel in the surrounding tissue until their energy reaches a value with sufficient annihilation cross-section. The annihilation with an electron becomes effective only at small kinetic energies. Lower energy decays are envisaged, but the energy spectra are not mono-energetic due to the 3-body reaction nature of the β decay. The 3-body decay energy distribution, the large straggling of electrons, and the limited knowledge of the exact stopping power of the surrounding tissue makes a backward calculation of the movement impossible.

Electro-magnetic fields change the movement of the positrons, hence the projected range reduces due to the circular movement in a static magnetic field. New approaches for combining the PET analysis with the magnetic resonance tomography (MRT) allow for a suppression of this physical range limit. At the same time these technologies smartly open up synergies: pH sensitive PET-MRT drugs which attach different amounts of O–H groups to the vector depending on the surrounding pH level allow for local pH measurement in a combined PET-MRT analysis. The MRT with its sensitivity to hydrogen atoms detects the hydrogen in the O-H groups, while PET enables separating the natural hydrogen from the drug related amount. In the vicinity of tumours the pH level changes due to the metabolism of the tumour.

6.2 Radiation Therapy

Fight the enemy where they aren't.

A tumour by itself is not a critical illness, think of the term "benign tumour". Cancer, as a sub-class of tumours, grows and spreads in the body (metastases). This is what makes it a critical disease and at the same time a challenge to effectively treat/cure it. Cutting out or destroying the main tumour is the first and most important step, but only a strategy which is able to take up the battle even where the enemy is apparently not and fight on each length scale will provide a universal cure for cancer.

It appears to be a contradiction: Getting healed by radiation. Not only few people fear nuclear technology because of the carcinogenic effects of radiation on the human body. As we learned in Sect. 2.7.1 this is true, but already the ancient Greeks knew: the dose makes the poison. This quotation directly leads to the goals and technological challenges of radiation therapy and nuclear medicine in general: Reducing the dose to healthy tissue to the absolute minimum required and focussing the dose to unhealthy tissue as strongly as possible. The technological limit reduces to the term selectivity: Reducing the dose to healthy tissue. The body can repair damaged DNA and recover from radiation effects within certain limits, providing a certain tolerance to technological imperfections. After the therapy the life-quality and life-expectancy of the patient have to be improved, medical science speaks of survival rates or disease-free survival. Radiation finally kills any biological organism, therefore only killing the tumour is not a challenge, but the patient needs to survive the therapy with a benefit. This survival rate depends on the selectivity of the therapy. Luckily, accelerators provide a remarkable set of technological options and physics to address the selectivity.

Currently, nuclear therapies focus mostly on cancer treatment. Besides this main block of cancer treatment, also non-oncology applications exist, since the concepts of nuclear therapy are not limited to cancer cells and e.g. bacterial inflammations can also be treated via radiation. Cancer involves cells with reproduction errors, typically fast and uncontrolled reproduction. Cancer requires an initial cell, which, by bad luck, mutates, survives, and manages to reproduce. After reaching a certain cell conglomerate size or if the cell properties are too similar to the original regular cell type, the immune system of the body cannot recognize the cancer cells as malignant therefore the natural defensive safety mechanisms protecting the body from malfunctioning cells will not destroy it. All cancer cells emanate from a certain type of original tissue by an unlucky mutation, maintaining most properties of the original tissue. This defines the type of cancer and the primary tumour location. Thyroid cancer for example originates from thyroid cells. Even though its metastases might be found in other parts of the body it is still possible to identify them as thyroid cells. Furthermore, since all cancer cells originate from the same cell zero, they duplicate genetic information and cell features of it.

The first step of a selective treatment lies in the exact 3D localisation of the tumour. In Sect. 6.1 we discussed accelerator based diagnostic methods available to locate tumours without sectioning the whole body. If the tumour cannot be removed surgically, for example when it lies deep in a vital organ or infiltrates it on the microscale, radiation therapy provides a solution. The diagnosis represents the first step required for external radiation therapy discussed in Sect. 6.2.1, but also for the surgical sub-class of internal methods. In this aspect, the physics of beam-matter interaction requires a deeper understanding of the tumour location, since energy-loss connects to the amount of passed atoms and their nuclear charge not to the passed distance. Consequently, a diagnostic method needs to clarify the stopping power of the surrounding tissue for a precise spatial positioning of the treatment, strongly connecting accelerator based diagnostics and therapy. Ideally, a biological mechanism would automatically take over this delicate task of dose delivery. This internal biological delivery of radioactive isotopes via so-called vectors forms the idea of the internal radiotherapy discussed in Sect. 6.2.2.

Accelerators provide a wide range of options for both therapy approaches. Photons, α, β, ion, and neutron radiation can all be generated with accelerators in broad energy ranges from keV to GeV. The range in connection with the tumour location and depth in the body represent the first selection criterion for the particle species and energy. Deep tumours require a high range and energy-loss physics favouring the energy deposition in depth with minor deposition in the entrance channel (Bragg-Peak behaviour) while surface near tumours require short ranges to protect the underlying tissue. Furthermore, accelerator based production of radioactive sources (see Sect. 5.1) provide the isotopes for vector based therapy but also for external therapy via decay radiation (e.g. 1.33 MeV photons from ^{60}Co).

The molecular mechanisms of the destructive effect of radiation on cells in general and cancer in particular are not perfectly understood and a topic of active research. As far as we know, the effect of radiation therapy relies mostly on ionisation of the atoms in the tissue via scattering reactions (see Sects. 3.1, Fig. 3.5 and 3.2). This transfer of energy to the electrons in the material is often termed Linear Energy Transfer (LET). LET and stopping power are connected, but the LET limits the view of the energy-loss to the short-ranged secondary electrons inflicting the highest energy deposition density close to the particle track. In radiation material science (see Sect. 7.4) it could be understood as the damage cascade density, or the mean primary knock on atom (PKA) energy. The ionisation directly changes/destroys the complex biomolecules breaking cellular molecules. In particular the damage to the DNA helix represents the major cause of death to the cell since it controls the cell working, but the DNA has a damage threshold: While single DNA defects might be repaired by cell internal mechanisms, multiple defects strongly increase the probability of cell death, yielding a lower limit for the short-term energy deposition required for an effective radiation treatment. Higher LET increases the probability for multiple damage events per cell/molecule, a situation hardly repaired by the cell internal system. The statistical nature of beam-matter interactions blurs this limit, since the cells in the irradiation region will receive a statistical distribution of doses. Consequently, higher single doses potentially improve the therapeutic success, but at the same time increase the dose to the surrounding healthy tissue, the so-called overkill. An optimal ratio has to be found defining the overkill threshold. This lack of understanding and control requires empirical quantities for the overkill and a stepwise irradiation with intermediate success control of the remaining tumour mass. Several applications with a few weeks to month distance compensate for this statistical effect by targeting the surviving cancer cells of the last application after the metabolism sorted out the dead cells.

A secondary, indirect effect of treatment relates to the presence of oxygen in the affected cells. The ionisation produces free radicals for example by breaking H_2O present in all cells into H plus the HO radical. The radicals chemically attack the cells, in particular the vulnerable DNA, increasing the destructive potential in the low LET case. Cells lacking oxygen supply are less prone to the destructive effect of free radicals due to an intermediate process of radicals with oxygen. Unfortunately, tumour cells often lack oxygen (hypoxia) due to their destructive nature on the metabolic system. This reduces radiation therapy efficiency of hypoxic tumour cells up to a factor 3 in the low LET case. Furthermore, hypoxia promotes spreading of the tumour to overcome the nutritive deprivation making this tumour type a particularly dangerous and hard to treat case (Vaupel and Harrison 2004).

The strength of cancer hiding it from the immune system could also be its major weakness. Cancer cells survive in the human body because of their genetic code being similar to their corresponding host/origin cell type that the immune system cannot identify them as an enemy. Little changes of their DNA could already change this situation, amplifying the therapeutic effect of radiation therapy by demasking tumour cells which subsequently activates the immune system to destroy them. The ionisation of a water molecule hydrating the DNA and the subsequent recombination of electrons from other parts of the DNA molecule potentially breaks bonds in the large DNA molecules, already changing it with only a single ionisation reaction (Ren et al. 2018). A cell with such a modified DNA would be targeted by the immune system. A contribution of the immune system would dramatically enhance the therapy effect and at the same time allow for reduced doses applied to the patient, also reducing adverse side effects.

6.2.1 External Beam Therapy

External radiation therapy, also external beam therapy, applies an external radiation source (the accelerator or an isotope source) firing particles through/into the body. Due to the mass and size of the human this requires ion energies above about 100 MeV, electron energies of some 10 MeV, or photon in the 1 MeV range. The choice of ions has a certain influence on the dose localisation and, besides technical aspects of the accelerators favouring protons. Helium and carbon ions have potential treatment advantages over protons with the disadvantage of requiring $\geq$GeV energies.

Particle beams could be seen as a disembodied scalpel. Like every scalpel, they require precise application to cut away only the diseased tissue. Instead of the surgeon's hand, particle beams require computers, beam optics, and precise knowledge of the target. All technological effort aims at increasing the precision, which is

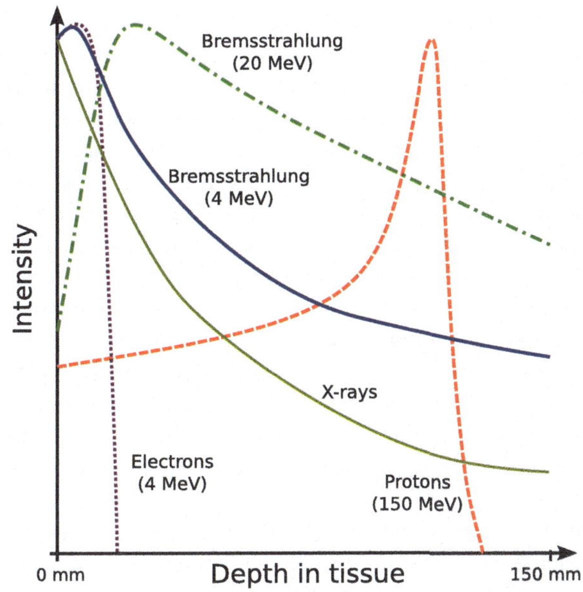

Fig. 6.11 Comparison of depth effect of beam irradiation in tissue. Electron exhibit also a peaked behaviour, but the strong bremsstrahlung emitted by the electron stopping leads to an unfavourable depth distribution. Protons on the other hand purely damage by their stopping behaviour. Photon with their exponential decay cannot offer a depth variation. *Source* MBq Disk Bew; by Cepheiden, CC BY-SA 3.0, commons.wikimedia.org/w/index.php?curid=7509488

equivalent to reducing beam induced damage to healthy tissue, which would increase the risk of producing secondary tumours and losing organ functions important for health and quality of life of the patient. Here the physics of energy-loss discussed in Chap. 3 with the strong differences between photons and charged particles comes into play. Figure 6.11 demonstrates the damage intensity over depth induced by photons, electrons, and ions, pointing out the fundamental fuzziness limiting the scalpels precision.

Photons with their energy-loss by reduction in intensity feature an exponential decay of dose with depth with a decay constant μ reducing with photon energy. The strongest dose occurs at the entrance surface, making them optimal for treating surface-near tumours. For deeper tumours, the situation becomes worse since the part of dose deposited in a deep tumour exponentially reduces with its depth. Higher photon energies slightly improve the situation to a more flat dose distribution, but physics limits the part of dose in the tumour to about the tumour size in depth divided by the photon range. Up to a few 100 keV X-ray energy, DC X-ray tube are used, while for higher energies either collimated radioactive sources such as ^{60}Co or MeV LinACs based X-ray tubes with solid targets emitting electron Bremsstrahlung are used. The rotation of the X-ray source around the patient enables using several entrance channels for deep tumours for a reduction of the peak healthy tissue dose in each entrance channel, but the potential gain of a 4π rotation remains limited.

Electrons already offer an improved situation with concurrently higher required particle energy for a given range. Figure 6.11 demonstrates the electrons localised energy deposition with a fixed and relatively sharp end of dose in depth due to the limited particle range. Electron beam therapy improves upon dose localisation due to the fixed range of electrons. Extensive straggling reduces the spatial confinement and with this the selectivity of the radiation dose. Even worse, the stopping of electrons produces a broad spectrum of photons up to the electron energy via Bremsstrahlung. This spectrum adds to the electron dose, extending the dose range beyond the electron range (+straggling) limit. The physical aspects favour photons and electrons for surface near tumours. Lastly, ions, in particular protons, demonstrate a completely different behaviour but also require the highest beam energies. Their different stopping power evolution with energy, see Fig. 3.6, induces the Bragg-Peak, an energy deposition maximum close to the range limit. For deep tumours this reduces the surface near dose drastically, since the stopping power is lower at higher energies, and, by clever placement of the Bragg-Peak in the tumour, concentrates most of the dose to the target tissue.

This fundamental physics favours the use of ions for deep lying tumours. Cyclotrons produce the required ion-beams in the range of 70–250 MeV for protons ($\approx$365 mm range) and up to a few GeV for heavier ions. Protons require the lowest beam energies due to the atomic charge dependence of stopping power and offer technical advantages with respect to ion sources (as discussed in Sect. 2.4) and beam optics (Sect. 2.3.2). Cyclotrons with variable extraction radius allow for a continuous energy modulation during treatment for adjusting the depth range. Electrons require LinACs to reach the 10 MeV range. MeV photons require adding a bound electron X-ray source (see Sect. 4.3.1) to the electron LinAC, while regular DC X-ray tubes (Fig. 4.14) suffice for smaller photon energies. Due to the constructive aspects of these devices they all have dimensions in the order of 5 m, except for X-ray tubes.

Any beam targeting can only be as good as the actual target location is known. The term "irradiation planning" describes the process of localisation and determination of doses to be induced in every point in the body. For this we need to localise the target/tumour by using diagnostic methods such as PET, MRT, or CT described in Sect. 6.1. Unfortunately, placement of beams does not work in a spatial dimension, but energy-loss of beam, the stopping power, depends on the amount of passed atoms/area and their nuclear charge and mass. Since the body consists of about 60% water, water phantoms/reference bodies allow for a first assessment of the local dose distribution experimentally. Most tissue types feature densities close to water (1 g/cm^3), but compositional and density variations and the deviations from Braggs mixture rule (see Sect. 3.2) require readjustment of this first order information using information of the actual patient. For example thyroid tissue features a density of 1.05 ($\pm$0.05) g/cm^3 with stoichiometry $H_{10}C_{12}N_2O_{75}Na_{0.2}Cl_{0.2}K_{0.2}I_{0.1}$ (Woodard and White 1986) resulting in a 7% smaller length specific stopping power compared to water for 100 MeV protons. In this example the range would differ by about 7 mm for both tissue types, a critical value due to the influence on the Bragg-Peak position on the dose distribution. Interestingly, X-ray attenuation coefficients are quite well understood and, like ion stopping power, they scale with the nuclear charge of the target matter. Unfortunately, density and nuclear composition are not fully exchangeable due to different scaling of the stopping power and photon absorption with these quantities. Information from 3D X-ray analysis (=computer tomography, CT) yield a 3D distribution of the mixed absorption coefficients, providing a way to obtain the necessary information. Adding information at several photon energies adds more equations to enable solving the ambiguity in composition and density, but the differences in scaling are small for photons, partly preserving the uncertainties.

With all information on patient and target, the possibility for an inverse radiotherapy planning arises. This implies first identifying the target area and setting the required dose distribution. Based on this, the required beam and scanning parameters will be calculated using physical models for the energy deposition and biological models for the resulting dose distribution. Figure 6.12 shows such a calculated "Line of fire" with the flat maximum dose zone around the tumour and a low dose in the entrance channel. In order to fully catch the tumour, in spite of uncertainties in positioning, the high dose region needs to be extended beyond the actual tumour by the magnitude of the positioning uncertainty (typically in the mm range) or tumour cells might remain, ready to rebuild the tumour. At the same time, the local dose at the unhealthy tissue has to exceed a killing threshold but stay below an upper limit.

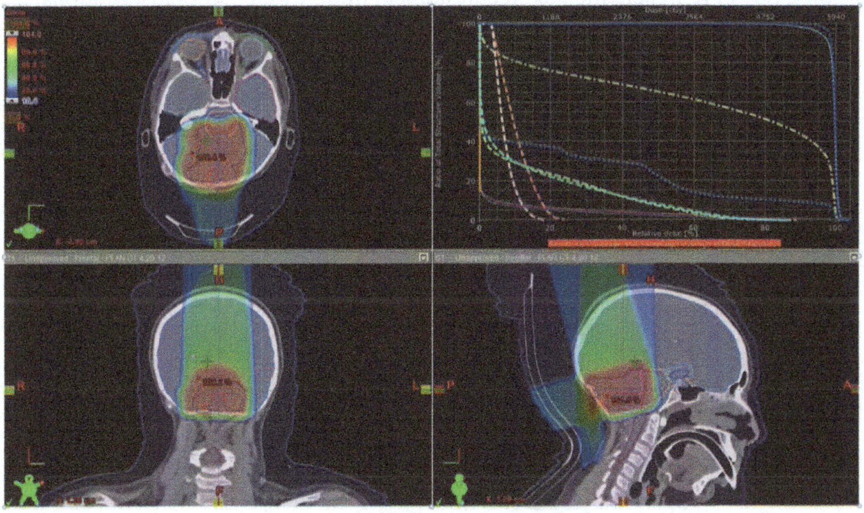

Fig. 6.12 Example of a computer assisted proton irradiation therapy planning of a delicate brain tumour. Blue marks low dose regions and red the maximum dose region. The target tumour is depicted by a red circumference within the maximum dose region. The entrance channels of the beam at the back and the top of the head deliver relatively small doses to the healthy tissue due to the Bragg peak effect. Several entrance channels focussed on the same region reduce the healthy tissue dose, but require more complex beam systems. The stopping power in the entrance channel determines the depth of the dose maximum. Reprinted from Buchsbaum et al. (2014) with permission by Springer

Doses above this limit would induce no additional therapeutic effect (overkill), but additionally damage the healthy tissue in the entrance canal.

In this step, the available instrument options for positioning the Bragg-Peak become relevant. The optimal therapy moves the Bragg-Peak laterally via beam optical systems and in depth via a change of beam energy on the accelerator side. Older systems use thin and thick masks specifically manufactured to shape the beam and reduce the energy from a fixed projectile energy. The optimisation of the dose delivery mechanism represents a current technological development path with several advantages in positioning and practical aspects. Small beam spots with variable range deliver the best solution from a physical point of view. Technical realisations apply active beam scanning via ion optics (dipoles), called pencil beam scanning (Fig. 6.13) together with variable beam energy. The rotation of the beam around the patient enables using multiple entrance channels. Moving the patient implies risks of losing the spatial calibration obtained for irradiation planning due to the elasticity of the body. For this reason, the beam has to move around the patient. Rotation of the complete accelerator system including its beam optics are technically challenging, but beam optical solutions are limited to a few discrete directions, potentially having sensitive organs in the specific entrance channel.

With perfect beam control, the uncertainty in the target/patient geometry and stopping power represent the next technological limit. Compositional variations, e.g. in water content, different densities, and the exact positioning of the patient induce uncertainties for the stopping power. The high variability and mechanical flexibility of tissue changes its beam-matter interaction properties on a short time-scale (e.g.

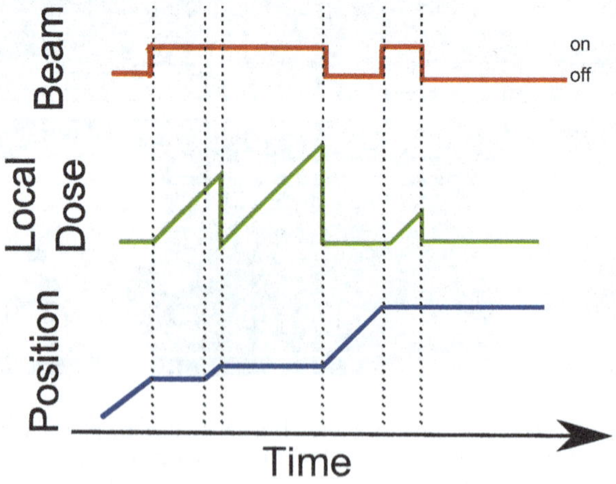

Fig. 6.13 Schematic representation of the beam scanning irradiation. During movement, the beam is off, limiting the tissue dose. Upon reaching a treatment position, the beam stays constant until the pre-planned dose is reached. Together with beam-energy variation a 3D dose pattern can be induced

via breathing) but also on a longer time-scale via metabolism, water content, build-up of new blood-vessels, and so on. An ex-situ characterisation can therefore only provide a limited accuracy in any case. New approaches integrate analytical tools, such as X-ray sources, into the beam irradiation setup to minimize the time between characterisation and irradiation of the patient. A true in-situ approach has to exploit the radiation emitted during treatment for a live correction of the ex-situ data, further increasing the positioning accuracy and reducing the required dose halo. Similar to ion-beam analytics (Sect. 7.1.5) this exploits the secondary radiation emitted by beam-matter interaction. Promising approaches exploit the formation of positron emitting isotopes such as ^{11}C or ^{15}O with short half-lives in the order of minutes. While these isotopes are challenging for external PET as discussed in Sect. 6.1.2, their short half-life is optimal for delivering high count-rates for in-situ analysis. The 3D tomographic information generated by the integration of a 3D PET scanner into the irradiation enables a measurement of the local beam energy. Nuclear excitations such as $^{12}C(p, p\gamma_{1\text{-}0})^{12}C$ or $^{12}C(p,\gamma_0)^{13}$ N (similar to the PIGE analysis method) add another degree of information. The prompt emission of the MeV scale photons disables all transport effects moving emitting radioisotopes, but their high energy represents technical challenges for detection efficiency and collimation required for a precise 3D localisation of origin. Any secondary radiation requires precise knowledge of nuclear reaction or production cross-sections to convert the local emission density of the secondary particle to a physical quantity such as beam energy or Bragg-Peak position via depth dependent calculations (Sect. 3.4). The more of this information comes together the better the result due to larger equation systems enabling for example the cancellation of mass density and elemental stoichiometry.

The clinical efficiency and the adverse effects such as secondary tumours induced by the therapy strongly depend on the accuracy and with this, the exact setup applied. The ongoing technological advances of ion beam therapy complicate the clinical assessment of the method, since not all of the devices in use actually apply the latest developments. In addition, the economical aspect improves due to optimisation of the accelerator constructions and the advancement of software. Yet the irradiation with ion beams represents the most costly treatment option, already due to the high required beam energy and the connected costs (€/eV rule of thumb). Lower costs are realised with electrons and photons with the disadvantage of problematic dose-depth distributions.

In principle, neutrons allow for a treatment cancelling this contradiction. The trick is to exploit the extreme variations of nuclear reaction cross-sections of neutrons with mixed targets. While charged particle reaction cross-sections are relatively similar for most targets and the nuclear reaction related energy deposition is low compared to the energy-loss connected with the Coulomb interaction of the projectile with matter/electrons, neutrons do not interact via the electronic energy-loss mechanism but deposit their energy mostly via nuclear interactions. ^{10}B application together with neutron energies $\ll$100 keV exploits this aspect due to its high neutron $^{10}B(n, \alpha)^{7}Li$ cross-section leading to a local formation of MeV ions in the so-called boron neutron capture therapy (BNCT). The energy-loss due to the interaction with protons in tissue represents the major elastic loss component (also known as neutron moderation). The

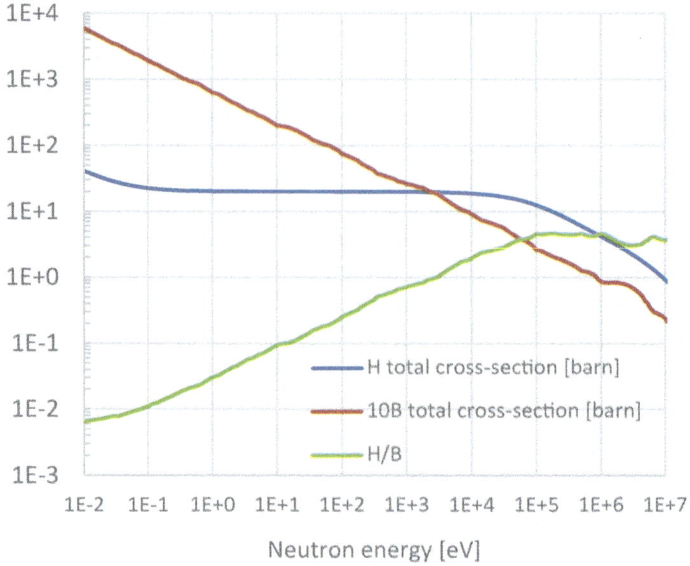

Fig. 6.14 Comparison of the $^{10}B(n, \alpha)^{7}Li$ to the elastic scattering cross-section of n–H (Koning et al. 2015). At 2.7 keV both cross-sections are equal, below $^{10}B(n, \alpha)^{7}Li$ dominates, above n–H dominates. As long as the ratio of cross-sections H/B $\ll$ ($Q = 2.79$ MeV)/E the boron doping of tumours provides improved dose selectivity, assuming the atomic density of H and ^{10}B are equal in the tissue. Neutron energies up to about 100 keV achieve good ratios

$^{10}B(n, \alpha)^{7}Li$ cross-section roughly equals the scattering cross-section with hydrogen, representative for all tissue including the healthy, in this energy range (OECD Nuclear Energy Agency (NEA) 2017). Figure 6.14 depicts these competing cross-sections for different neutron energies with the H cross-section being roughly representative for healthy tissue dose and the ^{10}B cross-section for tumour dose. The ratio of cross-sections is not the only factor, but the deposited energy of the nuclear reaction with its $Q = 2.79$ MeV exceeds the one deposited in elastic scattering (limited by the neutron kinetic energy in the keV range) by orders of magnitude up to the MeV range. Besides ^{10}B also ^{155}Gd and ^{157}Gd are currently under investigation for offering this advantage.

A particularly neat example of BNCT features a liver cancer with several metastases, which was treated by neutrons (Zimmermann 2006). Due to the long range of neutrons in matter, the irradiation dose would be homogeneously spread over the whole body, not focussing on the tumour tissue. Therefore, the liver was surgically removed from the patient, brought to a neutron source (fission reactor) for irradiation, and finally reimplanted into the patient. Removing the organ from the patient makes the treatment extremely selective. In order to further localise the radiation dose to the tumour cells within the liver, these metastases were doped with boron, an element with high neutron reaction cross-section. Neutrons see boron as a bull's eye leading to a strong localisation of the radiation dose to the tumour in this case. The liver can regenerate its mass, compensating both losses due to the tumour and the irradiation damage in healthy cells.

Quite recently accelerator based neutron sources opened a field of application of >10 MeV neutrons with sufficient flux density for therapy. The cyclotrons used for direct proton therapy hit a beryllium target close to the patient, which acts as a neutron source, see Sect. 4.2. Here the ^{10}B doping loses its advantage for selectivity (see Fig. 6.14) and the neutron works similar to a charged particle beam via collisional energy-transfer instead of the nuclear reaction mechanism of BNCT. This fast neutron therapy suffers from high costs and the difficulties of controlling the neutron beam which is not possible via classical beam optics. A potential advantage lies in the higher LET of neutrons. A higher LET reduces the effect of hypoxia specific radiation resistance mentioned above, a typical problem for photons, electrons, and protons with their lower LET.

6.2.2 Internal Metabolic Radionuclide Therapy

This section focusses on the so-called metabolic radiotherapy, a truly internal method since it exploits the internal metabolic system of body and malignant cells. Metabolic therapy offers a more general approach for cancer treatment by exploiting the physiology and the natural metabolic mechanisms of individual cells on a molecular level. In order to follow and target specific cell types, so-called *vectors* were invented. The vectors act as a crowd of radioactive-pizza delivery guys delivering small amounts of isotopes directly to the desired addresses. The term vector defines an entity which can trace the specific properties of a given cell type throughout the body, even if the cancer has spread in metastases. It exploits internal transport system such as blood vessels, just like the blood delivers oxygen to cells requiring it. All cells require numerous supplies and interact with hormones and other cells via receptors, each of them representing a potential entrance path of the vector to the cell, see Fig. 6.15. The biological technological challenge lies in identifying the options specific to the certain cancer cell type to be treated. Chemotherapy, as the classical cancer treatment, exploits the higher rate of cell division/mitosis of cancer cells compared to healthy cells. By blocking the mitosis it achieves a certain level of selectivity to cancer cells, but the strong side-effects of this therapy indicate its relatively weak selectivity. For ideal selectivity, no healthy cell in the body should feature the same receptor or at least have a substantially lower consumption/interaction rate with the vector than the malignant cells. In this case, a distribution of the vector through the blood system will naturally target the cancer selectively. The receptor selectivity represents a strength for the selectivity, but a weakness for the technology, since, in contrast to the external therapy, each cancer type requires the development of a specific vector.

Upon infusing the vector with a nuclear isotope it becomes a tool for nuclear imaging and therapy. The vector provides the transport module, while the radioactive isotopes contributes the treatment module. So far, three levels of vector complexity were applied: The direct use of an element used by the metabolism such as iodine

Fig. 6.15 Sketch of vectors with a receptor compatible with the specific cell. The vectors carry nuclear products, dock to the cell and transfer the nuclear product to the cell. Modified from svgrepo.com license CC-BY-4.0

from NaI (a salt) enriched with a radioisotope, ^{131}I in this case, and its selectivity for the consumer cells (thyroid cells). The human body makes use of Iodine only in the thyroid, with small amounts being present in the excretion tract, see Fig. 6.16. Thyroid cells store and consume the iodine for forming hormones. Consequently the radioactive iodine concentrates at this cell type and selectively irradiates the iodine storing cells via its decay. This first level is very thin, as basically only this one example exists, but it stands responsible for about 90% survival rate after this treatment (Eichhorn et al. 2004). The complete loss of the thyroid in the frame of this therapy is a downside, but can be compensated by artificial hormones.

The next level vectors comprise more complex metabolic molecules such as sugars, with the prominent example FDG marked by ^{18}F used in PET analysis. This level allows the use of a broader range of radioisotopes by enabling use of a few additional elements such as Fluor and their radioisotopes. The radioisotope's element has to be a building block of these metabolic molecule vectors, therefore a potentially metabolised molecule containing the radioisotope's element has to be found and synthesized using the radioisotope. This strongly limits the selection of isotopes, since not every element binds to biological molecules and the molecule has to interact in a relevant way with the metabolism. Furthermore, the synthesis of the vector has to coincide with the production of the radioisotope.

The, so far, highest complexity was achieved by carrier cells such as antibodies or viruses, which can be infused with any radioisotopes. The vector possesses a cargo space, which is chemically unspecific and hence able to carry any element. Nuclear treatment has the advantage of requiring only minute quantities of the nuclear isotopes and hence also of the vectors. Some vectors with good specificity but high toxic potential might be non-applicable for chemical treatment of cancer, but in the minute quantities required with nuclear treatment their chemical toxicity remains acceptable for the body. The radioactive isotopes act as a booster extending the range

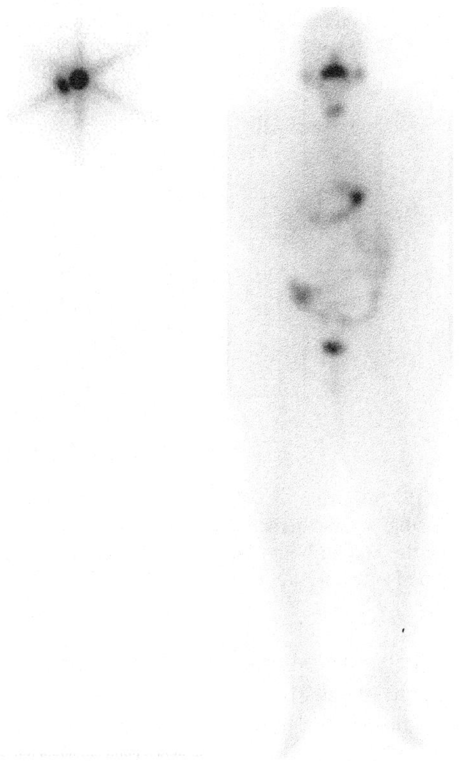

Fig. 6.16 Distribution of ^{131}I visualised by analysis of the characteristic γ-rays emitted upon decay. The left image shows the situation shortly after application. The right picture shows the situation after the third treatment about 6 month later. The picture uses a different scale and shows only little iodine uptake in the salivary gland, the bladder, and the gastrointestinal tract. The thyroid was completely disintegrated by the therapy. Original work by Drahreg01 on Wikipedia, license CC-BY-SA 3.0

of applicable vectors by weakening the vector toxicity normalised to the therapeutic effect.

The vector represents the first stage of selectivity and the carried nuclide the second step. Let us take a look at the example of radioactive iodine for thyroid cancer treatment. Figure 6.17 shows the radiations of ^{131}I, emitting photons in the range of $0 < E < 723$ keV and electrons in a similar range (together with electron-Bremsstrahlung). In 89.4% of the decays ^{131}I emits an electron via a 3-body decay with up to 606.3 keV. The electrons have a range up to about 2.2 mm in tissue (National Institute of Standards and Technology 2019). Since 606.3 keV lies below the electron stopping minimum, the highest ionisation effect lies at the end of the electron range. Consequently, the applied dose will not restrict to the iodine location, but a dose halo of, in the ^{131}I case, 2.2 mm thickness will form around it irradiating also healthy tissue.

In the energy-loss sections of Chap. 3 we learned about the low range of ions in matter, potentially reducing the dose halo. Electrons feature intermediate ranges. α-emitters with their typical 5 MeV discrete emission energy, such as the accelerator produced Po isotopes, achieve a range of about 40 μm in biological tissue, about the diameter of a medium sized cell or the dead layer of the skin. This range provides an optimal selectivity for the radiation application. Unfortunately, the half-lifes of

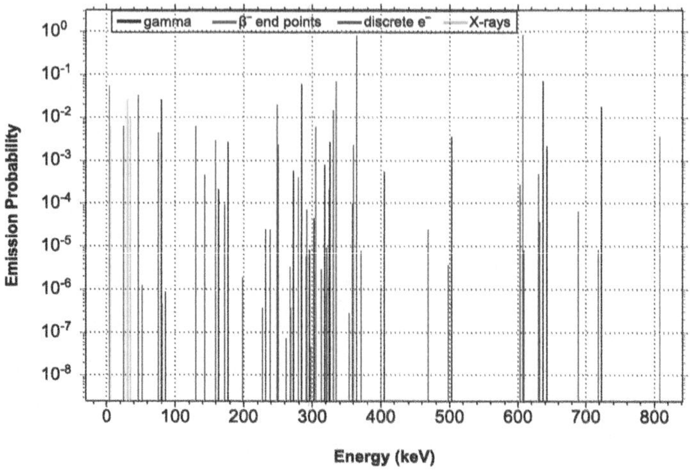

Fig. 6.17 Spectrum of ^{131}I decay radiations. The isotope strongly emits electrons/β^- radiation responsible for the therapeutic effect. The main γ-lines lie in between 300 and 700 keV and mostly escape the human body due to their high energy, leading to low dose levels for the patient and allowing for diagnostic imaging

Po and also the other accessible α-emitters in the uranium and thorium region lie in the range of mostly years and above, bad for fission power waste but also medical applications. The half-life of the radioisotopes plays another important role for the selectivity. At least a few hours are required in order to prepare the vector infusion and allow the vector to reach its target in the body. Long half-lifes, on the other hand, distribute the dose over too long time, limiting the treatments effectivity or requiring administration of extreme amounts of isotopes. Furthermore, besides the physical half-life, many elements also have a limited biological half-life by excretion and exhalation. ^{131}I with 8.02 days represents an optimal compromise.

^{131}I is actually a bad example for accelerator applications since it is a natural side product of nuclear fission. Interestingly, ^{131}I makes up one of the major risks of nuclear fission accidents and is the very reason iodine tablets are distributed for protection against its thyroid carcinogen effect after fission reactor accidents. In a way, ^{131}I is its own poison and antidote. The physical details of its decay demonstrate the typical problem of selectivity of internal radiotherapy and the corresponding development challenge for new radioisotopes. Even if the vector provides a perfect selectivity, the range of decay radiations worsens the selectivity. Figure 6.18 compares different emitter types regarding the localisation of their dose distribution. The 0.7 MeV electron roughly represents the endpoint energy of the ^{131}I case, demonstrating that even the lowest energy γ-rays emitted by ^{131}I barely contribute to its destructive effect compared to the 606.3 keV β^- decay. The β^- decay with its distribution of electron energies will further change the picture by increasing the low range dose.

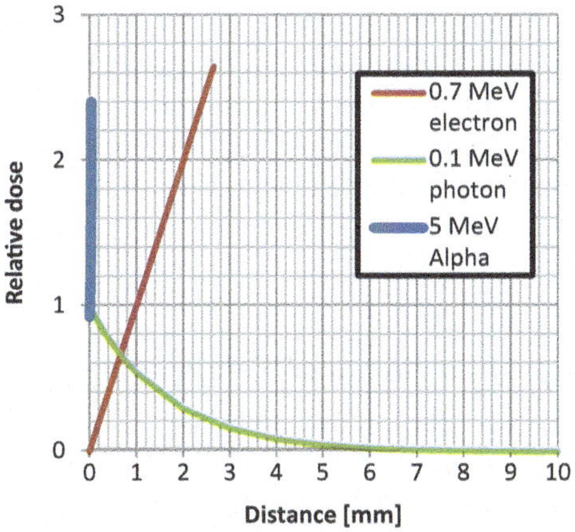

Fig. 6.18 Relative comparison of dose-distance distributions for a radioisotope point source located at $x = 0$ in striated muscular tissue (National Institute of Standards and Technology 2019). Photons feature the longest range with an exponential decay, electrons have a range on the mm scale with a peak dose at the end of range. α-particles reach only a few ten μm deep (36 μm in this case). The lower the range the more damage induced per length for a given particle energy

Figure 6.18 demonstrates a therapeutic radioisotope requires the right type and energy of emitted particles. Accelerators, or practically proton-beam irradiation, will generate different isotopes and sometimes elements compared to fission reactors, but these isotopes can in principle pertain similar properties compared to the "traditional" fission products. The generation of a selected isotope then requires also the right beam energy (and with this a specific accelerator) for sufficient cross-section without producing too much of other, unwanted isotopes via (p, xn) reactions. Figure 6.18 also clarifies why ^{18}F, a perfect isotope for imaging/diagnostics, has only little therapeutic effect. ^{18}F emits 1655 keV positrons and with this 511 keV photons. The positrons deposit energy similar to electrons, but their high energy prevents a selective dose application and furthermore limits the local dose density due to a distribution of the decay energy over a large volume. The dose required for tumor killing cannot be selectively applied to a controlled volume. Other isotopes such as ^{64}Cu with a β^- end point at 579 keV or the potentially interesting ^{182}Re with a zoo of photon and low energy electron emissions (typical for heavy elements) are required. Finding and producing these isotopes (Sect. 5.1) in connection with a suitable vector limits the technology of internal metabolic therapy. Every cancer type requires a repetition of this step due to the selectivity of the vectors, a clear disadvantage compared to the generalized treatment of external beams. On the other hand, the selectivity and transport mechanism of vectors offers a kind of auto-aim capability, killing even the smallest metastases not even visible to diagnostics.

Bringing a contained radiation source such as an X-ray tube or enclosed radioisotopes close to the tumour via surgery or natural openings (breathing, anus ...), the so-called brachytherapy, represents the last option combining aspects of both external and internal radiotherapy. It offers good solutions for certain types of cancer, but from a physical perspective it features the detection and placement problem of the external therapy discussed in Sect. 6.2.1 and the range selectivity problem of internal therapy. The full flexibility in isotope selection and the dramatically lower technical requirements/costs (no vector or large accelerator) are strong practical advantages. Since the main challenge lies in surgical placement it is usually not considered as part of nuclear medicine.

6.2.3 Selectivity from a Physical Perspective

The four methods (external photons or charged particles, metabolic vectors, brachytherapy) described in this chapter feature four completely different physical approaches and different particle types. X-ray therapy as a technically simple method allows for cost efficient treatment and avoids radiation protection issues, yet physics limits its selectivity due to the energy-loss mechanism of photons. Charged particles, and in particular protons, are more complex projectiles with a significant depth selectivity induced by the Bragg-Peak of energy-loss, allowing to efficiently reduce healthy tissue dose. This relatively new technology has not reached complete maturity and is still connected to the highest costs among all options. Both methods are applicable to any kind of cancer in any location inside the body. The fat patient problem or physically speaking the entrance "window" thickness of the body represents a practical restriction, though. On the downside, they require a localisation of the target tumour and the quantification of energy-loss properties of the surrounding tissue, strongly connecting their success to diagnostic results.

Internal therapy avoids these problems. The applied radioisotopes limit its selectivity. Here accelerators offer access to additional isotopes options not accessible for production by neutrons (from fission reactors or neutron sources). The major strength of the technology is also its major weakness: Every type of cancer requires developing a specific vector and these usually require specific isotopes chemically compatible with the vector.

Brachytherapy appears somewhat outdated in view of the other methods, but practical advantages should not be underestimated. Radiation therapy is in general expensive and requires advanced technology in hard- and software. Brachytherapy uses only little information on location, shape, and energy-loss properties of the tumour, yet in many cases it offers an optimal treatment. In the end, the complex nature of cancer requires considering all four options since a universal solution/nostrum was not yet found.

References

E. Bell, F. Grünwald, *Radiojodtherapie: bei benignen und malignen Schilddrüsenerkrankungen* (Springer, 1999). ISBN 978-3540659136

J.C. Buchsbaum, M.W. McDonald, P.A. Johnstone, T. Hoene, M. Mendonca, C.-W. Cheng, I.J. Das, K.P. McMullen, M.R. Wolanski, Range modulation in proton therapy planning: a simple method for mitigating effects of increased relative biological effectiveness at the end-of-range of clinical proton beams. Radiat. Oncol. **9** (2014). Article number: 2

S.R. Cherry, J.A. Sorenson, M.E. Phelps, *Physics in Nuclear Medicine* (Saunders, 2012)

R.E. Curtis, J.D. Boice, M. Stovall, L. Bernstein, R.S. Greenberg, J.T. Flannery, A.G. Schwartz, P. Weyer, R.N. Hoover, Risk of Leukemia after chemotherapy and radiation treatment for breast cancer. New Engl. J. Med. **326**, 1745–1751 (1992). https://doi.org/10.1056/NEJM199206253262605

W. Eichhorn, H. Tabler, R. Lippold, M. Lochmann, M. Schreckenberger, P. Bartenstein, Prognostic factors determining long-term survival in well-differentiated thyroid cancer: an analysis of four hundred eighty-four patients undergoing therapy and aftercare at the same institution. Thyroid **13**(10) (2004). https://doi.org/10.1089/105072503322511355.

A.H. Elgazzar, *Orthopedic Nuclear Medicine* (Springer, Berlin, Heidelberg, 2004) https://doi.org/10.1007/978-3-642-18790-2

A.B. González, M. Mahesh, K. Kim, M. Bhargavan, R. Lewis, F. Mettler, C. Land, Projected cancer risks from computed tomographic scans performed in the United States in 2007. Arch. Intern. Med. **169**(22), 2071–2077 (2009). https://doi.org/10.1001/archinternmed.2009.440

J.H. Hubbell, S.M. Seltzer, *NIST Standard Reference Database 126* (1996). Abgerufen am 2019 von https://www.nist.gov/pml/x-ray-mass-attenuation-coefficients

M.L. Iglesias, A. Schmidt, A.A. Ghuzlan, L. Lacroix, F.D. Vathaire, S. Chevillard, M. Schlumberger, Radiation exposure and thyroid cancer: a review. Arch. Endocrinol. Metab. **61**(2) (2017). https://doi.org/10.1590/2359-3997000000257 .

A. Koning, et al., *TENDL-2015: TALYS-Based Evaluated Nuclear Data Library*. Von (2015). Abgerufen https://tendl.web.psi.ch/tendl_2015/tendl2015.html

P. Miller, N. Long, R. Vilar, A.D. Gee, Synthesis of 11C, 18F, 15O, and 13N radiolabels for positron emission tomography. Angew. Chem., 8998–9033 (2008). https://doi.org/10.1002/anie.200800222

National Institute of Standards and Technology, in *NIST: Introduction of e-star, p-star, and a-star*. Von (2019). Abgerufen https://physics.nist.gov/PhysRefData/Star/Text/intro.html

OECD Nuclear Energy Agency (NEA), in *JANIS*. Von (2017). Abgerufen https://www.oecd-nea.org/janis/

A. Rahmima, H. Zaidib, PET versus SPECT: strengths, limitations and challenges. Nucl. Med. Commun. **29**, 193–207 (2008)

X. Ren, E. Wang, A.D. Skitnevskaya, A.B. Trofimov, K. Gokhberg, A. Dorn, Experimental evidence for ultrafast intermolecular relaxation processes in hydrated biomolecules. Nat. Phys. **14**, 1062–1066 (2018). https://doi.org/10.1038/s41567-018-0214-9

M. Rickhey, O. Koelbl, C. Eilles, L. Bogner, A biologically adapted dose-escalation approach demonstrated for 18F-FET-PET in brain tumors. Strahlenther. Onkol. **184**, 536–542 (2008). https://doi.org/10.1007/s00066-008-1883-6.

P. Vaupel, L. Harrison, Tumor hypoxia: causative factors, compensatory mechanisms, and cellular response. Oncologist **9**(Suppl 5), 4–9 (2004)

T. Weitkamp, A. Diaz, C. David, F. Pfeiffer, M. Stampanoni, P. Cloetens, E. Ziegler, X-ray phase imaging with a grating interferometer. Opt. Exp. **13**(16), 6296–6304 (2005). https://doi.org/10.1364/OPEX.13.006296

D.R. White, J. Booz, R.V. Griffith, J.J. Spokas, I.J., Wilson, Report 44, tissue substitutes in radiation dosimetry and measurement. J. Int. Comm. Radiat. Units Meas. **23**(1) (1989). https://doi.org/10.1093/jicru/os23.1.Report44

H.Q. Woodard, DR White, The composition of body tissues. Br. J. Radiol. **59**(708), 1209–1218 (1986). https://doi.org/10.1259/0007-1285-59-708-1209

R. Zimmermann, *Nuclear Medicine—Radioactivity for Diagnosis and Therapy* (EDP Science, 2006)

Chapter 7
Material Analysis and Testing

Abstract Material analysis and testing add information and systematics to development processes. This boosts the technological development speed and aids in generating new ideas and solutions. As such, science and industry try improving their accuracy, resolution, and contrast mechanisms to enable extracting more knowledge from processes and samples. Accelerator based methods exploit photons, electrons, neutrons, and ions in probably >100 individual methods. All of them exploit similar physics and devices discussed earlier in this book. This chapter discusses about 30 methods, highlight their information properties to enable the reader understand and select a complementary set of methods for a given analysis problem.

Progress is only possible through death and retirement (H. W. Möller)

This chapter concentrates on applying accelerators to gain deeper insight into materials and components in order to proof this saying of my father wrong. These samples can originate from industrial context for quality assurance, from scientific developments, or whatever can be imagined. The electron microscope represents a straightforward example of the similarity of these two worlds. Originally developed in a university its importance and value soon became apparent in fundamental research. The extreme magnifications possible due to the short wavelength of electrons compared to visible light opened up the nano-scale world. Not only physics became interested in the technique to study phenomena in solid state or material physics, but also biology and other disciplines applied it. Finally, when the devices became more mature and cheaper they found their way into quality assurance and development laboratories of metal, electronics, and many other industries.

The three main tasks of material analysis are the *identification* of chemical and isotopic constituents, their *quantification*, and their *localisation* in spatial and structural respect. These three analysis qualities translate to a decision diagram when seeking for a suitable analysis for an application. Figure 7.1 aids in identifying the actual requirements when thinking about an analytical method. Not every problem requires localising, identifying, and quantifying all elements with isotopic resolution on a nm scale throughout m^3 volumes with ppm accuracy. Strength usually leads to drawbacks in other aspects. For example, nm resolution will hardly be possible in a m^3 sized bulk sample, but rather on a mm sized surface layer sample. The strength

S. Möller, *Accelerator Technology*, Particle Acceleration and Detection,
https://doi.org/10.1007/978-3-030-62308-1_7

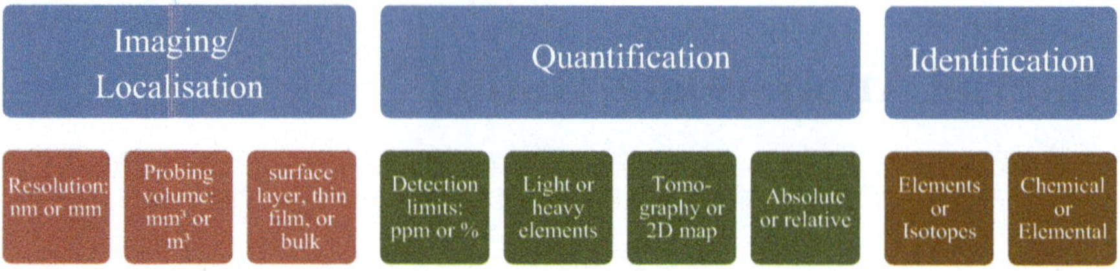

Fig. 7.1 Decision diagram for comparison of photon, electron, ion, and neutron beams in analytics. All methods have their specific strength and weaknesses given by the "or". No single method can satisfy all optima of each of the three categories

of each method will be given in the following sections. Combining several methods cancels weaknesses of the individual methods at the expense of additional time and costs.

An example situation: Synthesis of materials by physical (e.g. sputtering) or chemical (e.g. solid-state reactions) means is a delicate business. The input materials or sputtering targets can be easily characterised and their composition is usually known with sub-percent accuracy, but the final product not necessarily follows this composition. Besides these raw materials, the injection of reactive process gases such as air may be required for certain products. Vessel materials tend to mix with the reagents. Several reaction pathways or steps will connect the initial raw materials with the final product. In the case of sputter-deposition, these are the sputtering process, a gas phase transfer, the sticking onto the substrate and finally a solid state reaction on the substrate. In all steps non-ideal processes take place and the actual skill of producing the targeted material lies in the smart optimisation of the input parameters. For example synthesizing a NCM lithium cathode material with this technique involves mixing Li, Co, Mn, Ni, and O. A set of different stable compounds exist with the given ingredients and even the targeted material has, by intention of charging and discharging it, the possibility for varying lithium stoichiometry or content. A single compound is of interest, the possible outcomes have no use for a battery. Careful optimisation of reaction temperature and time, gas composition, and non-stoichiometric mixture of the ingredients (e.g. a bit of excess lithium) will lead to a good product. This way of optimisation, or calibration of the production method, required for obtaining an optimal and reproducible material cannot be taken blind. Tomographic information on the elements distributed over the some 100 μm thick cell with sub-% absolute accuracy is required. Relevant methods will be discussed in the following sections, but a full solution to the requirement is actually not available, so far. In conclusion, analytics play an important role in scientific research and industrial production by being the eye of development.

After production, the role of analysis continues in the use of products. Over the lifetime of a product deterioration and fatigue take place. These abstract terms connect directly to physical and chemical changes of the product. This could be oxidation of an electric contact, the formation of inactive phases inside a battery or lattice-defects and micro-fractures in a reactor vessel. A sensitive detection of these initially minute changes and extrapolation of the trends represents an economical means for shortening testing times and understanding failure mechanisms for improving component lifetime in an economical fashion. Precise analytics allow identifying problems and limits before they become practically relevant. The more precise and complete the analytical methods reveal these changes, the better and the more economical the final product will become and the less probable unforeseen situations and regress will occur.

Today's analytical methodical park includes methods for detecting single atoms (usually nuclear methods as in the medical diagnostics) up to macroscopic structures and quantities of regions deep in a material bulk. For the quantification counting statistics, signal noise, and background represent the main limitations, while for the spatial resolution of the localisation the wavelength of the probing particle constitutes a natural limit. As we know from quantum physics everything in nature, and in particular sub-atomic particles, can be considered as either a wave or a particle. The De-Broglie wavelength λ of a particle beam of momentum p is given with Planck's constant h by

$$\lambda_{\text{particle}} = \frac{h}{p} \approx \frac{h}{mv} \tag{7.1}$$

The wavelength describes the so-called diffraction resolution limit, with visible light wavelengths in the range of 380–700 nm limiting the possible resolution of optical analytics to rather large structures. Physicists definitely found tricks to circumvent these optical limits, but these tricks lack in universal applicability, see for example the 2014 Nobel prize on super-resolved fluorescence microscopy. To achieve better resolution the electron microscope was invented, leading to a reduced wavelength/diffraction limit of e.g. 12.2 pm for a 10 keV electron beam. Even shorter wavelength can be reached with ions and neutrons at high energies which is the physical basis of probing quarks and their substructure using particle physics accelerators such as LHC at CERN.

This extension of physical limits beyond optical and chemical analytics represents the fundamental strength of accelerator based analytical technologies. Practical solutions or devices, respectively, hardly reach these ideal limits, but it constitutes an ongoing challenge. The extent to which these technical challenges were solved defines the quality of an instrument. We will discuss these technical solutions and the advantages of different methods for different applications. Some challenges are more difficult to solve for certain probing particles and hence for each analysis task a different particle type might be advantageous. Commonly specific tasks require specific optimisations, leading to a diverse situation of similar, but in detail different analytical devices for a single method.

Some technological challenges reoccur throughout single methods or even groups of methods, which is the very reason Chap. 2 belongs into this book. The detection of the products and secondary particles emitted in the beam-matter interactions exploited for the analytical method represent one of these challenges where technological limits of integral technical components limit the analytical technology.

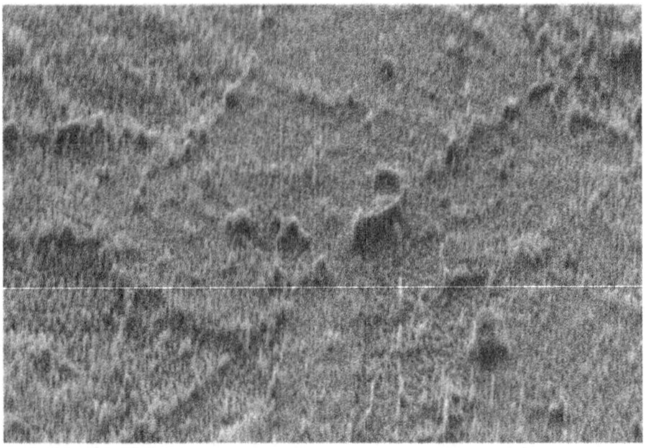

Fig. 7.2 3 kV scanning electron microscopy image of a polished stainless steel surface after 50 min of hydrogen plasma exposure. The imaging reveals the development of a nano-structure with precipitates and grain boundary rifts. Courtesy of Marcin Rasinski

The detection of electrons and ions with silicon-based electronic detectors reaches an efficiency of 100% as soon as the sensitive/active detector thickness reaches the stopping range of typically 100 μm to a few mm, see Fig. 2.43. For photons and neutrons, detection efficiencies quickly drop for energies above a few keV. This reduces the detection limits and requires frequent detector calibration for an absolute quantification of the results.

The calibration transforms a qualitative to a quantitative measurement. For ion based analysis this step usually drops out due to the usually 100% detection efficiency, leading to a so-called calibration-free method. Imagine an electron microscopy image showing for example a metal surface as the one in Fig. 7.2. Imaging represents a qualitative analysis, meaning the image shows us something structured. It is impossible to derive dimensions, particle thickness, composition of the structures or for example the form of cracks from the image, although this information is encoded in the image. The information remains hidden. Quantification of such an image requires many steps from the physical connection between a certain difference to the measured signal, the incoming flux and fluence, over the analysis geometry and optics up to the detectors and statistical image analysis. Although acquisition of quantified data requires significantly higher effort, the quantified results provide more information often crucial to find trends in datasets or exclude false conclusions due to too many assumptions on what should be seen in the image compared to what the image actually shows.

An analysis apparatus needs certain equipment and design aspects in order to allow for a calibration or even a calibration-free analysis, but also the understanding of the underlying physics requires a certain maturity in the form of theoretical understanding. In the end, even the best calibration-free method relies on a calibration/measurement of angles, sizes, and distances of the actual apparatus with all the tolerances and alignment involved.

Table 7.1 Cost comparison for analysing samples for their elemental composition. Costs were estimated with 2018 prices without claiming knowing the absolute truth or certain device prices

Method	Accuracy (%)	Detection limit (ppm)	Resolution (nm)	Upkeep (k€/year)	Invest (k€)	Cost/sample
XRF	1	10	10^6	115	300	30
EDX	10	1000	10	120	300	45
PIXE	1	1	1000	150	1500	75

Investments assumed to depreciate over 20 years. 3000 (EDX, PIXE) or 4000 (XRF) samples per year assuming 200 days with 8 working hours with one expert working full time on the device. Upkeep includes room, power, maintenance, and 1 expert (100 k€/year labour costs)

Technology is always advancing in our modern society (which is definitely not self-evident). As a result, new devices come onto the market tackling the challenges of the different methods and extending the range of the technically feasible. This should encourage the reader to consider new technologies and combinations of these, if existing components are insufficient. This term technical realisation describes the point of a certain device/instrument/apparatus in this multi-dimensional space of methodical and technological challenges. German language has very precise and convenient words for describing technical devices and would use the term "apparative Umsetzung". In English language the term intends to describe how a machine implements a certain technical challenge within the technological limits. The apparatus represents one of the main building blocks of any analytics. Having a good idea is another building block, but it must be possible to implement it technically. Furthermore, having a first implementation, a prototype, often leads to the insight of how to improve the apparatus power. The following generations of this device will then implement better resolution, economics, precision, and so on, gradually approaching the technological limits. This prototyping approach is typical and very practical for complex problems and hence found its way not only into analytics, but also into many organisational and business processes such as agile management.

Randomly picking out an example allows for a calculation of analysis costs for three different methods available for laboratory applications. The task: Analyse the elemental composition of a material. The methods: Photons using the x-ray fluorescence spectroscopy method (XRF), electrons using the energy dispersive x-ray analysis (EDX), or protons using Particle induced x-ray emission analysis (PIXE). These methods are all based on the emission of characteristic x-rays but exploit different projectile types. Table 7.1 compares a few important aspects. Personnel costs dominate the upkeep for all devices. The large device required for PIXE consumes more power and space, leading to higher costs compared to the table sized devices for XRF and EDX. High investment costs of accelerator based analytical tools lead to a strong scaling with degree of capacity utilisation. The assumption of full utilisation leads to comparable costs for all three methods, but with lower utilisation, in particular PIXE specific costs will increase rapidly due to the higher investment costs. On the positive side, PIXE offers the best detection properties.

This chapter lists numerous examples of accelerator based methods, pushed as far as it can go without loss of generality. Numerous other methods exist not even mentioned here, but listing all methods would go beyond the scope at least of this edition. For all this, there's only one thing you should know: In the end it doesn't even matter how many methods we list. It is important to understand the basics, then we get them all. Look down the list and what's there marks examples of what ought to be. (freely from: Linkin Park—In the end).

7.1 Ion-, Electron-, and Photon Beam Analysis

- You will never become one of the best without accepting your feedback, even the worst of it

Just like for your personal development, technical developments require feedback. Imagine for example the task of synthesizing a glass, but it comes out dull. What to do? Trying again until is an option, but if it's not working after the third approach you scientifically qualify as stupid. Analysis methods will provide an insight into the reasons and the points for optimisation of the production process. The analysis using accelerated ion- and electron-beams and secondary photons produced from accelerators features several distinct qualities for material analysis. Numerous competitive methods based on light, chemical, or physical mechanism exist. Usually these are even less expensive, but accelerator based methods offer the strong combination of quantified accuracy, spatial resolution, and atomic or nuclear sensitivity.

Figure 7.3 depicts the geometry of the interaction underlying an analysis. Geometry represents the main aspect and also challenge of the corresponding analysis devices. In the following, the angle between the incident beam and the normal of the target surface will be depicted by α. The reaction angle θ depicts the angle between the detector and the beam direction. By this definition, backscattering depicts particles

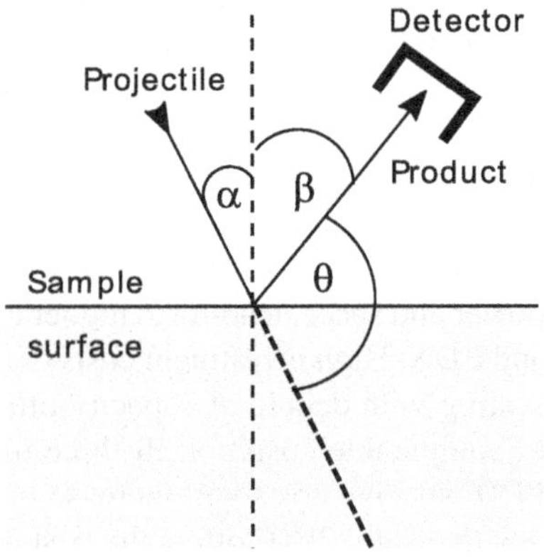

Fig. 7.3 The fundamental geometry of surface analysis methods

scattered towards the half-space of the incident beam ($\theta > 90°$), while forward scattering depicts the particles scattered in beam direction ($\theta < 90°$). Due to the limited range of charged particles in matter, described in Sect. 3, forward scattering restricts to targets of μm thicknesses or gracing incidence geometries ($45° < \alpha < 90°$). Since the beam carries a forward momentum its direction has a special meaning for the reaction geometry in particular via the conservation of momentum before and after the reaction. The angle θ represents a circle around the incident beam. The detector position on this circle is arbitrary with respect to the reaction, but in the case of $\alpha \neq 0°$ the path length inside the sample of the particles exiting the sample depends on the angle between sample surface normal and detector β. The angle β becomes relevant if stopping and absorption are considered by the fact of significantly larger stopping and absorption in solids.

$$\rho = \frac{I}{D * \sigma} \tag{7.2}$$

The quantification of a target material per investigated surface area ρ using massive particles (not photons) from an infinitely thin target slice with the amount of incident projectiles on the investigated surface D, the amount of the received events I of an interaction given by a cross-section σ. Equation (7.2) states a proportionality between concentration of the sought material and the detector signal I. Together with counting statistics (Sect. 7.1.1) this represents a technological limitation for detecting minute quantities, a detection limit. D represents the integrated beam current received by the sample. The more beam current and the more time spent for each measurement, the smaller the concentration ρ the method can detect. Thinking of a matrix of measurement pixels acquired for a 2D sample mapping such as Fig. 7.2 this directly defines the amount of points we can measure on a given sample area in a given analysis time. Therefore (7.2) also connects lateral resolution, beam current, detector properties, detection limits and accuracy.

In this section we discuss several methods based on photon (the types originating from accelerators), electron, and ions beams. These three types have their distinct properties regarding technological availability, such as the outstanding quality of electron sources (Sect. 2.4), their continuous matter interaction (stopping or loss of quantity), and their nuclear properties regarding nuclear and electronic interactions. Furthermore, the particle energy distinguishes the analytical methods with a particular difference between energies below and above about 1 MeV. Below this energy mostly only interactions with the electrons of the sample take place, while for higher energies the relevance of the electrons reduces while the importance of nuclear interactions increases. These six general accelerator based analysis regimes translate into the methods discussed here. Due to technical feasibility, the high energy photon and electron regimes are hardly commercially available, while the high energy ion analysis is just on the edge of becoming commercially available. As this book focusses on established technology the following sections will restrict to the commercially available subset.

7.1.1 Physical Concepts, Detection Limits, Accuracy and Uncertainties

- Results *wouldn't be nothing, nothing without an error or uncertainty (It's a results world)*

Material analysis by accelerators relies on the interaction of photon, electron, ion, or neutron beams with matter (see Chap. 3). The only other degree of freedom is the particle energy (which is limited by technical and economic constraints). These statements make the physical world of the analysis smaller and easier to understand. The analytical information itself relies on the change of the properties of these projectile beams and/or the emission of secondary particles originating from interactions with the target/sample and their subsequent detection.

These interactions take place on different physical levels, depending on projectile type and energy. Three distinct levels describe these interactions with higher energies probing smaller structures. On the low energy range up to some keV, chemical bindings are probed. Increasing energy opens access to the electronic structure of atoms connected to the elements, but also structures such as crystalline order become visible. In the MeV range, the Coulomb barriers begin to fall and nuclear information becomes accessible, adding isotopic sensitivity. Usually technical limits of energy resolution blur the information of lower energy levels to invisibility for higher energy analysis beams. Consequently we have to decide which type of information (elemental, isotopic, chemical) the analysis needs to yield in order to decide on its details. Neutrons differ from the other projectiles due to their extremely weak electro-magnetic interaction with targets, making them more universal but also the technically most challenging projectiles.

Not all data interpretation is as easy as imaging. Even imaging often contains layers of information, think of a map displaying not only land mass outlines but also altitude, ground type, or borders. With some physical understanding we could even derive the regions of intense rainfall or the movement of continental plates via their connection to mountains and altitude variations. Many of the methods discussed in this chapter are indirect methods. Unlike for example a voltage measurement with a multimeter, the sought after value is not directly measured, but only its impact on a detected secondary particle spectrum. This type of measurement requires a model for data analysis, see 3.5. Models such as stopping power and kinematics allow for a more or less unambiguous evaluation of the measured spectra, extracting the analysis result from the raw data. For this interpretation a model can employ a forward or a backward calculation ansatz. A forward ansatz varies the input parameters (e.g. the sample structure) of a given model until the resulting spectra match the measured ones. A backward ansatz applies mathematical inversion of given analytical models. Material analysis requires those physical and mathematical models for understanding the processes responsible for the obtained data. As such the model also introduces additional systematic uncertainties by the limited amount of cases and physics considered for constructing the model. A simple example: The face

Fig. 7.4 The face-swap app as a model for the limited understanding of an algorithm. The limited knowledge and situations implemented in the algorithm let it identify the authors face and a drawing of Kofi Annans face, although the latter is just a good painting not a real face (the model yields a false positive)

recognition software of a face-swap app tries to find human faces in a picture via the typical eye-nose-mouth combination. Figure 7.4 shows an example where this model leads to a misinterpretation of data since not only the human face features this structure. Data analysis by models potentially induces such an ambiguity into results, therefore stay sceptical about modelling outcomes.

Detectors, see Sect. 2.5, form an important technical aspect in this distinction as any result can only be as good as the data used for it. The detector dynamic range, the ratio of minimum to maximum detected signal level, typically lies in the order of some thousands. The larger the dynamic range, the more information and complex interpretations become available. In the digital world this originates from typically 8 ($2^8 = 256$ levels) to 16 bits ($2^{16} = 65{,}536$ levels) of signal sampling. Even if the dynamic range is not limited by digital sampling accuracy, it remains difficult to reach relative accuracies below 10^{-5}. In fact many of the methods presented here feature about $1\% = 10^{-2}$ of dynamic range. Most physical measurements suffer from the same technological limits leading to a stacking of problems. This starts at fundamental constants or cross-sections required for the data analysis, but hardly known with better accuracy than the applying method itself (since it is the same they are determined with).

Technically a shift of sensitivity enables circumventing the limitations of dynamic range. In photography modern cameras do this automatically by adjusting exposure time, aperture, and amplification (the ISO value). All analysis methods have mostly identical means for manoeuvring between large and small signal intensities. Unfortunately, this also has limits, resulting in the detection limit of the method, the smallest detectable signal. The elemental composition analysis described in Sect. 7.1.5 for example typically allows for detecting parts-per-million (ppm $= 10^{-6}$) of a single element in a sample. Background effects but also simple statistical limits of count rates induce the detection limit.

Due to these difficulties, physicists and mathematicians developed mathematical formalisms for dealing with errors/uncertainties. Both terms, error and uncertainty, are often used as synonyms. The word "error" sounds like a problem of the data, therefore today the more precise term "uncertainties" is common as it describes a natural part of the measurement world. In this formalism uncertainties are categorized into two groups: Systematic and statistical (or random). This distinction is very important for comparing data to each other. Data acquired on the same device may all show the same systematic deviation from the same data acquired on a different device. Device A always delivers "systematically" say 20% higher values than device B, but they could still show the identical trend of their data. When measuring a certain value under the same conditions repetitively the systematic uncertainty will not appear in between the data points. Here the statistical uncertainty remains. This difference relates to the absolute uncertainty to measure the true value and the relative uncertainty between two measurements following a certain trend. Statistically means the results will distribute around the true expected value of the measurement. An example: The dice throw counting statistic experiment. Every number on the dice should appear equally often, but after say 6 throws, 2 numbers probably didn't even appear once. The more often you throw the clearer the measurement indicates the equal probability of each number. Statistical uncertainties shrink with every repeated measurement and mathematically disappear in the limit of infinite measurements. Systematic uncertainties exist independent of the data statistics (loaded dices remain loaded no matter how often you throw them).

In any case, the uncertainty represents a confidence interval in the sense that statistically not all results must lie within the specified uncertainty corridor. For example specifying a 3σ uncertainty says 99.73% of the data points lie within the uncertainty corridor. The remaining data points may be well outside this range without disproving the result. Practically several uncertainties occur simultaneously, such as counting uncertainties, device drifts, input data uncertainty and so on. Summation rules allow for calculating the resulting total uncertainty. The result strongly depends on the connection between uncertainties. Independent uncertainties originate from independent sources, for example the counting errors of two detectors. Dependent uncertainties are related to common effects, for example temperature drifts of two devices next to each other. The two devices measure the impact of the temperature drift independently, therefore a correlation value has to be subtracted from the sum of uncertainties in this case. Dependent uncertainties are practically often negligible so they will not be discussed here. For a quantity A derived from several independently measured values x_i the error propagation rule for independent x_i reads:

$$\Delta A = \sqrt{\sum_i \left(\frac{\mathrm{d}A}{\mathrm{d}x_i}\Delta x_i\right)^2} = \sqrt{\left(\frac{\mathrm{d}A}{\mathrm{d}x_1}\Delta x_1\right)^2 + \left(\frac{\mathrm{d}A}{\mathrm{d}x_2}\Delta x_2\right)^2 + \left(\frac{\mathrm{d}A}{\mathrm{d}x_3}\Delta A\right)^2 + \cdots} \tag{7.3}$$

Calculating ΔA requires the equation deriving A from the x_i. The derivatives of this equation enable calculating the individual errors via the change (=first derivative) of A towards the variations in x_i. Multiplying this sensitivity towards x_i with the uncertainty range Δx_i yields the contribution to ΔA. $\Delta x_i/x_i$ has to be small or higher order derivatives are required for an exact solution, but if this is violated the measurement has anyways more severe problems than correct uncertainty propagation. We evaluate this at the example of (7.2) with $A \equiv \rho$ where we assume I, D and σ to be uncertain.

$$\Delta\rho = \sqrt{\left(\frac{\mathrm{d}\rho}{\mathrm{d}I}\Delta \mathrm{I}\right)^2 + \left(\frac{\mathrm{d}\rho}{\mathrm{d}D}\Delta D\right)^2 + \left(\frac{\mathrm{d}\rho}{\mathrm{d}\sigma}\Delta\sigma\right)^2}$$
$$= \sqrt{\left(\frac{e}{D\sigma}\Delta \mathrm{I}\right)^2 + \left(\frac{-eI}{D^2\sigma}\Delta D\right)^2 + \left(\frac{-eI}{D\sigma^2}\Delta\sigma\right)^2} \tag{7.4}$$

Equation (7.4) demonstrates the calculation of total uncertainty of ρ. The reader may insert some exemplary numbers to check how the relative uncertainty $\Delta\rho/\rho$ depends on the uncertainties of the other values. Figure 7.5 provides an example of two different experimental counting situations of ΔI as an inspiration. Counting statistical uncertainties are the prime example of uncertainties in analytical methods

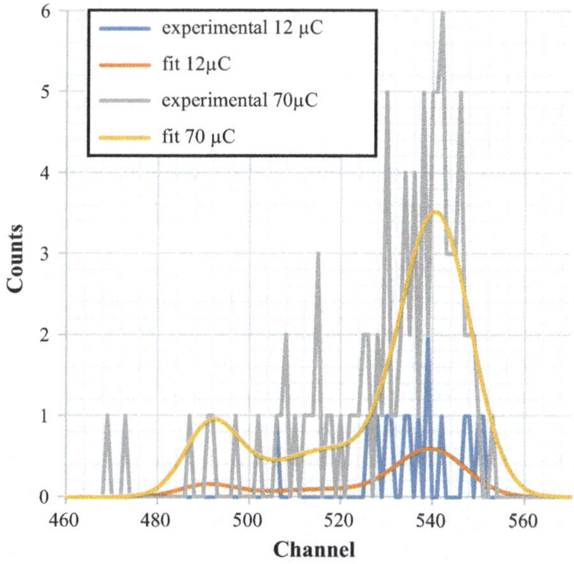

Fig. 7.5 NRA measurement with small and large integral signal level of the $^{10}B(\alpha, p_0)^{13}C$ on the same NCM811 battery material sample after 12 and 70 μC of α-ion dose. The number of counts increases proportionally. Evaluation of the boron content becomes increasingly accurate with more counts. In the 12 μC case a part of the interpretation becomes questionable due to detection limit with 1 ± 1 counts while the 70 μC case offers a large signal-to-noise ratio around channel 540

as many of them rely on counting particles. Derived from the Poisson distribution, (7.5) describes the uncertainty of the detected events/counts ΔI as its square-root.

$$\Delta I = \sqrt{I} \tag{7.5}$$

The largest partial uncertainty dominates the total uncertainty of ρ, while three identical uncertainties result in a square-root increase. In this special example ρ depends linearly on all three input quantities, but in cases where powers of higher order or exponential relations connect the derived with the measured quantities, some quantities affect the result stronger than others and small uncertainties in a quantity with high scaling result in large uncertainties of the derived value. Complex models potentially even feature several incompatible outcomes within larger uncertainties, if the dependence is not strictly monotonically (e.g. sine function).

Understanding errors allows improving an analysis technique on the technical side by identifying the critical values yielding the highest benefit for the cost of reducing their uncertainties. Finally, uncertainties have a certain notation standard

$$\begin{aligned} A &= 5.130^{+0}_{-0.15} \\ B &= (1.23 \pm 0.1) * 10^8;\ B = 1.23(10) * 10^8 \\ C &= 1050 \pm 65(\text{sys}) \pm 123(\text{stat}) \end{aligned} \tag{7.6}$$

In the example (7.6) we have three uncertain quantities *A*, *B*, and *C*. *A* has a value of 5.130 without a positive uncertainty (it's an upper limit), but only a negative uncertainty of 0.15, a notation often seen in technical drawings where this type of uncertainty is called tolerance ensuring e.g. a pin fits into a hole. Quantity *B* has a symmetric uncertainty of $0.1 * 10^8$. Both notations of *B* have the same meaning with the bracket version giving the uncertainty of the last digits of the preceding number. Quantity C features two separable symmetric errors, the systematic error of 65 and the statistical error of 123.

Besides the accuracy on the single data-point level many applications require mapping information in 2D or 3D. Accelerator based methods implement the lateral resolution via the beam spot size. A smaller spot size leads to smaller/better lateral resolution, but in general also reduces the beam current (=signal intensity). Beam optics and special particle sources reduce the spot sizes with typical demagnifications in the order of 100. The limits of spot size are mostly of technical nature with currently achievable spot sizes down to about 1nm. Typically the methods are prefixed with a μ- or nano- in order to indicate a small spot size (e.g. SIMS vs. nano-SIMS) version of the regular method. Despite the spot size, these methods show no difference to the un-prefixed versions. Smaller spot sizes are not always desirable, not because it's good to hide information (its not!), but it depends on the sought information. Smaller spots are prone to irregularities such as dust or remainders of mechanical polishing on the sample while larger spots average out these aspects resulting in reduced uncertainties. Both types, local and mean result, contain different kinds of information. Finally yet importantly, the selection of spot size, mapping area, and point density define the amount of required measurement time and processing resources.

The holy grail of analytics, a full tomographic (3D) mapping requires additional depth information in every measured point of the 2D map. Essentially four ways for depth resolved information exist. The enery-loss of charged particles equals a connection between projectile energy, product energy and depth (S in units of $\mathrm{d}E/\mathrm{d}x$). Consequently, energy resolved detectors provide depth information (see Sect. 3.4), but straggling and energy resolution effects limit the depth resolution. This type of depth resolution represents the strong point of ions due to their lower straggling compared to electrons. The second option exploits sample rotation in transmission mode, similar to the medical computer tomography. This mode requires beam transmission through the sample, generally requiring the range inherent to photons. A more general option exploits the variation of range and reaction cross-sections with beam energy. The change ($\mathrm{d}/\mathrm{d}E$) of these parameters with energy defines the resulting depth resolution. Increasing projectile energy increases the range. Adding up results at several energies allows for defolding the results according to the depth dependent sensitivity at each energy. Lastly, the sample can also be mechanically sectioned and analysed in a cross-cut geometry. In addition to classical mechanical methods, sputtering based methods enable a removal of surface layers, revealing deeper layers with the progress of removal, see Sect. 7.1.4. Sputtering can also remove slices on the nm scale as described for example in Sect. 7.1.3. Inhomogeneous sputtering and the surface modifications induced by the sputtering introduce uncertainties and roughness to the surface.

Roughness generally challenges material analysis methods, in particular the ones on the micro- and nanoscale. Technical surfaces are rough. This has several implications in particular for inhomogeneous media and layered structure analysis as depicted in Fig. 7.6. An analysis beam larger than the characteristic lateral roughness scale will yield a different result than a beam smaller than the lateral roughness.

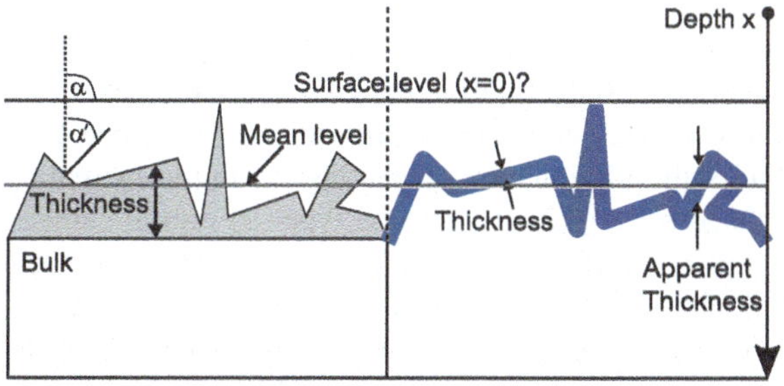

Fig. 7.6 Rough surface can mean a lot of things. The left part shows an example of a rough layer on a smooth surface while on the right part the layer is smooth and the surface rough. Correspondingly impact and exit angle and the definition of depth can change. The mean level indicates the surface as assumed for R_a. It has to be clear what the term surface and depth mean. Is the rough surface the flat projection area or the true local inclined surface? How does a large and a small beam-spot change the result?

Angular variations of the local impact angle from the macroscopic impact angle alter the sample description as seen along the projectile path to what we would expect from a flat surface, in particular with grazing incidence methods. Exact surface profiles determined via profilometers or optical methods are often impractical as input for result interpretation. Several technical parameters exist to describe roughness in a reduced way based on statistical distributions. The value R_a, states the average deviation of the local surface from a mean level. Milling, drilling, and turning typically results in R_a in the order of μm, while fine grinding and polising reach values down to $R_a \approx 10$ nm. The smaller R_a, the larger the range and the spatial resolution, the less impact the roughness will have. Consequently, microscopic analysis often requires mirror finished samples for scientific result quality.

7.1.2 X-ray Absorption and Scattering Analysis

X-rays in the range of keV to about 100 keV offer a large variety of analytical methods categorized in three methodical groups: Imaging, spectroscopy, and scattering. Among the numerous methods in these categories this book can only discuss a few selected examples. The x-ray production for these applications to the largest extent relies on the bound electron type sources, namely of x-ray tubes (Sect. 4.3.1). Above a certain voltage threshold (in Germany 30 kV) many countries require special radiation protection measures and/or trained personnel, due to the increasing penetration depth of the x-rays resulting in relevant outside doses. Devices with lower voltages and appropriate shielding can usually operate without these costly additions and therefore provide the working horses of analytical x-ray applications. A few applications make use of free electron sources due to their higher brightness and wavelength variability opening up additional experimental optimisation dimensions, e.g. a kind of x-ray "colour", and improved result quality to the analysis. For mobile and handheld applications also radioisotope x-ray and γ-ray source are applied. The analytical methods physics are independent of the x-ray source, but the bound electron source variants have to deal with reduced signal intensity, broad x-ray spectra, and worse spatial resolution compared to the free electron source variants.

Imaging exploits the photons for projecting a sample in real space onto a 2D detector. These detectors are usually special types of digital camera chips, but also could be imaging plates, arrays of individual silicon x-ray detectors, or scintillators can be applied. Analytical imaging probes on all scales from macro- to micro- to nano-scale. For imaging x-rays offer a different contrast than visible light imaging due to the different absorption mechanisms in matter. More importantly, x-rays penetrate material enabling a 2D see-through and 3D tomographic imaging impossible with visible light. The physical difficulties of x-ray optics prevents using classical imaging techniques for most applications. Instead, imaging offers a purely geometric option for magnifying objects as depicted in Fig. 7.7. The spatial resolution using this technique mostly depends on the detector pixel pitch and the x-ray source dimension.

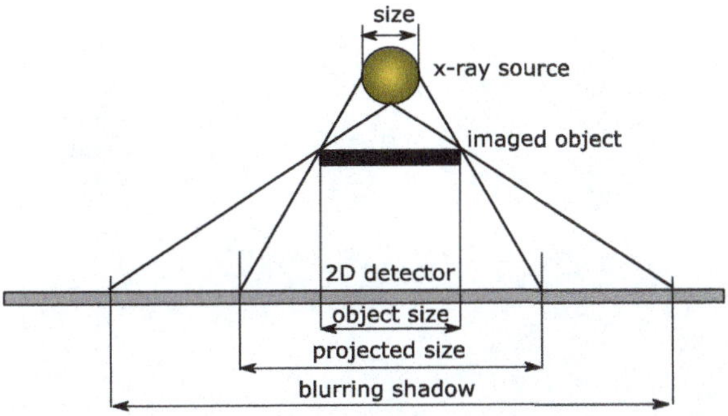

Fig. 7.7 Geometry results in a larger projected size of objects close to the x-ray source and distant to the detector. Via these distances a form of magnification can be set. Larger magnification goes hand in hand with an increase in image blurring, but smaller x-ray sources reduce this effect

The method of x-ray absorption imaging, regularly applied not only in medical diagnostics, works in a similar fashion also in industry and material analysis. In this context it is called radiography and is part of several industrial norms (e.g. DIN EN ISO 19232, 17636, 5579, 11699). The contrast derives from a contrast of different material (e.g. plastic vs. metal) and from the material thickness between source and detector. Defects, cracks, and porosity in a material reduce the effective thickness inducing a contrast for x-ray inspection, even when they are hidden inside the material or when they are smaller than the spatial resolution. The penetration property of x-rays allows seeing into hollow objects such as tubes or chip packages (see Fig. 7.8) or compounds for locating problems without dismantling the objects.

Practically the same method can deliver tomographic information by acquiring several images when rotating detector and source around the sample, similar to medical tomography (Sect. 6.1.1). In the technical context smaller scales are of interest compared to the medical application. Consequently, x-ray micro-tomography in its latest form reaches spatial resolutions down to a few 10 nm (Maire and Withers 2014). Improving the spatial resolution below the 1 μm requires x-ray focussing optics. These optics are not as compact as visible light optics, but several options such as Fresnel zone plates, gratings, or so-called K-B mirrors exist (Maire and Withers 2014). Computerized post-processing for extracting size distributions of for example inclusions or volume fractions of different contrast zones (≈materials) produces also quantitative information in addition to the qualitative imaging. Photon beams with narrow spectral distributions, such as monochromated synchrotron light, enable high contrasts for specific elements when spectrally placed on the K-absorption edge of the corresponding element. Using energy-resolving pixel detectors adds compositional information to the image using the spectroscopic methods discussed below.

X-ray imaging allows for fast image acquisition in the order of several 10 ms to hours, depending on required contrast, spatial resolution, and imaged volume. This time-scale enables in-situ analysis of processes such as crack formation in mechanical testing. Figure 7.9 shows an example from biological science of analysing the effect of drugs on a parasite infection using a combination of x-ray imaging and spectroscopy

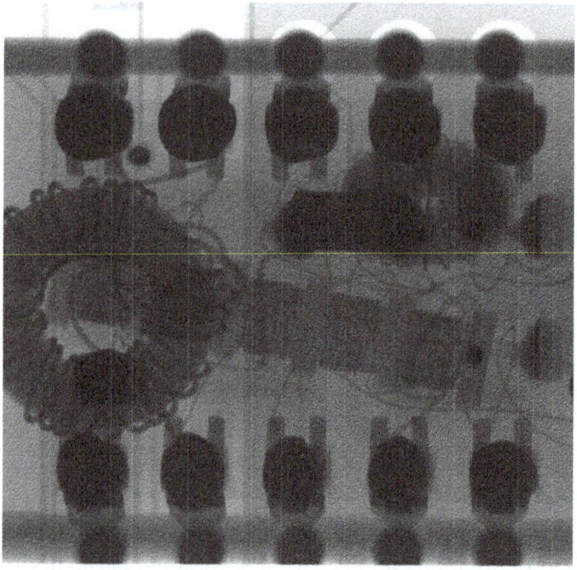

Fig. 7.8 Example x-ray imaging enabling inspecting the wiring and components in an integrated circuit chip without opening the structure. Reproduced from X-Ray_Circuit_Board_Zoom.jpg: SecretDisc derivative work: Emdee under CC-BY-SA-3.0

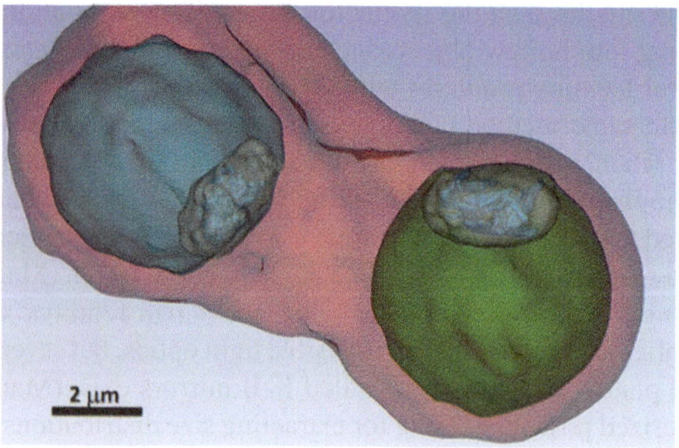

Fig. 7.9 This x-ray image shows details such as the vacuole of the two Plasmodium parasites (artificially coloured in blue and green) inside an infected blood cell (red) acquired using multiple wavelength synchrotron x-ray tomography with sub-µm resolution and a combination of imaging and XRF. Reproduced from (Kapishnikov et al. 2019) published under CC-BY 4.0

methods (Kapishnikov et al. 2019). Mapping of certain elemental components such as Fe or Br contained in the cells and the drug in combination with highly resolved imaging enabled following the action of the drug and its metabolisation.

This brings us directly to the spectroscopic methods. In addition to the straightforward analysis of x-ray absorption, spectroscopy analyses the energy spectra of secondary particles emitted by x-ray absorption, providing additional information on the chemical composition of the sample. These spectroscopic methods exploit the energy of the photons and secondary particles released by them, requiring energy resolving detectors. These energies and their changes connect to chemical and elemental composition via the interaction of photons and bound electrons resulting in an elemental contrast.

The x-ray fluorescence (XRF) method sends x-rays into the sample to produce secondary x-rays of lower energy. The primary x-rays ionise and excite the atoms as depicted in Fig. 4.16, similar to the working of EDX (7.1.3) and PIXE (Sect. 7.1.5) mentioned later on. Among this triumvirate of secondary x-ray based elemental analysis methods XRF represents the compromise in terms of spatial resolution and accuracy but with the lowest overall costs. EDX offers the best lateral resolution. PIXE provides the lowest detection limits and best accuracy. The electrons recombine emitting again x-rays of an energy characteristic to each element. Free electron light sources enable improved contrast and lateral resolution compared to bound electron sources through higher brilliance allowing for smaller beam spots and exact excitation wavelength positioning on the elemental absorption edges (Fig. 3.4) resulting in a relatively stronger contribution of the absorption edge to the total signal than with broad excitation spectra. Figure 7.10 shows an XRF spectrum of a stainless steel. Each peak represents a certain atomic transition specific to the elements present in the target. The signal integrates over the penetration depth of the x-rays in the sample, hence a depth information can only be acquired using tomographic methods. The typically good energy resolution of about 130 eV of modern silicon drift detectors enables separate quantification even of neighbouring elements. The secondary x-ray yields monotonically increase with incident x-ray energy, requiring a certain minimum x-ray energy for practically relevant detection limits.

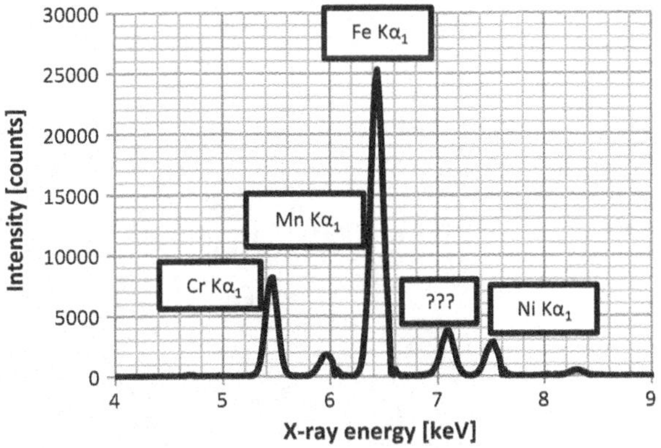

Fig. 7.10 XRF spectrum of a 1.4301 stainless steel using the handheld device depicted in Fig. 7.43 with 40 kV, 10 μA and a Rhodium x-ray source target. XRF detects the main elements Fe, Cr, Mn, Ni via their Kα1 lines. Can you identify the unknown (???) peak using Fig. 3.4? Permission granted by Bruker Nano GmbH. (Fe Kβ1)

X-ray photo-electron spectroscopy (XPS) exploits the information stored in secondary electrons emitted upon x-ray absorption, the so-called photo-electrons. The x-ray energies of typically around 10 keV emitted from K-α line emission, see Fig. 3.4, allow releasing electrons from the atoms. The photons transfer their enery to the electrons bound in the target atoms. The resulting electron energies are characteristic for the specific atomic energy levels. The electron kinetic energy results from the photon energy minus the electrons atomic binding energy. This results in keV range electron kinetic energy. This electron energy restricts the information depth to a few nanometres below the surface. Figure 3.4 shows us some of these binding energies, but it displays only part of the truth. The displayed values represent the ideal situation of an isolated atom, but in molecules and compounds, e.g. H_2O, the values change by some eV due to the binding situation. This chemical binding energy shift is the very reason we have stable molecules and compounds and provides the energy released for example by burning hydrocarbons. XPS allows determining these values.

Different binding types with different energy shifts exist for the different molecules, see Fig. 7.12. Sometimes, a single stoichiometric composition can arrange in different binding configurations with different chemical and physical properties, enabling an identification and quantification of the chemical state via XPS. The technical realisation of XPS with a crystal monochromator analyser attached to a bound electron x-ray source, and sample handling and manipulation is shown in Fig. 7.11.

The figure shows a XPS device with hemispherical electron energy analyser for increasing the secondary electron energy resolution to sub-eV. The high energy-resolution in combination with the finite dynamic range of the detector requires setting it to a certain atomic binding level for each measurement, but the close energies

Fig. 7.11 Technical implementation of XPS with the typical hemispherical electron energy analyser attached to a larger system with sample loading and other techniques. Copyright Forschungszentrum Jülich/Tobias Wegener

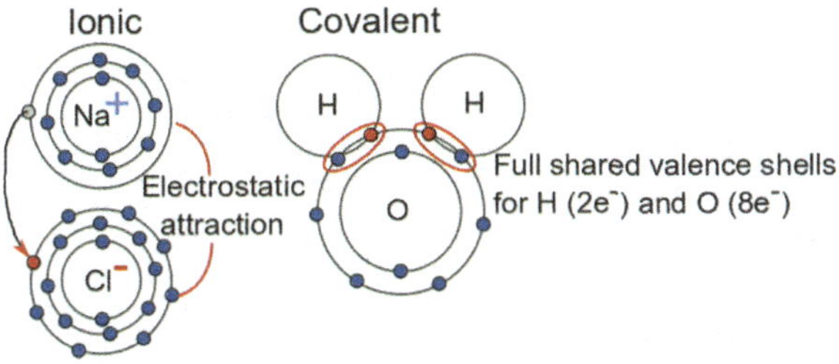

Fig. 7.12 Molecular bindings change the electron orbitals or electronic structures, respectively. The total amount of electrons and quasi-neutrality remain conserved, but the binding energy changes from the mono-atomic value. This is the very basis for the binding and chemistry. The exact differences depend on the binding type and the binding partners. The differences are accessible by XPS

of chemical binding states require this energy resolution level. A background signal level arises in the detector due to reactions of the ejected secondary electrons with the sample on their way towards the detector. Due to the numerous required calibrations and the background, XPS usually achieves a statistical uncertainty in quantification in the order of 15%.

XPS easily implements into in-situ experiments for parameter studies. The method can even run in a few 10 mbar ambient pressures for studying the influence of the interaction of surfaces with atmosphere. Accessing this information and testing the influence factors for the formation of a certain binding configuration or its impact on processes reveals important details of the underlying chemical reactions. Figure 7.13 demonstrates such a study investigating the formation of carbohydrates in the Fischer–Tropsch process on a cobalt surface. XPS reveals not only the presence of carbon, but it can quantify the occurrence of distinct binding state in relation to process parameters.

Scattering based methods represent a type of imaging accessing the reciprocal space via diffraction of x-rays on periodic structures in the sample material. These periodic structures have to be in the order of the x-ray wavelength, namely in the order of nano-metres. The reciprocal space working of scattering methods originates from larger x-ray scattering angles being induced by smaller sample structures, a reciprocal connection. These structures can be crystal lattice spacings, atomic positions in molecules, or microscopic precipitates/inhomogeneities of materials (e.g. pores in a solid). Figure 7.14 demonstrates the physics of diffraction. In this case, a lattice spacing d results in a constructive interference of a parallel wave front if the path length difference 2δ equals the photon wavelength multiplied by a positive integer. This so-called Bragg condition connects the reflected intensity with the discrete structural features in the sample. Tuning of wavelength and incident angle θ allows scanning for these features.

With this said the method of X-Ray Diffraction analysis (XRD) is defined. XRD enables investigation mostly of crystallographic properties. Each element, compound, and phase (e.g. α vs. γ iron) features a specific set of distances in its

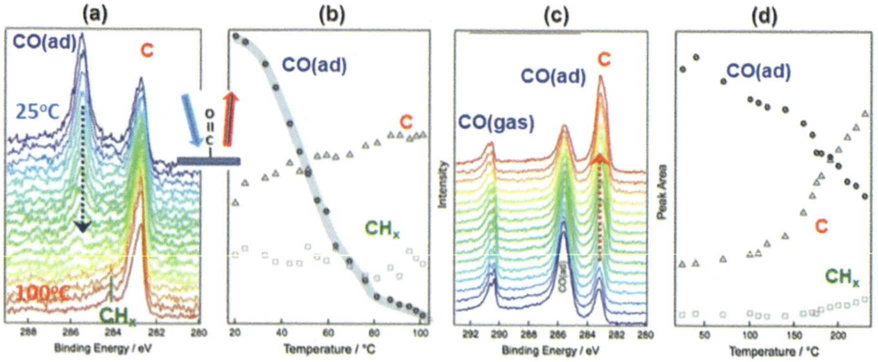

Fig. 7.13 Carbon in-situ XPS spectra acquired with synchrotron light on a cobalt foil first loaded with CO gas then heated in vacuum **a**, **b** and in 0.133 mbar CO, **c**, **d**. **a** shows the electron binding in three different states ascribed to adsorbed CO molecules, C impurities and CH_x molecules. **b** Summarises the signal levels, showing a reduction of adsorbed CO with temperature. **c** Demonstrates the separated quantification of adsorbed and gaseous CO and C impurities due to state dependent electron binding to the C atom. **d** Shows the decrease of adsorbed CO with temperature and a corresponding increase of CH_x and C. Reprinted from (Salmeron 2018) with permission by Springer

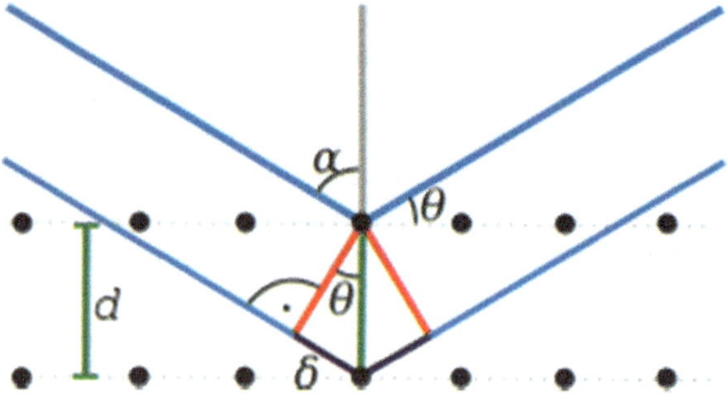

Fig. 7.14 A typical diffraction geometry on a regular lattice of spacing d results in constructive and destructive interference of reflections from different layers depending on impact and exit angle θ

lattice due to the specific atomic arrangements. With a given wavelength, typically generated with an x-ray tube and a crystal monochromator, the Bragg peaks can be scanned by mechanically rotating the sample against the x-ray beam and the detector via the angle θ. These so-called 2θ spectra show several peaks as demonstrated in Fig. 7.15. Knowing θ and the x-ray wavelength, Bragg's condition allows calculating the lattice spacing d.

Not only the fundamental distances matter in a crystal as the situation in Fig. 7.14 depicts. Think of the crystal as an artificially planted forest where the trees have fixed distances. If you walk along the forest you will see all trees behind each other from several positions. For crystals this works similar and a pattern of Bragg peaks show up

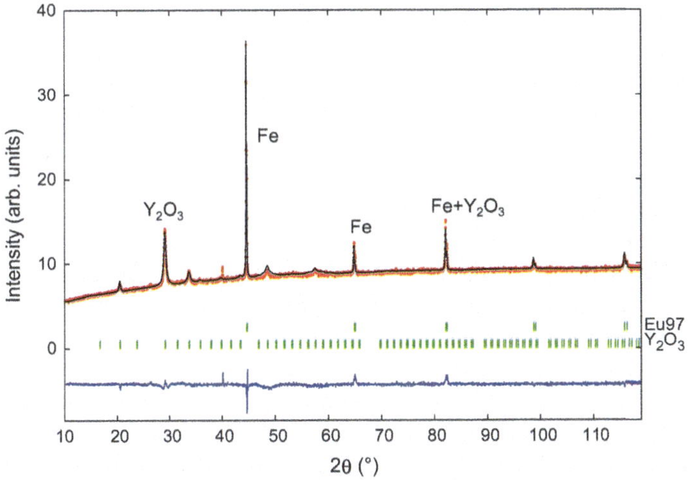

Fig. 7.15 XRD spectrum of Eurofer-97 stainless steel coated with a thin Y_2O_3 layer using a Copper x-ray spectrum. The lines below the spectrum indicate where the reflexes of each compound can be found. From the steel only Fe contributions are visible. The blue line at the bottom represens the difference between fit and experimental data. Courtesy of Anne Houben

in the 2θ spectrum. These peaks allow identifying the contributing structures in mixed compounds via fitting spacing databases for elemental and compound materials to the measured data. When walking through a natural forest where all trees grew randomly we cannot see such a long distance order, at best a limited order on short distances induced by a minimum distance between two trees required for them to be able to grow. In XRD this situation equals analysing an amorphous material. The disorder of amorphous materials makes them invisible to XRD, the method works only for ordered structures.

XRD distinguishes whether a compounds constituents are present as separate phases (e.g. precipitates in a matrix) or as a single mixed/solute phase (e.g. intermetallic phases). Separate phases will show both individual reflexes in reduced intensities. Mixed phases typically show lattice spacings in between the pure elemental values of their components. The semi-empirical Vegards law states a linear scaling of the resulting spacing with the mixture ratio.

Vegards law allows for a rough quantification of mixed phases, but generally technical limitations prevent a clear quantification of compositional aspects using XRD. Inhomogeneity, roughness, and surface layers (e.g. oxides) influence the XRD result. Together with the varying beam spot-size and penetration depth when varying θ, a sound quantification becomes nearly impossible. Some compounds such as CrO_2 and FeO_2 feature very similar Bragg peak patterns, preventing a separation of both.

For the analysis of non-crystallographic structures Small Angle X-ray Scattering (SAXS) can be applied. Specialised literature for the application of small angle scattering for different tasks in material science and biology (Chaudhuri et al. 2017; Schnablegger and Singh 2017). The website (Brookhaven National Laboratory 2018) provides some basic information and evaluation software links. SAXS detects nm sized features through scattering of x-rays at electron density contrasts. For example a pore in a metal represents a maximum reduction of the electron charge density to zero. This changes the local refractive index for the x-rays. SAXS allows measuring for example porosity or precipitate size distributions, their shape, and orientation (for example elongated cylinders or spheres). Biological small angle scattering, a variant of the material science SAXS, extends this towards structural analysis of biological macro-molecules, polymers, and nano-composites.

Figure 7.16 shows a synthetic SAXS result comparing three different sizes of pores present in a metal. The q coordinate represents the scattering angle. Due to the reciprocal nature of the measurement larger q equal smaller structure sizes. In other words, small objects induce large angle deflections while large objects induce small angle deflections. The graph has three distinct regions. At low reciprocal dimension q a more or less constant intensity limit occurs. This is followed by an exponential decay in the so-called Guinier region. The slope of this decay depends on the particle dimensions. We can see the Guinier region shifting towards higher q for smaller porosity scales. Towards higher q the Porod region with its hill and valley structure follows. The Porod region represents x-rays scattering on the inclusion to matrix interface. The slope of this region has a fixed value of −4 for smooth particle to matrix interfaces, but differs from this value for irregular or fractal inclusions. The absolute intensity in this region depends on the surface area per unit scattering volume of the inclusions. A small pore has less surface than a larger pore, resulting in an increase of the signal level with pore size for a constant pore number density. Together

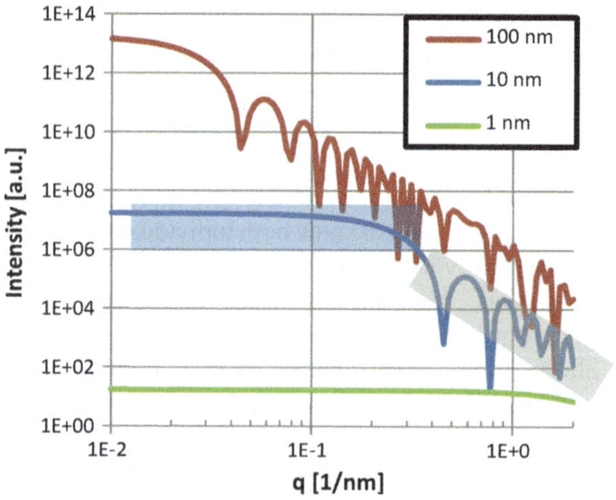

Fig. 7.16 SASfit 0.94.11 simulation of a SAXS spectrum for three different pore sizes with Gaussian distribution functions and constant number density. The graphs show a clear difference in the scattering pattern allowing analysing measured data by fitting. The left box marks the Guinier region, the right box the Porod region

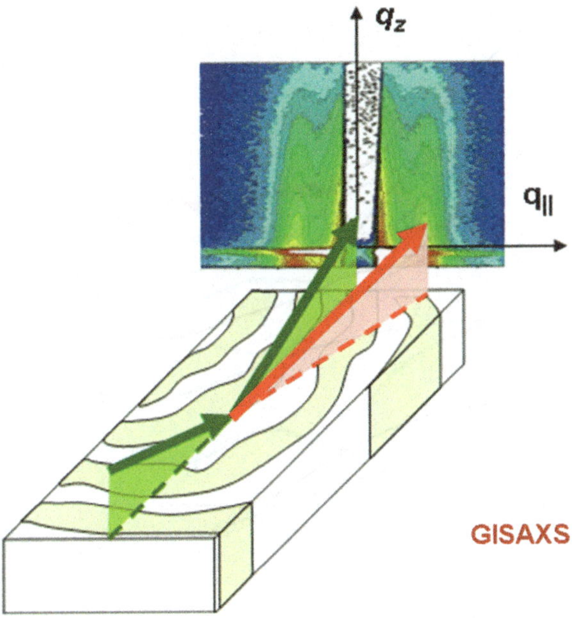

Fig. 7.17 GI-SAXS measurement principle. The x-ray beam reflects on the surface (green arrow) resulting in the specular peak. Scattering at electron density fluctuations in the material diverts the photons by a few degree (red arrow), resulting in a pattern representative for the size and property distribution of the scattering centres. Larger q equals larger scattering angle. From Smilgies, Public domain, via Wikimedia Commons

with information on the inclusion shape, this enables determining their size and form. At even higher q than displayed, the Bragg-region starts as discussed above.

SAXS has several technical variants depending on the details of the geometry. SAXS requires a 2D detector behind the sample in beam direction. Figure 7.17 depicts the geometry of the grazing incidence (GI-SAXS) variant which enables analysis of thick samples (=non-transparent for the x-rays). Classical SAXS works in transmission, but the transmission geometry requires thin samples since the sample has to transmit a certain fraction of the photons. In the Wide Angle X-ray Scattering (WAXS) the detector is positioned very close to the sample for measuring larger scattering angles equal to smaller structures. Increasing detector to sample distance reduces the maximum detected scattering angle due to the geometrical reduction of solid angle. Detectors mounted on rotating and translating stages enable extending the range of detectable scattering angles, enabling selecting between the different variants in a single setup.

In biological sciences, SAXS can provide insight into structural information on molecular assemblies. The structure of assembly of proteins has a strong influence on their function. The large amount of existing proteins and the development of new forms and applications makes structural analysis and with this understanding of a proteins working and function an ongoing and extensive task. The study (Hura et al. 2009) determined several protein structures using a high-throughput SAXS setup with samples consisting of the proteins of interest in liquid solutions of only 12 μl. The measurements used synchrotron light due to the higher brilliance requiring less sample material, yet providing more photon energy flexibility. Figure 7.18 shows

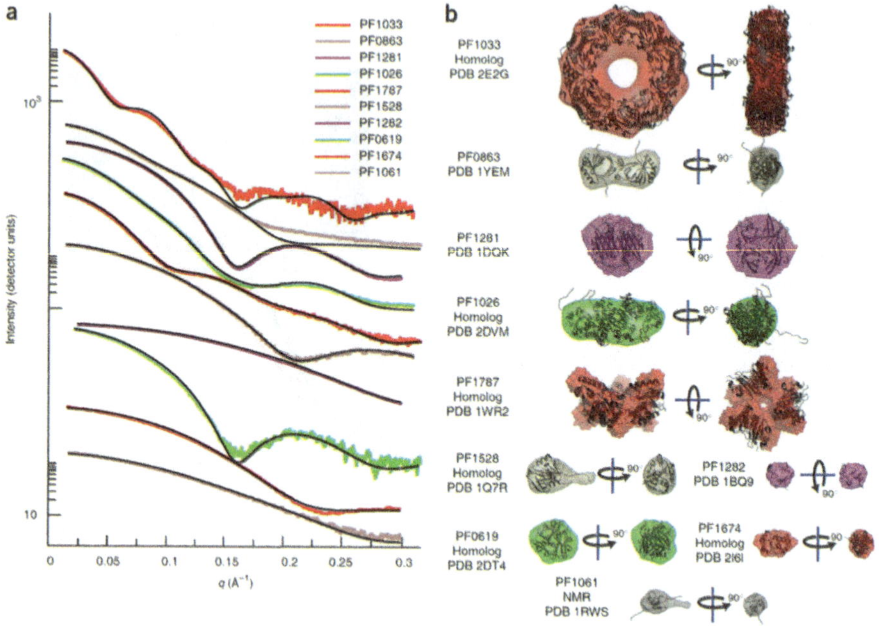

Fig. 7.18 SAXS provides information on shape and assembly of proteins in solution. **a** Experimental scattering data (coloured lines) compare well to SAXS theoretical calculations (black) of the known structures of 10 proteins. **b** The measured protein envelopes derived from the SAXS data overlaid with the existing molecular structures (ribbons) from two perspectives with 90° molecule rotations as indicated by the arrows. Monodisperse samples were used. All monomeric units had a 9-amino-acid His tag attached. Protein colouring according to curve colouring in (**a**). Reprinted from (Hura et al. 2009) with permission by Springer

the comparison of the structures of 9 reference proteins analysed by SAXS in this study. Data analysis and molecular structure codes allow defolding the SAXS data to obtain structural envelopes of unknown molecules to a spatial resolution <1.5 nm.

Generally, analytical science seeks to not only reach the highest spatial resolution, but also the highest time resolution. Some know the technological challenge in this time resolution from daily photography experience, may it be a smartphone or a professional camera. The pictures we take become brighter the longer the exposure time set in the camera. At night there is only little light and the camera software increases the exposure time in order to produce a visible image. The long exposure time blurs the image due to the camera shaking in our hands. Consequently, shorter exposure time, equivalent to shorter/better time-resolution, requires a brighter illumination of the scene. Photons are the ideal projectile for this task, since they can be produce with high intensity and detected with high efficiency. Time-resolved measurements exploit ultra-short photon pulses with pulse length down to attoseconds (10^{-18} s). Now comes the second challenge: How to make a shutter/exposure time this short? The answer is simple, it is impossible. Mechanical shutters reach their limits in the 10 μs range. Electronic switching times come down to picoseconds at best (think about how small capacity switches this requires).

Scientists exploit a trick: A virtual time-resolution can be generated via observation of numerous similar processes each observed at a different time index relative to a starting condition. The illuminating light pulse length then defines the time-resolution. Ultra-short pulses generate an ultra-short time resolution. Since technically the pulses cannot be repeated in such high rates, the virtual time-resolution requires the observation of another instance of the same process. Taking the image requires the same number of photons, independent of the pulse length. Consequently, ultra-short pulses have to provide ultra-high intensity beams to maintain the detector signal intensity. These intense beams destroy the target on impact, but in the ultra-short exposure time this will not become visible. The movement of the inspected atoms cannot be resolved if the pulse-length is shorter than the spatial resolution divided by the explosion velocity. The correlation and time-wise sorting of the individual measurements requires a trigger and delay setup or just brute force measurements connected to an intelligent filter algorithm. Technically, ultra-short and intense light pulses are the specialty of synchrotron light sources and in particular of FELs, see Sect. 4.3.3. The additional feature of variable wavelength of accelerator light sources enables adding additional information to the analysis.

7.1.3 Electron Beam Microscopy

Electron microscopy was originally invented about 90 years ago to extend the imaging resolution of optical microscopy via the better diffraction limit of electron beams as discussed with (7.1). Nowadays electron microscopy reached a high level of technological maturity and commercialisation with numerous manufacturers on the market. Commercial devices can be bought like cars with a long list of extras mounted to the basic setup. New types of electron sources with low emittance (see Sect. 2.4) enable significant reductions of beam spot sizes, increasing spatial resolution, compared to the original designs. Even more importantly, the capabilities of electron microscopy nowadays extend beyond pure imaging/localisation on the nanometre scale. Many analytical and quantitative methods based on different geometries and secondary particle types offer identifying and quantifying methods, making electron beam microscopy a valuable and versatile tool for analytical applications in science, medicine, and industry. As always the reader is referred to specialised literature for more details, e.g. (Williams and Carter 2009; Hawkes and Spence 2019).

The basis of all methods is the Scanning Electron Microscope (SEM) named setup. Scanning describes the methodology of image acquisition with an electron microscope. In contrast to the eye or a camera, the SEM methods feature only a single point detector, not a 2D matrix detector. Generating a 2D map using a point detector requires moving the bright spot (=the electron beam) across the imaged object, "scanning" it. The beam follows a meander pattern to minimize the travelled distance for covering typically a rectangular area. Knowledge of the beam position via the settings of the beam optics provides the 2D localisation of the detector signal. Figure 2.30 depicts a typical beam optical design of an electron microscope. Thermionic LaB_6 emitters generate a low emittance electron beam. Top-level devices use field emission sources instead of the LaB_6, providing even better spatial resolution. Variable DC voltages in the order of 1–100 kV allow for an optimisation of the projectile energy for different applications. A set of magnetic lenses focusses and steers the beam on the sample. The low mass of electrons requires only small optical elements compared to ions resulting in table-top devices. The distance between last beam optical lens and sample is typically in the order of a few millimetres. A multi-axis manipulator enables sample positioning and rotation. Compact devices in combination with the high range of electrons allows for the sample analysis even outside UHV conditions, an important aspect for biological samples, at the expense of analytical performance.

The physics of contrast and information of SEM rests on the interaction of electrons with matter. Electrons require a conducting surface layer or the surface will charge up and repel the projectiles. The electron stopping power in the range of 1–100 keV allows for ranges of a few 10 nm up to a few 10 μm. Higher energy improves the image resolution and opens up more analysis channels, but the longer range shifts the contrast towards larger depth, away from the surface. Within this range the beam impact induces multiple processes as depicted in Fig. 7.19, each resulting in its own SEM based sub-method. The figure should not be misunderstood: All emitted particles are produced starting from the surface, the picture states the terminal range of these particles exiting the sample. The analysis range stays below the projectile range due to thresholds of reactions and the necessity of the secondary particles leaving the sample to be detected. In particular, methods based on secondary electrons have the smallest depth range of only a few nanometres (due to typically a few 10 eV kinetic

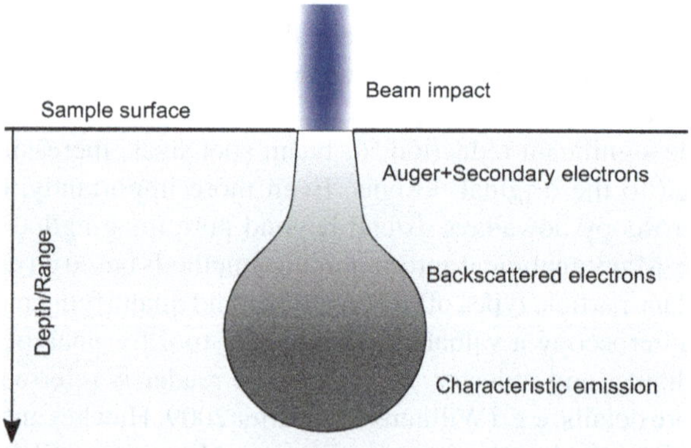

Fig. 7.19 The analysis cone of electron microscopes. The electron beam passing through the analysed matter loses energy and broadens up via the scattering with target electrons (straggling). The analysis relies on secondary particles emitted from the beam interaction, consisting mostly of lower energy electrons and x-rays, each with its own range to leave the sample. The range of x-rays allows them to leave the sample even from the largest depth. Low energy electrons can only leave the material from shallow depth as indicated by the grey gradient

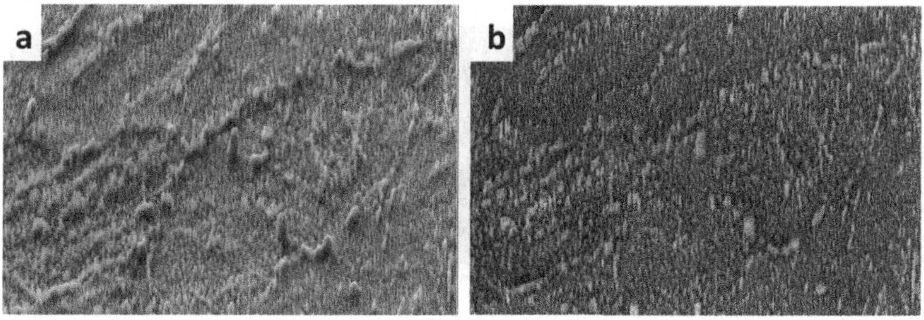

Fig. 7.20 5 keV secondary electron image of a plasma etched steel acquired with two different detectors on the same $10 \times 7\ \mu m^2$ sample area. Sample tilted by 54°. The plasma developed nanostructures depending on grains and plasma resistant precipitates. Secondary electrons **a** and backscattered electrons detected in-lens **b** lead to different contrast highlighting different features. Copyright Forschungszentrum Jülich/Marcin Rasinski

energy). Variation of projectile energy allows for a depth resolved analysis, but the strong straggling of electrons limits the depth resolution of the SEM.

Imaging clearly represents the reference result of SEMs. For imaging Secondary electrons (SE) are detected. SE are emitted depending on the electronic structure of the sample, but more importantly edges and thin parts of rough surface morphology result in increased SE yields, while cracks or shadowed areas release fewer SE. This morphological contrast gives the SE images a natural 3D appearance as shown in Fig. 7.20a. The reflection of projectiles results in higher energy products, the Back-Scattered Electrons (BSE). BSEs result in a stronger signal for heavier constituents, giving a qualitative information on the elemental composition. Therefore, brighter spots represent heavier elements and darker spots lighter elements, Auger and photo-electrons. The angle of product detection has a strong impact on the results, since the BSE cross-section decreases strongly with angle (Jablonski et al. 2016). Therefore, BSE detector has to be positioned coaxial with the projectile beam (in-lens detector) for sufficient signal. The SE detector is installed typically up to 90° away from the beam, suppressing the BSEs by several orders of magnitude. SE and BSE result in different images as demonstrated in Fig. 7.20.

The qualitative results of imaging can be clarified by an overlay with elemental quantification methods. The energy-dispersive x-ray analysis (EDX) method relys on interaction with electrons bound to the atoms to produce x-rays according to Fig. 3.4. The emission of characteristic x-rays requires at least the x-ray energy to be provided by the projectile energy, but even for heavy elements such as tungsten lines in the keV range exist. Figure 7.21 shows a case of Mo deposition on a polished W sample. The x-ray detector yields a spectrum for each point on the sample map. The evaluation of the spectrum results in a local composition. The 2D map of this compositional result shown in Fig. 7.21b, c enables identifying the nature of the surface morphology. The

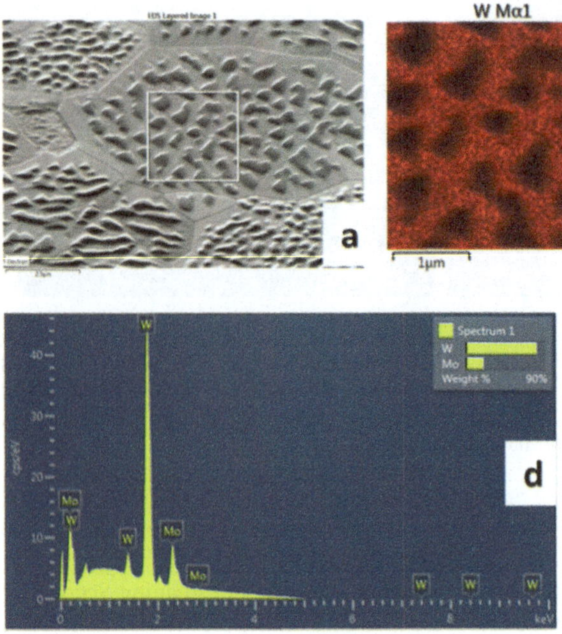

Fig. 7.21 EDX map of a tungsten surface where Mo features were deposited by physical means. The SEM image **a** shows the structural features. The EDX maps taken in the centre of (**a**) show the W $M_{\alpha 1}$ x-ray line map (**b**) and the Mo $L_{\alpha 1}$ map (**c**), adding compositional information to the imaging. The underlying x-ray spectrum (**d**) enables elemental identification and compositional quantification. Courtesy of Marcin Rasinski

Bremsstrahlung background, visible in Fig. 7.21d, limits the accuracy to about 1%. Consequently, EDX is not suitable for tracer analysis.

The Electron Backscatter Diffraction (EBSD) method reveals local crystallographic orientation. For more details regarding the aspects of crystallographic structure the reader is referred to solid state physics literature. A fluorescent screen converts the scattered electrons to an image, the so-called Kikuchi pattern. A 2D camera records the image for each point on the sample. Automated software analysis converts the image to the crystallographic map displayed in Fig. 7.22. The orientation map clearly visualises the grain boundaries and grain sizes at the surface. Post analysis software enables determination of grain size distribution. In combination with the other SEM methods, the influence of crystallographic structure on certain processes can be studied, as for example Fig. 7.21a depicts a grain orientation dependent deposition pattern.

So far the SEM methods were strictly bound to the sample surface. The product emission properties place a clear physical limit to the SEM technology, but the combination with sputtering methods allows extending this limit. For sputtering an ion beam of similar size as the electron beam is required. The so-called Focussed Ion Beam (FIB) module enables removing material on the scale of a few 10 μm. Figure 7.23

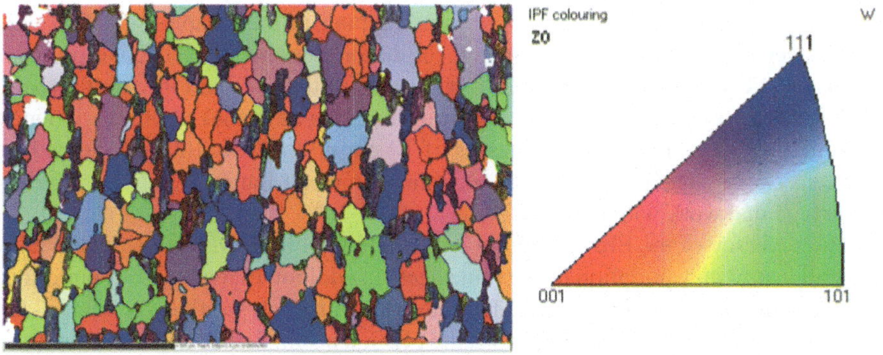

Fig. 7.22 EBSD map of a polished tungsten sample showing the mixture of grain orientations in a poly-crystal. The crystal has 3 fundamental orientations, 001, 101, and 111. Each orientation receives one of the RGB colours, colours in between RBG represent grains where all directions have an angle against the electron impact angle. Courtesy of Marcin Rasinski

Fig. 7.23 3 keV electrons SEM picture of a FIB cut in a plasma exposed steel sample at 5000× magnification. The trench has a width of 6 μm. Grain boundaries apparently affect the nano-structure as they remain slightly visible. The cut allows measuring the height of the nano-structures and reveals their substrate. Courtesy of Marcin Rasinski

shows a result of a FIB process. The FIB beam was scanned over the sample surface, removing material in a triangular cross-section. Corresponding sample alignment gives access to the sample depth via a cross-cut view. The cross-cut enables investigation of the thickness of the surface morphology in this case. Continued removal would allow generating a 3D video via slice-wise erosion. All of the formerly discussed methods can be combined with the FIB method, extending their analytical potential. The range of electrons again limits the spatial resolution, since small structures will transmit electrons, mixing the information (e.g. of EDX) with the material behind.

This transmission brings us to the last idea. The term SEM usually depicts the analysis of thick targets. The meaningful analysis of thin targets would require removing the material behind. This analysis method is called transmission electron microscopy (TEM). The separate analysis of nanostructures is not the only aspect of TEM. The analysis of thick samples (complete beam stopping in the sample, see Sect. 3.2) implies certain limitations to the spatial resolution, see the analysis cone problem discussed in Fig. 7.19. Transmission allows restricting the information to the thin neck of the analysis cone, removing the physical limitation of spatial resolution induced by straggling. This requires thicknesses of 10 nm up to a few 100 nm, depending on beam energy. The preparation of such thin samples from bulk samples requires skilled manual work with the FIB.

The accuracy of the FIB method in combination with the SEM imaging enables separation and transportation of the thin slices. In addition to removal by sputtering, the beam-matter interaction with a platinum containing gas, specifically injected for deposition, enables deposition of Pt via the FIB ion beam. The Pt deposition method aids as a protective coating for FIB cutting, but enables also soldering to attach these slices to sample holders. Figure 7.24 depicts the four main steps. Firstly, FIB prepares a slice by removing material in front and behind the sample. The Pt soldering FIB technique attaches this slice to a transfer needle. The needle brings the slice to a macroscopic TEM holder, usually a special disk of 3 mm diameter. Pt soldering attaches the slice to this holder. A low-energy polishing FIB mode then removes

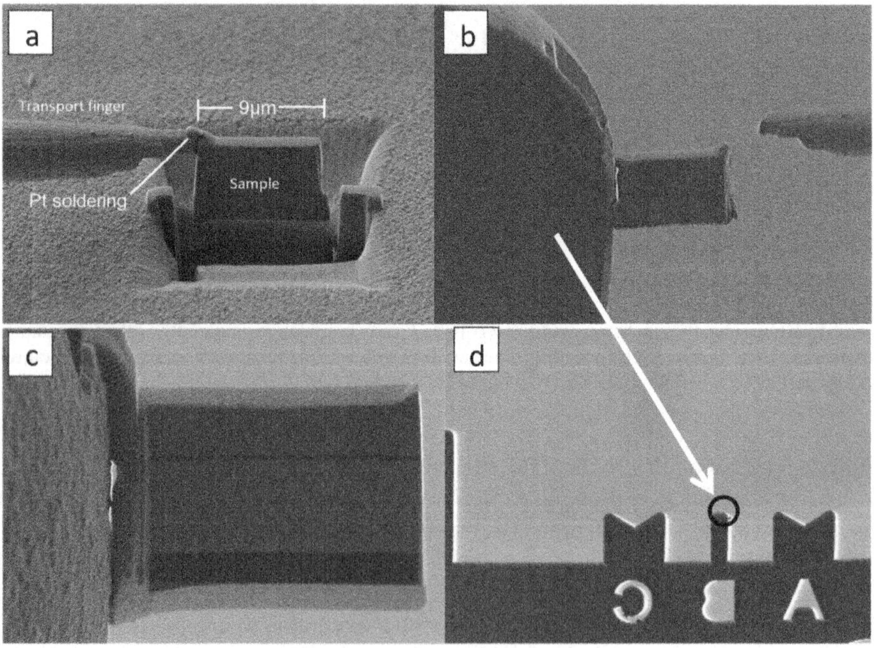

Fig. 7.24 Preparation of a TEM lamella. **a** First the lamella is cut out of the sample via FIB and Pt-soldered to a transport needle. **b** The needle transfers the sample to a 3 mm supporting holder disc. **c** Before TEM analysis the surface is polished down via FIB, removing the Pt and yielding the final analysis thickness. This process also modifies the sample, leaving for example displacement damage in the material. **d** The 3 mm supporting disc with "ABC" positions with the sample on the top right of position B (circle). Courtesy of Marcin Rasinski

material from the slice back and front, restoring the original sample structure and providing the final slice thickness. This part leaves also damage to the sample surface due to the ion impact, an effect visible in TEM analysis. The holder with the tiny sample attached is then brought to the TEM device for analysis.

A TEM analysis generates a 2D image of the sample with sub-nm resolution. All of the methods above can be applied like in a regular SEM. Extreme resolution is the domain of TEM. Cross-section analysis and nm scale imaging can be done in a regular FIB-SEM device without the complicated preparation process and the expensive TEM device. Extreme scientific setups such as PICO installed in Forschungszentrum Jülich reach imaging resolutions down to 0.05 nm, the atomic level. With the technological limit of straggling removed, reaching these extreme resolutions requires special beam optics correcting the higher order aberrations (spherical and chromatic) and a high level of vibration damping of the laboratory. Even without atomic level resolution, TEM provides additional information to SEM. Figure 7.25 shows an example where TEM reveals nanometre sized subsurface gas bubbles induced by gas implantation. TEM also reveals defect structure, e.g. dislocation loops and voids, hardly accessible by other methods.

The SEM technology seeks for increasing resolution and additional contrasts. One of the upcoming technological developments are ion beam based options. Ions feature shallower information depth and less beam spot blurring due to the increased

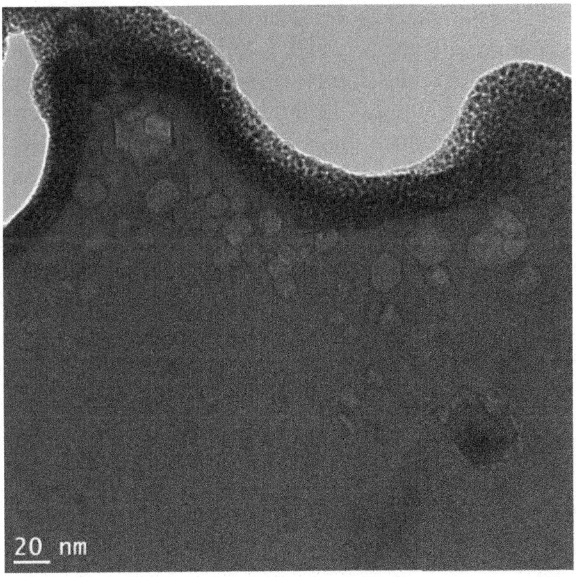

Fig. 7.25 TEM image of a stainless steel irradiated by deuterium and helium plasma. The lamella has a thickness of 80–100 nm. The gas implantation formed gas bubbles just below the surface. A 20 nm platinum layer (dark part) on the sample surface was deposited for practical aspects of preparation. Courtesy of Marcin Rasinski

Fig. 7.26 HIM image with 30 keV He beam (left) versus its 1 keV electron beam equivalent (SEM) of gold platelets on carbon. The ions provide better contrast and more information due to their higher stopping power localising the information closer to the surface. Reprinted from (Postek et al. 2007), with the permission of AIP Publishing

stopping power and reduced straggling of ions compared to electrons. The beam energies in the range of a few ten keV are slightly higher compared to SEMs, but similar spot sizes below 1 nm are technically possible (Tondare 2005). On the downside, in particular heavy ion beams induce significant surface sputtering in the keV range, altering the sample during analysis. Focused light ion beams, in particular Helium ions, reduce the sputtering since it connects to projectile mass. Low emittance ion sources similar to the Field-emission type used for high-resolution electron beam microscopes became available only in the last 10 years. The secondary electrons emitted upon ion impact produce images similar to electron microscopy imaging discussed above. The contrast of this so-called secondary electron imaging (SEI) arises from differences in surface binding of the electrons. The principle is identical to the electron beam imaging technique SEM, just with a different projectile. Competitive lateral resolutions are so far only achieved with He ion beams of a few keV using special field emission sources (Postek et al. 2007), but first devices of this type are commercially available. Figure 7.26 shows a comparison of this Helium-Ion-Microscopy (HIM) and SEM, indicating a high potential of improved imaging with ion beams.

7.1.4 Secondary Ion Mass-Spectrometry

- Nature is fair: Everything valuable comes with a price

Secondary Ion Mass Spectrometry (SIMS) exploits the release of secondary ions and ionised molecules via surface sputtering with an analysing projectile ion beam in the 0.1–50 keV range. As always the reader is referred to specialised literature on the field, e.g. (van der Heide 2014). The release of target material by sputtering defines SIMS as a destructive method, the sputtered areas are small though. Depending on the setup sputtering craters extend over 10 nm–100 μm in diameter, due to the typically atomic sensitivity of the applied detectors requiring only minute quantities of material. A detector catches the released material and analyses the particles for their mass either via ion optical separation via electro-magnetic deflection or a time-of-flight (ToF) tube with a chopping element. ToF detectors offer high transmission together with a possible acquisition of all masses at the expense of a discontinuous operation induced by the chopping process. The complete SIMS setup consists of two independent parts for the primary ions (projectiles) and the secondary ions (products) coupled via the sample as shown in Fig. 7.27.

SIMS reached a high level of industrialisation with many highly integrated commercial devices available on the market. The analytical strengths of SIMS responsible for this success lie in its complete elemental and isotopic sensitivity (H to U), low detection limits down to ppb (10^{-9}) and typically high statistical accuracy (count rates). ToF detectors allow separating even ions of close masses such as O and CH_4 (mass = 16 amu) or ^{184}W and ^{184}Re with mass resolution $\Delta m/m \approx 10^4$ up to masses in the order of 10,000 amu. Technically, mass resolution and detection limits are related in an opposing manner, therefore a given device layout always represents an application specific optimisation, but the flexibility of device layout enables solutions for most applications.

The keV energy range of projectile energies connects the analytical mechanism of SIMS with the electronic structure of the sample. This marks a significant difference to MeV particle beams which connect mostly to the nuclear structure of the sample, see e.g. Sect. 7.1.5. On the one hand, this difference in probing systematics eases the technical realisation due to smaller and cheaper devices (the idea of MeV/m Sect. 2.2). On the other hand the information becomes more difficult to interpret since it mixes compositional and structural aspects of the target. The primary physical effect responsible for the SIMS signal is sputtering induced by the keV projectiles, Sect. 4.1.

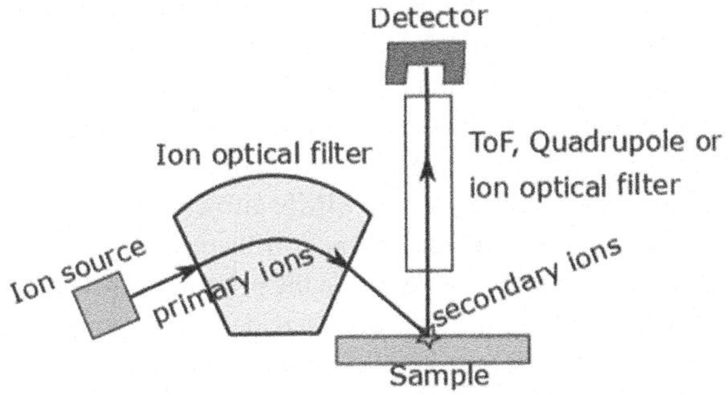

Fig. 7.27 Basic SIMS setup consisting of an ion source with filter optics (primary column) and a product ion filter with detector (secondary column). Different options for primary and secondary column layouts exist, defining the analytical capabilities of the device. The angles between projectiles, sample, and products also depend on device layout

Sputtering releases atoms from the target via a collisional cascade redirecting the momentum of the projectile to a backscattered product exiting the surface barrier (Fig. 4.2). Usually this emits atoms in the neutral state due to a decoupling of the projectile from the product via the multiple required intermediate collisions. Of course SIMS would not work if this was completely true. With a small chance, typically <1%, the emitted atoms exchange electrons with the surface directly after passing the surface barrier. The probability of this charge exchange process and whether the atom receives or loses an electron depends on the local potentials of surface and product. The understanding and quantification of these effects is a point of active research, but a general quantitative describtion was so far not found.

The ionisation induces a fatal weakness opposing the many analytical strength of SIMS. The sensitivity of SIMS depends on the electronic structure (molecules, phase, crystallographic orientation…) and composition (elements) due to a strong dependence of the secondary ion yields on these parameters, the so-called matrix effect. As a consequence, secondary ion yields span about 5 orders of magnitude inducing an unknown and varying sensitivity which in turn makes it practically impossible to provide fixed signal calibrations of sample concentration per detected ion. In other words, SIMS does not yield absolute compositional numbers (quantification) but only qualitative trends (identification). Figure 7.28 represents such a case where a thin-film $LiCoO_2$ battery was analysed after several charge and discharge cycles. The lithium signal clearly shows a decrease of the Li^- signal upon charging as expected. Chemically this depletion ends at about $Li_{0.5}CoO_2$ at the typical 4.2 V versus Li^+/Li charging voltage, the Li^- signal drops by a factor ≈ 5 in the discharge zone (first 200–300 nm), though. Charging altered the matrix, strongly influencing the negative ion yield.

Approaches on calibration with reference samples partially solve the matrix-effect problem for repetitive standard analysis situations by scaling the signal intensity and composition of the reference to the sample intensity. Dynamic situations such as the Li-ion battery resist this approach. Only in the tracer limit of a material of interest extremely dilute in an otherwise constant matrix, such as dopants in silicon, the reference calibration yields physically correct results. In general, the analysis of a sample means its composition differs from the reference (why analyse otherwise), therefore the comparison to a reference induces an uncertainty hardly quantifiable. Due to the chemical nature of the matrix effect, isotopic ratios will be correctly projected to their intensity ratios. A strength exploited in tracer studies for example in geological and nuclear sciences where SIMS allows for determination of (rare) isotope ratios revealing geological age or tracer transport.

SIMS' potential for localisation depends on the primary ion-beam spot size and the operational mode. The primary ion-beam spot size defines lateral spatial resolution as stated above. Scanning of the beam over the sample surface generates a 2D map. The depth resolution originates from the sputtering effect revealing deeper layers of the sample with ongoing removal of material. Here two modes exist: The dynamic mode relies on the sputtering effect of the primary beam. The surface recesses dynamically during analysis. This couples signal generation and depth progress. The static mode decouples both by adding a third ion column, the sputtering beam. This third

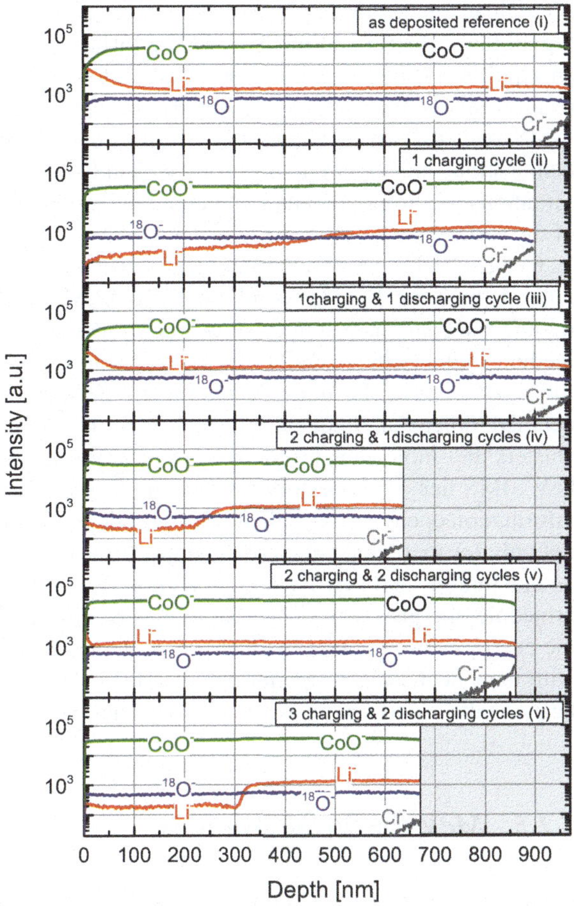

Fig. 7.28 Negative ion SIMS spectrum of a solid thin-film lithium ion battery at different charging states. SIMS analysis reveals the depth profile of Li and Co in the $LiCoO_2$. In the negative ion mode Co had to be detected as partial oxide. Charging depletes the Li while discharging refills it, but apparently not the whole $LiCoO_2$ layer contributes to the electro-chemical process. Reprinted with permission from (Dellen et al. 2016)

beam only removes materials from the surface, while the primary beam analyses the static surface after each layer removal step. The third column increases technical complexity, but an independent choice of analysing and sputtering beam energy and species generally enables more efficient sputtering and/or improved SIMS signal. Depth analysis down to about 50 μm becomes possible.

The choice of the primary ion strongly influences the yields for negative and positive ions. The technical implementations of the secondary column only allow for detection of either polarity. Most metals prefer to be in positive charge state while non-metals (O, C, H, F ...) tend to absorb extra electrons resulting in preferred emission of negatively charged ions. The released ions are not necessarily mono-atomic, but also oxides, hydrides, atomic clusters and all other imaginable combinations will be released, opening up tricks for intense signals of a given element in both modes, e.g. by detecting metals as oxides in the negative mode. The sputtering projectiles accumulate on the sample surface. This accumulation influences the electronic structure of the surface, coupling back on the surface emission properties. Consequently, an equilibrium of projectile deposition and removal sets in, but until the surface reaches this equilibrium the secondary ion emission is hardly interpretable. Static SIMS with its two incident beams further complicates the situation. Figure 7.28 demonstrates this effect in the strong Li and CoO signal changes in the first 10 nm. Situations requiring information on the first few nm require a sacrificial layer on top of the surface.

Current technical development routes for SIMS aim at higher integration with other techniques. Devices combining FIB-SEM with SIMS enable improved positioning and coherence of the results. The similar energy range and device size of both methods reduces the overall costs compared to two separate devices while allowing for faster analysis. From the technological perspective, inventions aim at revealing detailed molecular/chemical information of the sample via the detection of emitted clusters and molecules by MeV analysis beam impact. The difference of MeV to keV SIMS lies in not destroying these molecules upon projectile impact. Chemical information opens a new quality of information, imaging for example biological cells not only for their elemental composition but also for their constructive aspects such as membranes and cell nuclei. Protection of the fragile molecules requires lower projectile energies realised via large cluster projectiles (e.g. C-60 clusters sharing the energy between 60 individual projectiles) at keV energies or via MeV energies with their decreasing stopping power. MeV SIMS exploits the same setups as ion beam analysis and can be understood as a new type of detector in this context.

7.1.5 MeV Ion-Beam Analysis

Ion-Beam Analysis (IBA) constitutes a particularly large field with an exceptional amount of three to four letter abbreviations. On the one hand this demonstrates the complexity of ion-matter interactions and the required technologies, but on the other hand it relates to a lack of consolidation of the method due to the pending industrialisation of the technologies. Grouping a few interesting techniques using their analytical features and, in particular, the analysed particles makes life a bit easier and will be the preferred route in this book instead of elaborating one method after the other. Due to the scarce literature landscape on IBA the recommended reading for those interested in more details and physical aspects of the full set of analysis techniques is (Nastasi et al. 2014).

Already in the introduction, in Fig. 1.2, the physical situation was displayed. Ion-beam analysis exploits the full set of features and possible interactions of all incoming and outgoing particles in this scheme. The incoming projectiles can be any kind of ion. An individual method was defined for each physical interaction effect resulting in a certain type of products, leading to the many abbreviations. In contrast to electron microscopy, fundamentally new effects become possible with MeV ions, adding additional information aspects. More and different nuclear reactions, e.g. fusion reactions

($Q > 0$), between projectiles and targets are specific to the nuclear/nucleus aspect of ions. The ion mass allows for sputtering of the target, releasing material in the form of secondary ions and neutrals as discussed in Sect. 7.1.4. Different projectile ion species show qualitative differences in the interaction aspects with the targets such as nuclear reactions. Also several quantitative differences in the projectile stopping and the interaction with target electrons, e.g. regarding Bremsstrahlung, exist between different ions. The generally lower energy-loss straggling of ions compared to electrons allows for a depth-profiling using a recalculation of depth with the stopping power (similar to nano-machining discussed in Sect. 5.3.3) as discussed in Sect. 3.4. The drawbacks of IBA lie in the lower specific performance and higher cost of ion particle sources and optics compared to the electron equivalents, see Sects. 2.3 and 2.4. The probably even larger aspects of its lacking industrialisation lies in the complex analysis of the obtained data which contains significantly more information making it hard to disentangle the results.

MeV-IBA is a very flexible set of methods, meaning the methods can analyse nearly everything listed in the beginning of this chapter in Fig. 7.1, if properly optimized for the specific question. Each projectile species has its individual advantage (Table 7.2). Ion sources for protons (P) and deuterons (D) currently feature the highest brightness and lowest emissivity, leading to the smallest beam spots. D and ^{3}He ions offer many nuclear reactions with the highest Q-values physically possible, due to their excess/missing neutron, at the disadvantage of radiation protection issues with D due to strong neutron emission already at single MeV and the extreme costs of ^{3}He. P and D deliver a good range, ^{3}He and ^{4}He provide the better depth resolution at the expense of range. Heavy ions with their high stopping power provide the best depth resolution but nuclear reactions cannot be attained below a few 10 MeV, except for few rare isotopes such as ^{15} N. The physics of heavy ion sources working on the basis of emission of secondary ions of the respective species by plasma impact yield typically rather low heavy ion beam brightness. Lower emittance and higher brightness, respectively, practically result in smaller beam spots, which is equivalent to higher lateral resolution as the analysis only provides results averaged over the beam spot area.

The projectile energy influences the measurement in a similar fashion as the species. It influences the possible reactions via their cross-section and thresholds (Sect. 3.3), and it influences the stopping power and range (Sect. 3.2). Higher beam

Table 7.2 Comparison of different projectiles used for IBA. Each option has positive (+) and negative (−) aspects. Practical aspects sum up ion source properties, costs, and radiation safety

Projectile	Depth resolution	Range	Nuclear reactions	Practical aspects
P	−	+	+	++
D	−	+	++	−
^{3}He	+	−	++	−
^{4}He	+	−	−	+
Heavier ions	++	−−	−−	+

energies lead to a higher range and a lower depth resolution. In a strong simplification the analysis yields a given amount of points in depth, say 10, which distribute over the projectile range for a given energy, resulting, in this example, in a depth resolution of about 1 μm for a 10 μm range or 2 μm for 20 μm range and so on. The range defines the probing depth limit of the method, but since the projectiles have to overcome reaction barriers and the products have to leave the sample to be detected, the real probing depth depends on reaction and target and is usually significantly smaller than the projectile range.

IBA works in the MeV energy range covered by DC tandem accelerators (2.2.1) starting from a few 100 keV up to about 10 MeV for H and He and a few 10 MeV for multiply charged heavy ions. This accelerator type allows for variable energy and current. Typical IBA end-stations (Fig. 7.29) are similar to electron microscopes, but only few commercial products with limited integration exist. The analysis of ions introduces additional angle dependencies in most interactions, making a set of angles in the end-station setup very important. Considering the Rutherford cross-sections angular dependence already 0.2° of angular uncertainty introduce 2.5% of uncertainty in the cross-section. The beam angle of incidence with respect to the surface normal α, the product exit angle β (to surface normal) and the reaction angle Θ relative to the incident beam together with the detector to sample and the ion optics to sample distances define the end-station properties. The optimisation in this parameter space represents a technological challenge and limitation. It defines spatial resolution since smaller ion-optics to sample distances lead to smaller spot sizes (Sect. 2.3.2, equation (2.30)) but also larger geometrical straggling (a closer object appears larger = covers a larger solid angle) and less space for large (solid angle) detectors. Detecting the low fluxes of secondary particles for IBA usually requires high detector solid angles to provide sufficient count rates for acceptable statistical uncertainty, as this scales with the number of detected events. The combination of beam energy and beam current defines the sensitivity, accuracy, and detection limits. Each IBA method has its individual sensitivity given by the cross-sections and the solid-angle coverage with Table 7.3 giving some exemplary orders of magnitude.

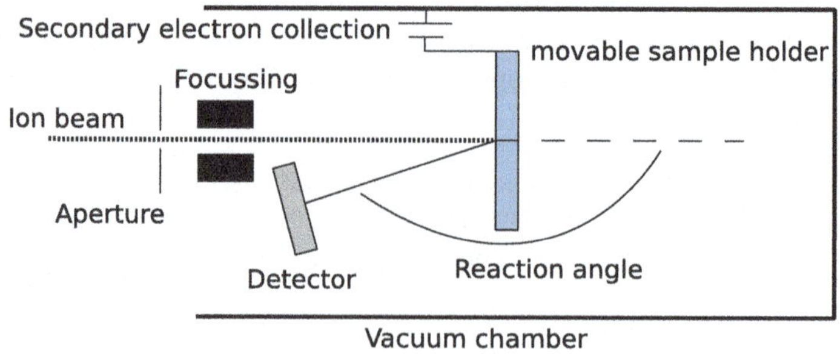

Fig. 7.29 Typical setup geometry for ion beam analysis. Ion optics form the beam coming from the left. Biasing the isolated sample holder enables secondary electron collection. The figure shows product detection in backscattering geometry. For transmission detection the detector is placed on the right side of the sample. Rotating the sample leads to a grazing incidence analysis

Table 7.3 Comparison of different IBA methods with respect to their fundamental sensitivity. An analytical method cannot detect all products, but its restriction to a certain detection angle reduces the fraction of detected products

Method	Products/projectile	Detector size (Sr)	Depth resolved
STIM/IBM	1	4π	No
RBS	10^{-5}	0.05	Yes
PIXE	10^{-5}	0.05	Partly
NRA/PIGE	10^{-7}	0.2	Yes

For mapping applications on the micro- and nano-scale, the ion beams produced by tandem accelerators cannot provide enough lateral spatial resolution. Ion optical beam focussing reduces the spot size by factors of 10–1000. The smallest spot sizes in the 10 nm range require additional apertures reducing the spot already before focussing to a few μm in size. Focussing increases the beam power density, potentially increasing the impact of the measurement on the sample by heating and radiation damage. If the measurement changes the sample, the result becomes meaningless. Low beam currents in combination with high detector solid angles allow mitigating this intrinsic technological disadvantage at the expense of counting statistics. Beam currents in the typical IBA range of pA to nA are tolerated by most materials, metals and inorganic compounds often tolerate even more.

The first class of analytical methods exploits the ion beam without classical detectors. Ion-beam-microscopy (IBM) counts the secondary electrons emitted per incident ion. Similar to electron microscopy, the secondary electron emission depends on the surface properties. Measuring the apparent ion-current while scanning the beam over the sample results in a 2D map of secondary electron emission values. Scanning-Transmission-Ion-Microscopy (STIM) exploits the projectile energy loss in the sample for imaging in transmission geometry. Increased energy-loss corresponds to thicker or increased stopping power (heavier elements) regions. Quantification of the energy loss yields depth-averaged mass density information as depicted in the first image of Fig. 7.30. Ion-Beam-Induced-Charge collection (IBIC) is a method for investigating semiconductors. It turns around the detector idea, by directing the projectile beam onto a semiconductor sample which is used itself as detector. Analysing the ion induced charge pulses results in charge collection properties and information regarding the response to displacement damage. Since the projectile beam energy is known, unknown properties of the semiconductor become visible. For example, defects in the semiconductor crystal alter the charge collection efficiency by hindering the charge mobility, a key property of semiconductors. All of these IBA methods feature the highest possible signal level, since every incident ion also induces a signal. In contrast to the methods relying on reaction cross-sections, this gives orders of magnitude higher signal intensities allowing for good image/data statistics even at low beam currents. Accepting low beam currents is equivalent to accepting small spot sizes/high lateral resolution.

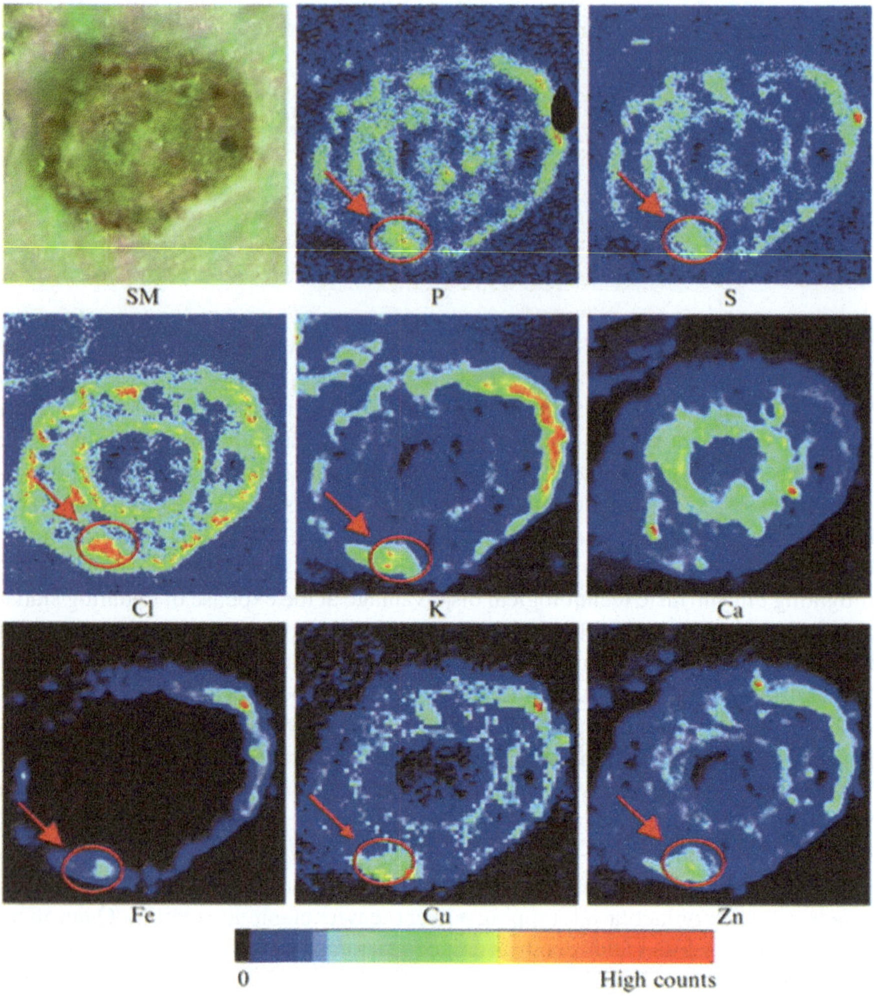

Fig. 7.30 Elemental mapping of a Viola lutea ssp. westfalica root sample enabling elemental localisation of its constituents via micro-PIXE (see below). The microscope picture (SM) STIM image shows a regular structure, but PIXE reveals an arbuscular mycorrhizal structure (arrow) in the elemental composition. Reprinted from (Vogel-Mikuš et al. 2009) with permission by Springer

The second class of methods recovers the information stored in released/backscattered ions. The class is further sub-sectioned by projectiles either above nuclear interaction thresholds or below them. Product ions contain particular high amounts of information since the similar mass to the projectiles leads to good contrasts due to the energy transfer. The detection of elastically scattered particles is named Rutherford-Backscattering-Spectrometry (RBS) and its non-Rutherford variant (complex cross-sections) is called Elastic-Backscattering-Spectrometry (EBS). The transition between the theoretically known RBS cross-sections and the

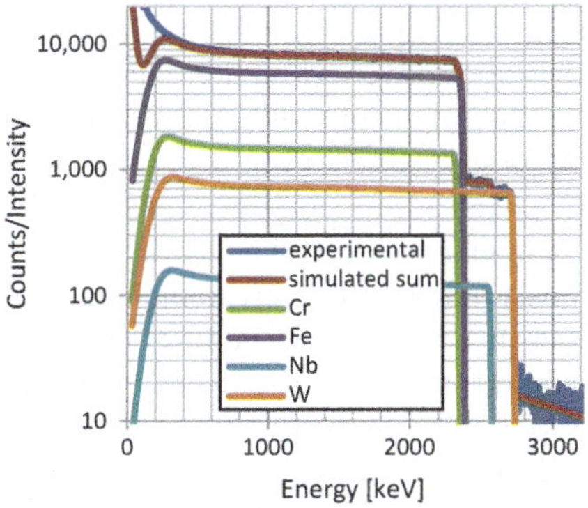

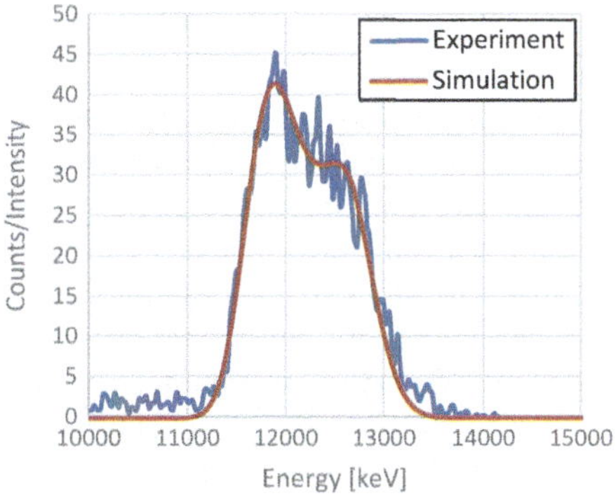

Fig. 7.31 Left: RBS spectrum of a ferritic steel consisting of Fe, Cr, Nb, and W. Right: NRA measurement of the D(^{3}He, p)^{4}He peak of deuterium implanted into the first 4 μm. The simulation fits the data by assuming a decrease of D density through the analysed depth. Both measured by 3 MeV ^{3}He ions. The simulated and elemental curves correspond to the data evaluation by SimNRA 7.02

more complex EBS cross-sections occurs roughly at half of the Coulomb barrier, see Sect. 3.3. With energies in excess of some MeV, nuclear reaction cross-sections become relevant. Nuclear reactions typically produce protons and ^{4}He with energy above the projectile energy. These products enable improved detection in particular of light elements with the so-called nuclear reaction analysis (NRA). Figure 7.31 compares RBS and NRA results from a steel sample. In principle both product types can be detected using a single detector, where the RBS part ($Q = 0$) can be found at scattering energies roughly up to the beam energy and the NRA part in particular sits at higher energies when $Q > 0$. In practice, different count rates due to the largely different cross-sections and significantly different product energies potentially require two specialised detectors for detecting the full energy of all particles.

Figure 7.31 demonstrates the main problem of RBS: The signals of bulk elements overlap. In the measurement we obtain only a single value per product energy, therefore we cannot disentangle the individual elements contributing to it. Every element starts with an edge, the so-called surface peak, with a slope given by the energy resolution of the detector setup. This edge represents the sample surface. From here on the projectiles enter the sample and continuously lose energy by stopping. If we consider the sample as a sandwich of depth slices, every slice would feature its own projectile energy according to the stopping of the beam. Lower beam energies also result in lower product energies, consequently extending the RBS signal to lower product energy with every layer of the sandwich. The heavier an element, the higher its Rutherford cross-section and the higher its surface peaks energies (due to the backscattering nature of the measurement). Consequently, lighter elements are buried in the statistical counting uncertainties of heavier elements making it impossible to detect for example C in W.

Elastic-Recoil-Detection (ERD), a variant of RBS in grazing incidence, improves mass and depth resolution to resolve this. Both Rutherford cross-section and stopping power increase with projectile proton number, resulting in improved detection limits (cross-section) and depth resolution (stopping-power) when using heavy projectile ions. By choosing incidence angles above 45°, the transmission direction moves to the sample front, allowing a detection of the light and heavy product particles which are otherwise stuck in thick samples due to conservation of forward momentum. The heavy product represents the target recoil in elastic scattering, hence ERD yields the best resolved compositional information. In contrast to RBS where the detector only receives particles of the projectile type and to NRA, where mostly protons and α's are received, ERD releases product particles of all elements and isotopes present in the sample. Identifying not only the energy but also the mass (or momentum $p = m * v$) of the product therefore yields additional information. Time-of-flight detector tubes with a thin transmission (timing) and a thick energy detector in series produce the required energy versus velocity maps as depicted in Fig. 7.32. Each element/isotope has a typical curve in this picture originating from the connection of velocity and kinetic energy through the mass ($E = mv^2/2$). The top-left represents the surface and the lower right edge the maximum analysis depth. The intensity is connected to the Rutherford elastic scattering, allowing for an absolute quantification similar to RBS.

The elastic scattering techniques are particular good in depth resolution, but several factors limit the depth resolution. The technological limit is given by the maximum stopping power (at the Bragg peak) and the energy-loss straggling. Technical limits arise from the dependence of the scattering energy on the scattering angle. The finite spatial extent of beam spot on the sample and the finite detector

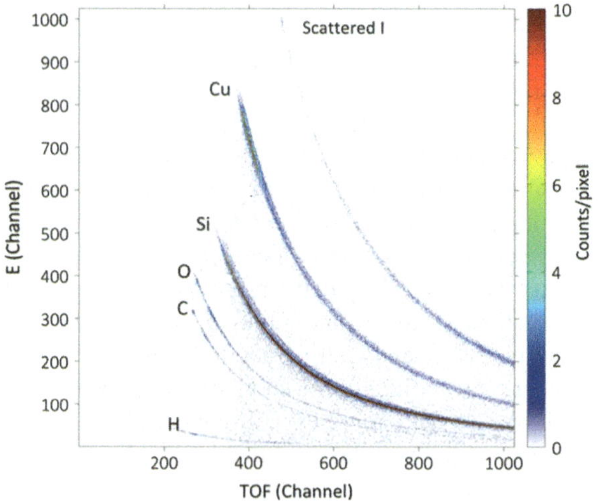

Fig. 7.32 Time-of-flight elastic recoil detection analysis (ERDA) with an iodine beam. The *Y*-axis shows the detected particle energy and the *X*-axis its time-of-flight, together spanning the whole energy–momentum space of the detected particles. Reprinted from (Siketić et al. 2018) licensed under CC-BY 4.0

aperture diameter directly lead to acceptance angle windows for the scattered particles. The wider the acceptance angle window the wider the detected particle energy width, since projectiles with different scattering angles (see kinematics in equations (3.16) and (3.17)) and path length through the sample mix in the detector signal. Beam intensity and detector counting statistics imply lower limits for these technical dimensions, since smaller beam diameter results in smaller beam current for a given beam current density and smaller detector apertures reduce the detector solid angle.

Steels are a good example for the limits of RBS. All steels contain C and H as critical components for their mechanical properties, but RBS will hardly be able to quantify these light elements. A better contrast in particular for light elements is achieved by Nuclear-reaction-analysis (NRA) of inelastic nuclear fusion reactions. The fusion reactions ($Q > 0$) introduce extra energy into the reaction, allowing a separation of NRA products from the elastically scattered particles which are limited in their kinetic energy by the beam energy and the elastic kinematics. Figure 7.31 demonstrates the proton peak of the $D(^3He, p)^4He$ reaction lying isolated at $\approx$12 MeV in a basically background free region, while RBS ends already at 3 MeV with many different reactions overlapping each other.

Light elements such as hydrogen and lithium often occur in energy related materials. Figure 7.33 shows a combined NRA+RBS analysis example of a Li-battery related material using a single passivated implanted silicon detector. The measurement contains quantity and depth distribution of all elements. The peak heights mostly

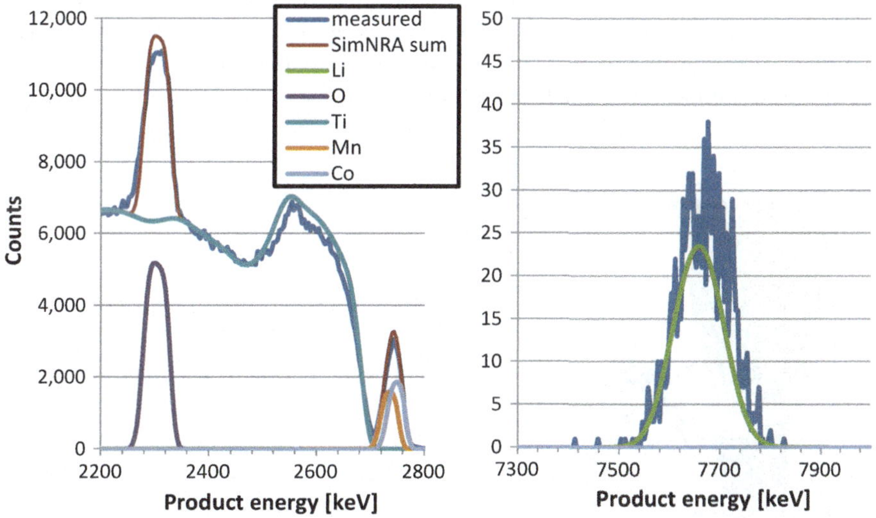

Fig. 7.33 3 MeV Proton beam based NRA and RBS analysis of a Li–Co–Mn–O ceramic thin film deposited on Ti with SimNRA 7.02 interpretation. The reduced slope of the Ti edge at 2700 keV and the width of the O peak at 2300 keV demonstrate the formation of a 150 nm TiO_2 film between substrate and the 500 nm thick ceramic layer. The element lines correspond to the perfect stoichiometry which slightly differs from the actual layer composition

yield the element concentrations, the peak width contain information on layer thickness (=energy loss), and peak height (also rising and falling edge) variations contain information on intermixture of layers and elements (e.g. by diffusion).

A critical factor for this high information content is the depth-resolution $R(x)$, measured in units of atoms/m^2. For charged particle products it depends on the scaling of the product energy $\Delta E_{\mathrm{product}}$ with depth in relation to the stopping power S.

$$R(x) = \Delta E_{\mathrm{product}}/S \tag{7.7}$$

Equation (3.6) quantifies the decrease of stopping power S with projectile energy/velocity, or in other words the decrease of depth resolution with increasing projectile energy. The product energy width $\Delta E_{\mathrm{product}}$ in (7.7) depends on several factors. Ideally, the mono-energetic projectiles result in a single product energy (Sect. 3.4), but the detector resolution, energy-loss straggling and geometrical aspects always lead to a window of possible projectile energies even for a mono-energetic beam. The depth resolution at the very surface depends on the ratio of stopping power to detector resolution plus geometrical broadening effects (here the energy-loss straggling is still zero). This so-called geometrical straggling originates from finite beam and detector sizes resulting in distributions of scattering angles and outgoing path length accepted by the detector, see Fig. 7.34. In Sect. 3.3.2 we discussed the impact of the reaction angle on the product energies (there E_3 and E_4). With increasing depth in the sample, energy-loss straggling induces a distribution of projectile energies further limiting the depth-resolution, see the right picture in Fig. 7.34. With increasing the projectile impact angle on the sample against the surface normal, the depth resolution increases due to the sine like increase of the projectile path length

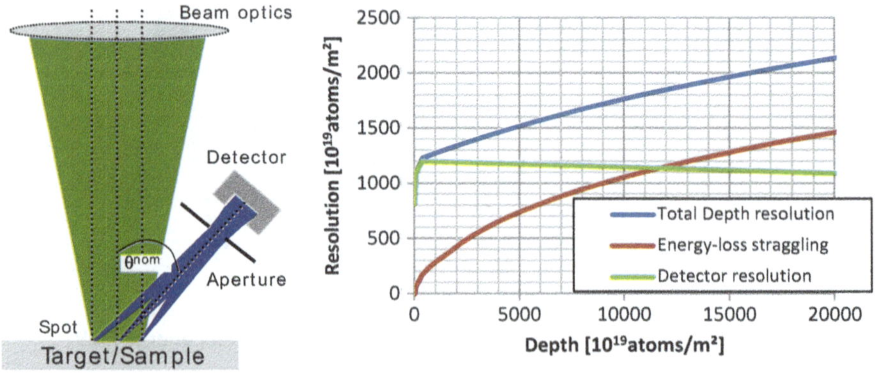

Fig. 7.34 Left: Geometrical straggling/broadening arises from finite spot size connected to different product origins and from finite detector apertures accepting a range of reaction angles around the design angle. Right: At the surface the detector resolution (here 15 keV FWHM) limits the depth resolution, but with increasing depth the energy-loss straggling dominates ($>1.2 \times 10^{23}$ atoms/m^2) the product energy-width

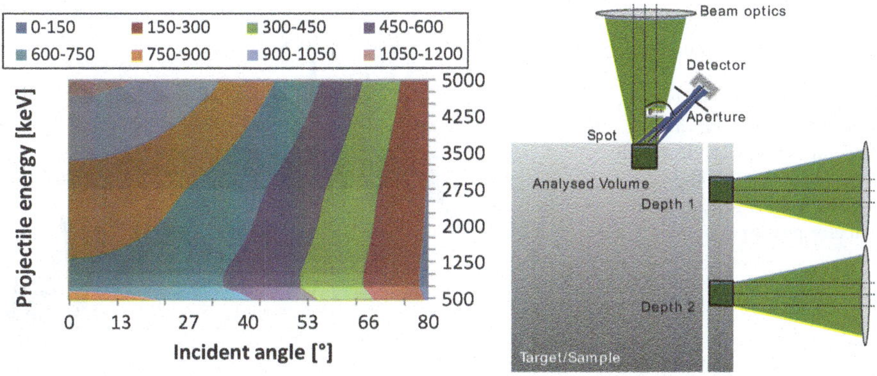

Fig. 7.35 The depth resolution changes with energy, impact angle, and depth at the example of the $^{12}C(^{3}He, p_0)^{14}$ N reaction. Left: best resolutions in units of 10^{19} atoms/m^2 are attained at low energies and large angles. Right: The same analysis from different sides of the sample or cross-cuts (white line) extends the tomographic range covering potentially the whole block

for reaching a certain depth (Fig. 3.15 in Sect. 3.4). With impact angle the beam spot size increases elliptically. Figure 7.35 left summarises the impact of projectile energy and impact angle on depth resolution. In the given example, the depth resolution improves by a factor 10 when changing from normal incidence with 5 MeV ions to 80° incidence with 1 MeV ions.

Analysing the sample from different sides increases the accessible volume. Short pathways are desirable for the analysis, since straggling continuously destroys the information retained in the products by broadening their energy distribution. Investigating the sample from the side or on a cross-cut allows for probing the whole sample volume (Fig. 7.35 right), but so far the technical limits of focussing render this alternative the option with worse spatial resolution. The analysis requires the projectile to reach the analysed depth, but also the products have to leave the sample to be detected. On the lower end (<1 MeV) high stopping powers lead to analysis ranges of typically a few 100 nm. On the higher end of about 30 MeV even a few 100 μm can be probed, but the lower stopping power decreases the depth resolution.

The third class of methods exploits photons emitted by the projectile-target interaction. Photons feature the best contrast, since they are not involved in the primary process, they are un-charged, and the peculiarities of the quantity loss mechanism (Sect. 3.1). This peculiarity lacks the depth resolution connected with the stopping power of massive particles, but sharp resonances in the production cross-section of gamma radiation can compensate this partially. Varying the projectile energy varies the depth of the resonance in the sample, enabling a sequential scanning of the sensitive depth through the sample.

PIXE is the ion equivalent of EDX, but with the higher mass of ions much less bremsstrahlung is emitted (see Sect. 4.3.2) and therefore the resonant peaks are on a smaller background and better detection limits can be achieved. Figure 7.36 demon-

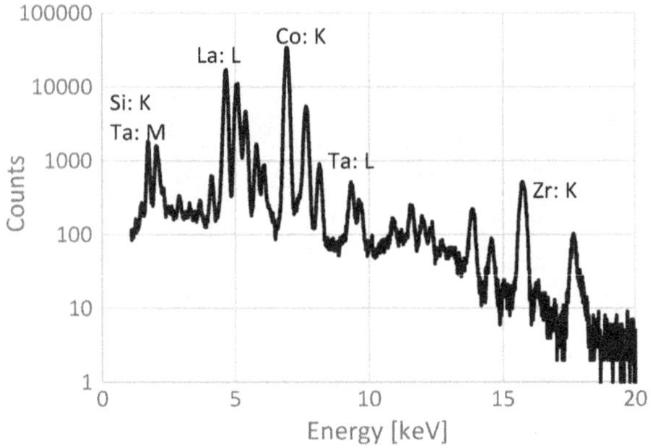

Fig. 7.36 PIXE spectrum of a solid state battery mixture of $LiCoO_2$ and $Li_7La_3Zr_2O_{12}$:Ta. The labels indicate the elements and their line groups (K, L, M) in the spectrum. The other, smaller lines originate from detector pulse-pileup

strates the capability of the method in detecting major elements with statistical uncertainties <1% and small impurities such as the silicon remaining on the sample from the SiC grinding paper. PIXE features high count-rates, but only limited depth information. Technologically respect PIXE represents the most advanced IBA method, due to the availability of complete analytical descriptions for its cross-sections and the high performance of modern integrated silicon-drift-detectors (SDD) used for x-ray detection compared to detectors for charged particles or γ-rays. State-of-the-art SDD detector energy resolutions of < 130 eV (@6 keV) enable separating practically all neighbouring K, L, and M lines, completely resolving the energy resolution as a technical limit of PIXE measurements.

Similar to NRA, the projectile can also induce nuclear excitations, forming the so-called Particle-induced-gamma-emission analysis (PIGE). These reaction could be simple excitations such as $^7Li(p, p_1)^7Li$ but also combined nuclear reactions such as $^7Li(p, n\gamma)^7Be$. The high energy $\gamma's$ can penetrate several mm of material, allowing a detection even through sample and vacuum chamber. Low count rates due to small cross-sections and detection efficiency in combination with limited number of reactions render a full analysis of a sample using only PIGE difficult. Depth profiling with PIGE follows a different mechanism than with charged particles. Naturally all γ's feature the same energy, as it is specific to a certain reaction and photons lose energy by reduction of intensity (Sect. 3.1). Deriving a depth information requires a sharp resonance in the reaction cross-section. Varying the depth of this resonance in the sample by changing the projectile energy results in a depth-profile by interpretation of the energy dependent count rates. In this case only the projectile energy-loss straggling deteriorates the depth resolution, but the counts of the non-resonant part of the cross-section induce an additional loss of depth resolution. With sufficient resonance peaking, the energy-width of the resonance can be considered as $\Delta E_{\text{product}}$ for calculating the depth resolution at the surface.

As a case study we plan a RBS+NRA measurement using MeV ion beams at the example task of analysing a lithium battery NMC ($LiCo_{1-2x}Mn_xNi_xO_2$) bulk cathode material nowadays used in most mobile applications (cars, tools, phones, etc.). The material contains an interesting mixture of light and heavy elements. The heavy elements are easily accessible via RBS, but their similar atomic masses make it hard to distinguish them via RBS. Adding PIXE data would allow for a clear determination of their concentration in depth with RBS adding surface near information. Lithium RBS will be hardly visible on the heavy element RBS background and its PIXE x-ray energy of 52 eV (Fig. 3.4) will be hardly detectable. Consequently an NRA reaction such as ^{7}Li(p, α)^{4}He yields the best detection properties. RBS and PIXE are largely independent of the projectile type allowing to conduct Li, Mn, Co, and Ni measurements in one run with an energy determined by the ^{7}Li(p, α)^{4}He threshold. The main isotopes ^{16}O of oxygen does not have a nuclear reaction with protons, but a resonant increase of the elastic scattering cross-section up to a factor 10 at 3–4 MeV. The Rutherford cross-section equation (3.1) scales with the square of the target nuclear charge, in principle yielding a ratio of the Ni to O cross-section of $(27/8)^2 = 11.4$. A factor 10 increase in O cross-section would therefore be sufficient for a clear O signal on the metals RBS background by compensating the difference in cross-sections. Alternatively, D ions offer nuclear reactions with O and Li isotopes. While protons at 2–4 MeV would allow for a localisation and quantification of all known elements, deuterons with their additional nuclear reactions would enable the detection of additional elements possibly present as impurities such as C and N. In a proton based analysis these elements are hardly detectable on the metal RBS background resulting in rather high limits of detection. Whether this additional information is relevant and worth the additional costs of deuteron operation remains to be decided considering the whole situation. The measurement will analyse down to a depth of 10–20 μm.

The challenge of analysing the data produced a vast set of analysis codes nicely compiled by the IAEA (https://nucleus.iaea.org/sites/accelerators/knowledgerepository/code/forms/allitems.aspx). The codes separate into four groups: Forward or backward interpretation and analytical or monte-carlo calculations. Most of the codes use analytical forward simulations. These codes try to find a situation which fits the experimental results best based on analytical equations. The codes therefore require an educated guess of the result and a good description of the experimental setup, requiring an experienced scientist or a trained neural network (Mayer et al. 2020). This input result is then varied until a minimum is found in the difference to the experimental data. Forward calculations have the disadvantage of possibly finding the wrong result even with a perfect mathematical match, if for example the measurement is not sensitive to a certain parameter or the experimental data exhibit significant statistical uncertainties. For RBS and NRA the code SIMNRA7 and NDF provide the most detailed implementations of analytical forward models. NDF also simulates photon based analysis in a consistent manner with charged particle spectra of the same sample. Specialised codes for PIXE such as GuPIX and Geopixe add additional features for quantitative mapping in certain wide-spread applications of PIXE such as geology. Monte-carlo code such as Corteo have particular strength in analysing complex geometrical situations such as 3D structure samples at the disadvantage of introducing computational noise due to their test particle nature. So far backward calculations were not implemented in a general purpose code due to the complexity and often ambiguity of IBA data.

In the beginning of ion beam analysis dedicated setups were used for each individual method. Over the years, a combination of methods has proven to provide superior results due to the weaknesses inherent to single methods. The combination of methods introduces new technical challenges, but the synergistic benefits of the so-called Total-Ion-Beam-Analysis (Jeynes et al. 2012) can eventually solve the inherent problem of ambiguities in IBA data. NRA detects light elements, RBS with its high cross-section reveals intermediate to heavy elements, PIXE adds the lowest detection limits for impurities and enables separation of intermediate elements, and STIM/IBM connect the results with other imaging methods.

Self-consistent interpretation through the knowledge of element identity and quantity from PIXE and PIGE combines with the depth information of RBS and NRA. Secondary gammas contain isotopic information in particular for light elements, while the x-rays provide the highest counting rates required for quantification. Secondary ions of RBS provide high count rates and excellent depth resolution, but the overlap of the low scattering cross-sections of light element on the in-depth signals of heavy elements prevents a statistically sound analysis of elements below carbon. NRA in contrast features attractive reactions with these elements often with high Q-values leading to a clear separation.

Not only combining different methods in a single measurement, but also different projectiles and energies mitigates the weaknesses even further. In Sect. 3.3.1, Fig. 3.13 demonstrates already slight changes in the energy potentially change nuclear reaction cross-sections by orders of magnitude and with that the detection properties change. The combination of different projectiles with different elemental and depth sensitivities even further increases the analysis result quality without need for extra equipment, see Table 7.2. Figure 7.33 demonstrated the quantification of ^{7}Li using 3 MeV protons. The same analysis with 3 MeV ^{3}He would add increased depth resolution in particular for C and O. Deuteron projectiles feature efficient reactions with sulphur and specific isotopes such as ^{6}Li and ^{18}O for studies of isotope tracers. ^{15}N ions reveal the hydrogen uptake via the H(^{15}N, α)^{12}C reaction.

In the future, current developments will culminate in highly standardised and automated analysis setups for nanometre-MeV-Ion-Beam-Tomography, if the reader allows for an educated guess. For a summary of some current technological challenges see (Mayer et al. 2019). New software and data analysis algorithms already now indicate the possibility for handling the data masses (Mayer et al. 2020). The advantage over electron microscopy lays in the depth information and two to three orders of magnitude higher sensitivity due to the absence of Bremsstrahlung and a broader spectrum of possible reactions. On the other hand the technological complexity makes it expensive and so far not developed to the point where an unexperienced user can quickly operate the analysis and consistently provide results as it is possible with modern electron microscopes.

7.1.6 Accelerator Mass Spectrometry

We saw many analytical methods in the preceding sections exploiting the idea of shooting accelerated particles onto a sample. Accelerator mass spectrometry (AMS) is the black sheep among the accelerator based analytics. It exploits the accelerator itself and attached beam optics for its analytical mechanism. The samples sit in a specially adapted sputter ion source producing a beam of sample ions. This process consumes the sample continuously, forming a multi-species beam consisting of all the elements and isotopes present in the sample. The integral beam current consists of the sum of the currents of its constituents and their different charge states. The more particles of a certain type present in the sample, the higher will be the ion current of this species. To be more accurate, the integral charge released of these ions will increase, since the sample will be completely evaporated after a certain measurement duration.

The topic of mass-spectrometry or sometimes also mass-spectroscopy combines numerous techniques of which AMS represents only a niche. This class of methods finds application in physics, chemistry, biology, geology, and even forensics by revealing elemental, isotopic, and even molecular composition of samples. This book cannot cover all the details of mass-spectrometry, but many of the methods rely on accelerator techniques such as ion sources and beam optics, but mostly working at eV to keV energies. For a deeper reading, the reader is referred to other books, such as (von De Hoffmann 2007; Gross 2006; Tuniz 1998).

In particular geological sciences have special interest in AMS due to its ability to quantify extremely scarce isotopes. The best AMS devices can detect these isotopes down to a fraction of 10^{-15} of the atoms in the sample. Examples include the isotopes with half-lifes of geological time scales of thousands to millions of years. ^{14}C (half-life of 5730 years) and in particular its ratio to ^{12}C (stable) marks the prime example for all AMS dating. Cosmic radiation generates ^{14}C on the earth surface for example via $^{14}N(n, p)^{14}C$ reactions in all carbon containing entities in minute amounts in the order of one ^{14}C atom per 10^{12} ^{12}C atoms. This ^{14}C attaches for example to CO_2 molecules. Upon death of, for example, plants they stop incorporating new ^{14}C nuclei, decoupling their isotopic composition from the atmospheric value (freezing). Via the known decay properties of ^{14}C and its surface isotopic ratio, the isotopic ratio of ^{14}C allows for dating of the death or, more accurately, the freezing point of the analysed sample. Figure 7.37 shows a calibration curve of this decay age, exhibiting relevant differences between the radiocarbon age derived from ^{14} to ^{12}C ratio and the real calendaric age for this calibration derived from other quantities. These systematic derivations originate from incomplete freezing of the ^{14}C content, variations of the natural atmospheric ^{14}C content over the years, and variations of the ^{14}C uptake due to consumption of non-atmospheric carbon during the living period. Further examples of interesting isotopes include ^{26}Al, ^{10}Be, and ^{36}Cl each with its own specialty in dating range and sample types.

The problem of these minute quantities lies in the overlap of masses, requiring extremely high suppression ratios. For example the masses of ^{14}C (14.0032 amu) and

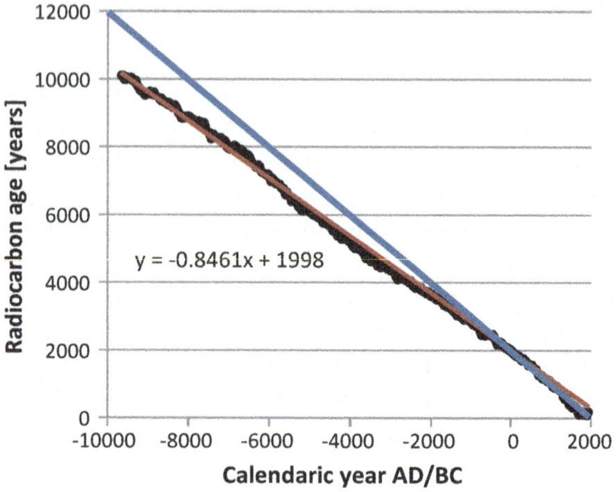

Fig. 7.37 The calibration of calendaric age and ^{14}C decay (black points and red fit) shows deviations from the ideal connection (blue line). Data compiled from (Stuiver et al. 1998)

^{14} N (14.0030 amu) differ only by 10^{-5}, while ^{14} N might comprise over 10^6 more atoms to the sample than ^{14}C as it chemically binds to many elements. Considering Gaussian broadening effects present in any device and statistical noise of counting experiments mentioned above (Sect. 7.1.1) we require highly resolving technologies. Calculating this example with two Gaussian beams of $x^1{}_{Max} = 1$ mm (1σ width) leads to a distance of 5 mm required in order for both intensities to be equal at the centre of the ^{14}C beam.

In order to develop such a technology the most important question is: In which quantities do the nuclei differ? We discussed already the tiny difference in mass. The difference in nuclear charge of 6–7 offers a significantly higher separation ratio. A technology exploiting this difference potentially offers higher separation efficiency, but it requires a full ionisation of the nuclei. We learned about several methods for ionisation, namely the plasma source and beam matter interaction in its stripping aspect. Plasma ionisation hardly reaches more than 2+ with the typical plasma temperatures. Stripping reaches higher ionization levels only >100 MeV, for example N^2 gas strips 99% of carbon ions to the 6+ state only at 100 MeV and above (Wittkower and Betz 1973). The stripping process always produces a mixture of charge states, prohibiting a clear separation along the maximum charge state, at least for the targeted concentration ratios of rare isotopes. In the stopping power itself, the nuclear charge yields a clear impact, with the limitation of broadened distributions by straggling. ^{14}C ions of 10 MeV feature a range of 7 μm (0.23 μm range straggling) in a carbon foil, while ^{14}N ions only reach 5.9 μm (0.2 μm range straggling) deep, potentially separating the elements. ^{12}C Ions reach 6.98 μm deep, a difference to ^{14}C ions well below the range straggling. The radioactive decay of ^{14}C also represents a difference to the stable ^{14}N, but with the minute quantities expected we will hardly see any statistically relevant amounts of decay products. A technology based on this aspect offers no prospect of success.

The technological challenge of mass spectrometry condenses to induce isolated beam path for each different mass to charge (m/z) value investigated in conjunction with a maximal and m/z independent transmission of particles from source to detector. Unknown transmission values make a calibration necessary. Usually electrostatic and magnetic dipoles select the mass of the accelerated ions by inducing energy/momentum dependent curvatures (see 2.3). We can derive the separation strength from a simple 90° dipole system. Mass resolution R in terms of mass m over the minimum resolved mass difference Δm derives from the ion curvature radius in the optical element r over the minimum curvature radius difference Δr where a projectile of different mass will not anymore be transmitted through the system. This derives from the ion optical deflection angle per mass difference φ, the drift-path length l, the aperture/detector width d in the bending direction after the drift element, and the angular beam spread ε via

$$R(E, m) = \frac{m}{\Delta m} = \frac{r}{\Delta r} \sim \frac{\phi * l}{\epsilon * d} \tag{7.8}$$

In the case of the magnetic dipole, the spatial separation depends equally on mass and particle energy. Consequently the beam energy spread due to fluctuations in E and the transversal emittance must be smaller than the resolved mass difference. Equation (7.8) demonstrates the importance of technical limitations of laboratory size. Large deflection angles and long beam tubes consume space. The ion source and accelerator define the emittance/beam spread with their emittance and acceleration voltage stability. The beam energy itself is of minor importance, as long as the beam loses negligible amount of energy in the beam lines. The aperture size defines the detection efficiency, as smaller apertures offer better resolution but also a certain loss of beam particles. Beam preparation in the form of the ion source efficiency and the accelerator transmission add up to this, giving the accuracy/statistical uncertainty of the systems measurements (different from the resolution). Detectors itself typically feature 100% detection efficiency. For a given beam species the accelerator transmission is independent of the beam current, at least for the extremely small currents down to fA considered by AMS. Therefore, it can be determined from reference samples with known isotopic ratios, enabling a quantitative calibration of the method.

Technical realisations exclusively rely on tandem DC accelerators. These devices readily provide the required beam energies in the order of a few MeV in combination with acceleration voltage stability $<10^{-5}$. This combination provides high resolution at acceptable costs together with the practical advantage of a grounded ion source and detector. Figure 7.38 shows an example AMS system. The system applies a 54° deflection electrostatic and a 90° deflection magnetic dipole on the low energy side and 180° magnetic filtering on the high energy side of the accelerator. The whole beam path has a length of about 40 m. Several quadrupole magnets along the path maintain the beam diameter. A switching magnet thereafter allow for reaching different end-stations with optimised detection properties for certain isotopes. With

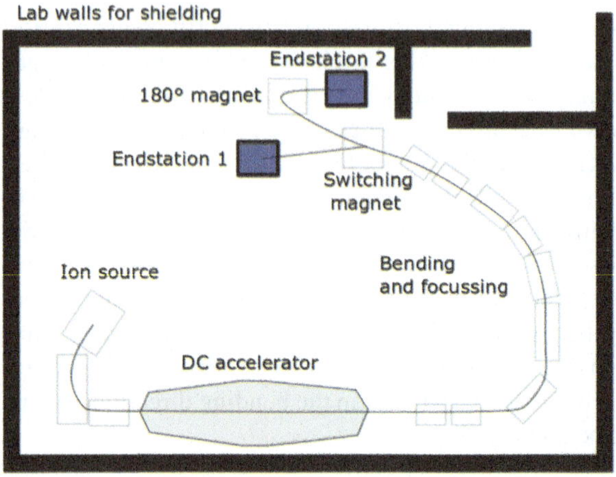

Fig. 7.38 Example of an AMS based on a 6 MV tandem DC accelerator. The sputter ion source on the lower left provides the beam of rare ions. Several magnets bend this beam after the accelerator by about 180° towards end-station (1) Lighter ions can be bend by another 180° towards end-station (2) for improved mass resolution. Quadrupole doublets keep the beam together on its way towards the end-stations

this beam optical system the AMS reaches an accuracy in the order of percent for ratios of the rare to the major isotope down to 10^{-12}.

In the example of the study (Friedrich et al. 2006)^{14}C dating was used for determining a huge volcanic eruption in the Bronze Age. For this dating, an olive tree found in the vicinity of the volcano buried alive during the eruption. The required sample mass for AMS lies <1 mg per data-point, allowing probing of several growth rings of the tree for improving the dating accuracy. In nuclear fusion research an experiment investigating ^{10}Be migration in a plasma environment (Bykov et al. 2016) was conducted. An enriched ^{10}Be sample was exposed to the JET Tokamak plasma, leading to a release of material by sputtering. From this original sample, the ^{10}Be migrates through the reactor and deposits on the >1000 times larger area of the reactor vessel. Only minute amounts of material migrate, but due to the extreme sensitivity and dynamic range of AMS these amounts can be followed. Extraction of surface near samples by grinding and subsequent chemical processing of the dust yielded several discs of material specific to certain locations in the Tokamak vessel. This allowed tracing not only the general migration of beryllium from the large beryllium surface present in JET, but also the migration from the specific spot of the enriched ^{10}Be sample in the large background of the natural beryllium migration.

7.2 Neutron Based Analysis

In addition to the neutral and massless photons and the charged and massive electrons and ions the neutron completes the analytical spectrum by being neutral and massive. As such, it offers unique properties, bearing the long range of a neutral particle and the interaction possibilities of a massive particle. Free neutrons decay with a half-life of 10.1 min, but even at low kinetic energies, the loss of beam intensity due to decay remains negligible. On the other hand, neutrons are difficult to handle in the sense of a beam due to the lack of electro-magnetic interaction neutron beam optics cannot work as for charged particles. This results in a lack of possibilities for beam focussing. Consequently, neutron beams remain large and microscopic imaging and analysis, as discussed for all other projectile types above, remains out of reach. This limits spatial resolutions usually to the mm scale. Neutron guide tubes (discussed in Sect. 4.2) represent the main technical option with some characteristics of beam optics, but these rather open up a certain technical freedom for placing analysis apparatuses than a dynamic control over the neutron beam phase-space. Furthermore, the generation of free neutrons is among the most costly options for analytical purposes, making neutrons somehow the last resort of analytical options.

Accelerator based neutron sources always exploit the neutron as a secondary particle from a charged particle nuclear reaction with matter, see neutron sources Sect. 4.1. These source potentially feature high brightness, but due to the fact that only a few percent of the incident charged particles induce nuclear reaction producing neutrons, the price per neutron projectile lies significantly above the price of charged particles and the achieved flux densities lie below those of charged particles. Only at charged particle energies above some 100 MeV each ion can produce several neutrons on average. Nevertheless, the neutrons physical properties lead to a very complimentary situation with charged particle analysis in terms of analysis range and contrasts. Neutrons feature high range, since electronic stopping powers or absorption have no meaning for them. This allows probing even meters of material, where charged particles can only probe up to a few 10 μm and photons a few 10 mm.

Neutron based analysis methods exploit the same fundamental concepts of scattering, imaging, and nuclear interactions as the charged particle and photon based analytics discussed above. This section will not elaborate all neutron methods in detail but show connections to equivalent ion or photon based methods. Besides these methods, where only the projectile is exchanged, neutrons offer also a few unique analytical options. The neutrons disadvantage of difficult control in the sense of beam optics represents also a strength. As a neutral particle it cannot annihilate with matter as charged particles, but it survives even a low kinetic energies or after many collisions with regular matter. Exploiting this feature allows for using not only fast (MeV) neutrons, which is the natural state after producing free neutrons via nuclear reactions, but they can be slowed down by collisions, usually with water, to thermal energy distributions or even to cryogenic temperatures. In the thermal (around room temperature or 0.025 eV) and cold (around liquid hydrogen temperature) kinetic energy range, the neutron de-Broglie wavelength (7.2) reaches the relevant nm-scale. The neutron can be treated as a wave, allowing accessing the methods similar to x-ray based analysis on similar scales. In the region of fast neutrons, on the other hand physical effects similar to MeV ion beam analysis take place.

Similar to other particles neutrons can be seen as individual particles or waves. Physically these two points of view are equivalent, the so-called wave-particle-dualism, but in practical analysis the mathematical formalisms and technical realisations differ. At thermal kinetic energy elastic and inelastic neutron wave scattering with the target becomes possible. Neutron scattering methods have a similar physical basis as the x-ray based scattering methods (e.g. SAXS) discussed in Sect. 7.1.2, only the contrast building physics is different. Photons see differences in the electron density, resulting in a scaling of the scattering cross-section with the target proton number Z or with density variations. Neutrons on the other hand have no perception for charges. The neutron scattering cross-sections depends on nuclear mass and nuclear interaction capability, see Fig. 7.39. The scaling with Z makes light elements difficult to detect using x-rays, in particular if mixed with heavy elements. The isotope dependent amplitude of neutron scattering does not have this overshadowing effect of heavy elements. Thermal neutrons have a special sensitivity for hydrogen, lithium, and boron. This sensitivity results in a completely new space of sample properties accessible by neutron scattering methods compared to their ion and x-ray equivalents.

For imaging, the different contrast-mechanism enables imaging in particular of light elements with an isotopic sensitivity. In addition to the different sensitivity

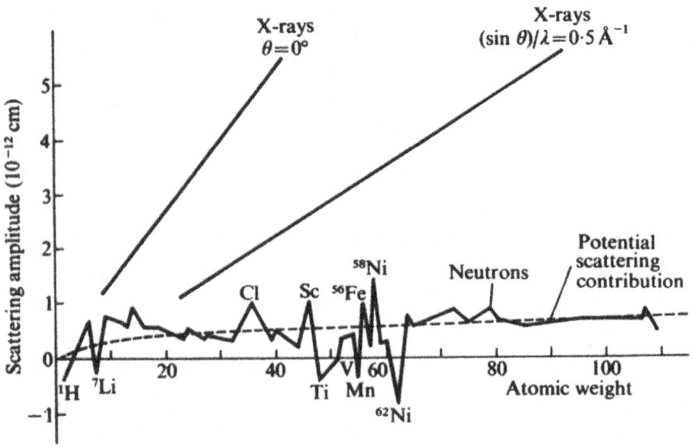

Fig. 7.39 Comparison of neutron and photon scattering amplitude. Neutrons exhibit no systematic dependence on elements or nuclear mass as x-rays do. The values even change by orders of magnitude for isotopes of a single element. The relatively constant scattering amplitude of neutrons represents an advantage over x-rays for detecting light elements. *From Research, London 7 (1954), 257)*

Fig. 7.40 Time-resolved 3D neutron tomography shows the rise of deuterated water (D_2O) in the root system of a lupine plant on the time-scale of minutes. The D_2O itself is invisible, but it pushes the visible H_2O (turquoise) bulk upwards. The deuterium as tracer enables a separation of the natural hydrogen present in biological systems from the investigated effect via neutron tomography. Reproduced from (Tötzke et al. 2017) published under CC-BY 4.0

compared to x-ray imaging, the neutron energies allows adjusting specific transparency of the target due to its impact on cross-sections (e.g. Fig. 4.7 and the connection between attenuation length and cross-section). Figure 7.40 shows an example of an in-situ thermal neutron tracer study of deuterium marked water in a plant (Tötzke et al. 2019; Tötzke et al. 2017). Rotating the sample in the neutron beam yields tomographic information.

In contrast to photons, neutrons have a magnetic moment. This induces a scattering contribution of the magnetic properties of the target nuclei. The scattering mechanism actually relates to the local magnetic field inside the sample. To implement this magnetic contrast the incident neutron beam has to be polarized, since neutron polarisations are naturally unordered. The magnetic spin polarisation direction can be selected by passing the neutron beam through a magnetic field.

Figure 7.41 shows a neat example of combining the neutron tomographic option with the magnetic spin polarisation of neutrons. This 3D magnetic field neutron imaging sample a simple copper coil in several different directions as a proof of principle (Hilger et al. 2018). A complex algorithm reconstructs the magnetic vector fields from the multiple magnetic points of view provided by the neutrons. The high penetration range of neutrons in matter enables probing the inside of bulk samples with spatial resolutions of 0.2 mm in this case. While the simple coil represents only an academic example, the same method was applied to investigate the magnetic flux inside a superconductor or the magnetic field structure in electric motors, battery cells (flowing current = magnetic fields), or magnetic materials to boost the understanding and inspire technical improvements and innovations. This method probes a quantity

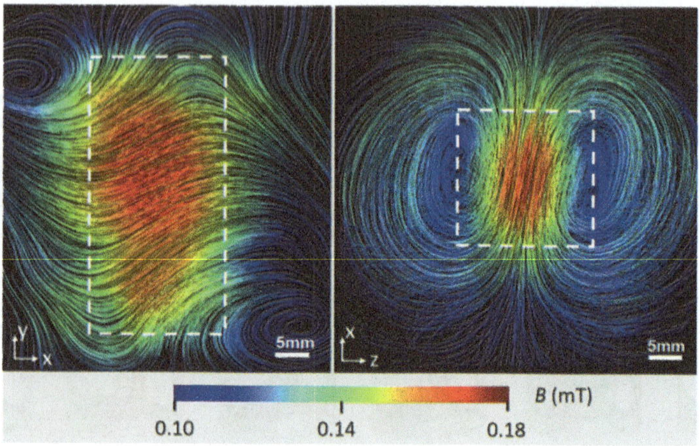

Fig. 7.41 Two different imaging direction using magnetic neutron imaging of the flux field lines inside a superconducting lead sample (dashed box). The exemplary measurement correlates well with analytical models of the expected field distribution. Reproduced from (Hilger et al. 2018), licence CC BY 4.0

inaccessible to all other projectiles discussed in this book, making it a valuable addition for science and industry.

The nuclear reaction of thermal neutrons with target nuclei results in most cases in (n, γ) reactions, but in a few cases also ions can be emitted. Nuclear reactions enable the three analysis methods discussed below: Analysis of the emitted γ, analysis of the decay of the heavy product, and analysis of the promptly emitted light products. The usage of thermal neutrons results from the widespread use of fission neutron sources, but with accelerators also fast/MeV neutrons become available opening up negative Q-value reactions such as $(n, 2n)$, and additional reactions resulting in light products.

Neutron-depth-profiling (NDP), the neutron equivalent of the ion based nuclear reaction analysis (NRA) discussed in Sect. 7.1.4 uses the light products of nuclear reactions. NDP reaches similar detection limits and analysis range as NRA, but due to the negligible projectile energy, the method achieves a better depth resolution with negligible heat load and radiation damage of the sample. This soft analysis aspect is important for samples sensitive to heat or radiation damage, for example in the biological context. The neutron beam can penetrate the sample from any direction. A charged particle detector positioned on the sample then collects the light products. With known neutron flux or a known reference sample NDP yields quantitative values for the surface near composition. Only a few naturally occurring isotopes emit light ion products with neutron interactions with a positive Q-value. Due to the negligible projectile energy thermal neutrons can only induce $Q \geq 0$ reactions. The possible isotopes/reactions are ^{3}He(n, p)t, ^{6}Li(n, t)^{4}He, ^{10}B(n, α)^{7}Li, ^{14}N(n, p)^{14}C, ^{17}O(n, α)^{14}C, ^{33}S(n, α)^{30}Si, ^{35}Cl(n, p)^{35}S, ^{40} K(n, p)^{40}Ar. Since the neutrons bring in no momentum into the reaction, the products will be emitted isotropically with an angle of 180° to each other. In the example of the $^{6}Li(n, t)^{4}He$ reaction with its Q-value= 4784 keV the ^{4}He product will receive 2050 keV and the triton 2734 keV.

Similar to NDP, Prompt-Gamma-Activation Analysis (PGAA) uses the photons emitted by the nuclear reactions of neutrons and target. For technical reasons these are both the (n, γ) photons but also the products of short-lived states. PGAA corresponds to the neutron equivalent of the ion-beam method PIGE (Sect. 7.1.5). The IAEA compiles information and data for PGAA on a specific website (www-nds.iaea.org/naa/portal.htmlx). Typically, PGAA uses thermal neutrons since these have the advantage of an easy, yet accurate absolute calibration of the result. A known flux monitor, usually a gold foil, placed in the neutron beam activates together with the actual sample. The flux monitor yields a so-called k_0 calibration factor relating neutron flux and activation. Science worked out tables of relative factors for all possible nuclides (De Corte and Simonits 2003). Using these factors, the activation efficiency of all nuclides can be derived from the flux monitor. This single proportionality factor holds even for varying thermal neutron energies. The effect of neutron energy cancels out, since practically all nuclides show the same scaling of cross-section with neutron energy as depicted in Fig. 4.7 in the thermal energy range. The reaction rate R of neutrons calculates as the product of the energy dependent reaction cross-section σ, the number of atoms N probed by the reaction behind σ, and the neutron flux F according to (7.9). With this calibration factor k_0, the measured activity R can be recalculated to a number of atoms of every nuclide present in the sample without knowing the individual σ. PGAA has detection limits down to a few 10 nano-grams for certain elements. PGAA features a mass related detection limit, since neutrons can penetrate large volumes homogeneously.

$$R = N * \sigma * F \tag{7.9}$$

Measuring the photons emitted after neutron irradiation represents the Delayed Gamma Activation Analysis also termed neutron activation analysis. This method integrates (7.9) resulting in the total amount of produced isotopes. In contrast to the PGAA method, the non-promt variant requires long integration times outside the irradiation facility due to long half-lifes of the produced isotopes, but it is technically less challenging.

7.3 Mobile Systems

All of the analytical methods discussed so far rely on sensitive equipment of table-top to synchrotron accelerator size placed in well defined laboratories. In many cases we discussed the problems of alignment issues and the development of ever increasing precision, but what if we cannot bring the sample to the lab? The adaptation of analytical tools onto production systems in the sense of an in-situ analysis already brings both closer together, but truly mobile applications go a step further in decoupling sample and analysis apparatus.

Semiconductor detector systems are usually quite robust and require only little power during operation. The main technological challenge lies in providing the required beams. Large accelerators with high beam energies already rule out due to limits of specific acceleration (MeV/m) discussed in Sect. 2.2. The available power further limits beam intensity and energy to the order of 100 kW for trailer-sized systems and the order of 10 W for handheld sized devices potentially operating battery powered. This power includes not only the accelerator but also auxiliary systems such as vacuum and data analysis computers. The main option for easing the power and voltage requirements (MeV/m limits) lies in the use of radioactive sources. Sources can provide α-particles of about 5 MeV, practically any photon and electron energy, positrons, and neutrons (with Be(α, n) converter). The weight of shielding and the potential exposure of surroundings represent a practical issue in particular for radioactive sources due to their isotropic emission properties. As mobile systems are potentially operated by untrained users radiation protection regulations become stricter.

Bringing the analysis system to the sample implies the analysis will happen in the classical outside, including air at 1 bar pressure. The extremely limited range of ions and electrons in air negates a decoupling of sample and analysis setup, but for photon analysis the use in air has only little drawbacks above a few keV. Higher energies using radioactive sources or the use of H_2 or He purge gas enable higher ranges. In addition, mobile systems suffer from reduced accuracy and spatial resolution due to less well defined particle beams and geometry. Beam focussing and high current density beams can hardly be realised under mobile space and power constraints. Finally yet importantly computational limits for complex analysis tools exist in mobile applications requiring special mobile CPUs or cloud based data analysis. The analysis of secondary photon spectra with their given peak positions requires only limited amount of computational analysis, but charged particle spectra are so far challenging even for desktop PCs.

Let us discuss a few examples. In several Mars rover missions a PIXE analysis system based on an α-isotope source was applied for elemental analysis of rocks (Rieder et al. 2003). The thin atmosphere on Mars of about 1 mbar just suffices for the 5 MeV α particles to travel a few 10 mm towards the Mars surface. PIXE allows for compositional analysis of rocks and soil, evidencing the existence of water on Mars. Since the decay emits also other particles, e.g. x-rays, source and detectors have to be shielded against each other in order to avoid excessive background signals. The setup of the Mars mission shown in Fig. 7.42 combines a compact x-ray detector behind an x-ray shield together with ion detectors and the ion source in a single tube of 52 mm diameter to be pressed onto the sample/Mars surface. The technical challenge of designing a precision analysis system sustaining a transport to Mars where it should provide high accuracy compositional data of numerous close elements are enormous, yet solvable using technologies discussed in this book.

More earthbound applications include the analysis of metal sheet thickness in steelworks using photon absorption. Processing of the sheets results in thickness variations, requiring a quick option of quality assurance without extracting samples. The absorption of photons is directly related to the metal thickness (see equation

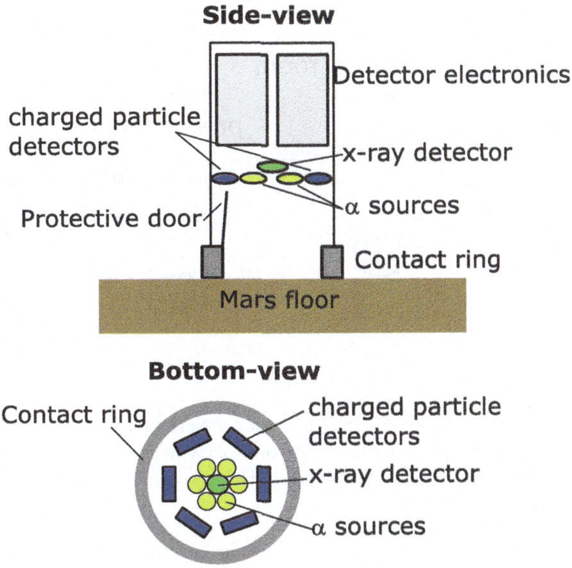

Fig. 7.42 Technical sketch of a mobile PIXE+RBS detector system with an isotopic α-particle source, similar to what was sent to Mars

(3.3)), since the metal type is mostly known in this application and a reference measurement without metal is also possible. The method uses a radioactive source in the handheld version or an x-ray tube in the stationary version. The same absorption idea can also be used to investigate filling levels of liquids or other materials in containers.

A more sophisticated analysis becomes possible with a mobile x-ray tube. Handheld XRF devices combine an x-ray tube operated at a few 10 kV with sub-mA current under the limitations of a battery-powered device. Figure 7.43 shows such a device used for on-site investigation of art. The devices allow for a multitude of analytical applications from ores over jewellery gold alloy composition to questionable materials containing hazards such as lead. Typical integration times are in the order of minutes. A standard silicon detector enables detection properties similar to stationary devices, but the limited x-ray source strength and the absorption of

Fig. 7.43 A handheld XRF device mounted for analysis of art pieces. Permission granted by Bruker Nano GmbH

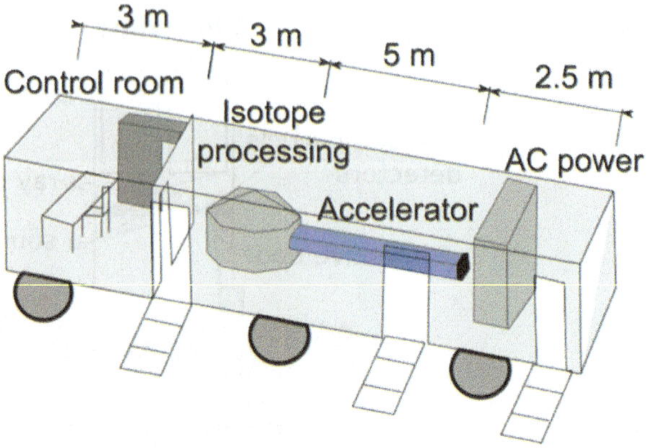

Fig. 7.44 Sketch of a trailer sized accelerator for neutron and isotope production. The system employs a LinAC for a few MeV protons with power supplys on the right end and production target and control room on the left end. The complete system fits to a german standard semi-trailer or shipping container

surrounding air lead to worse detection limits for light elements. In the keV range, corresponding to elements in the order of Mg (see Fig. 3.4) the absorption becomes relevant, first increasing the detection limits from typically a few 10 ppm towards % and rendering lighter elements completely invisible.

In principle, also regular accelerators could be made mobile, but the power consumption, weight, and the requirements on stability and vacuum quickly lead to trailer sizes as minimum space requirement, see Fig. 7.44. The specific acceleration achieved with modern LinACs, see Sect. 2.2, and improved detectors render potential in this market with energy limits in the order of 10 MeV. This energy suffices for producing various isotopes such as ^{18}F directly on-site, yielding clear advantages regarding isotope costs (see Sect. 5.1.4) or a backup solution for critical applications. So far, trailer-based neutron sources are the domain of miniature fission reactors, but accelerator based neutron sources could substitute these in aid of on-site medical (Chung and Lin 2001) or technical neutron based analysis. In conclusion, we have some technological starting points, but the vision of a multi-method tricorder like handset remains far away.

7.4 Radiation Damage

Radiation damage of materials is inseparably connected with accelerators and nuclear technology. In the preceding chapters we discussed numerous positive aspects of irradiation for material modification and analysis. The term *damage* on the other describes a negative and unwanted effect. Radiation damage has no clear definition, but the detrimental effects connected with it mostly relate to the negative impact of irradiation on engineering parameters of a material. An initial engineering design layout will, after a certain amount of radiation damage, become invalid. The damage effects have a large span ranging from a reduction in ductility of metals over drastic changes of resistance (see semiconductor doping Sect. 5.3.1) up to a reduction of optical transmission. A few of these aspects were already covered in Sect. 2.7.1 hence this section focusses on the aspect of engineering properties.

In order to understand the alteration of engineering properties a basic understand of these parameters and their meaning for an engineering layout are required. Let us consider an accelerator target as discussed in Sect. 2.6. This target consists of a 5 mm wall-thickness copper piece and receives an ion beam power $P = 1000$ W. It is connected to a 15 °C water-cooling pipe via an FKM (special vacuum rubber). According to equation (2.46), this yields a front temperature of 127 K above the coolant temperature. The beam parameters are fixed, hence the only variable for the resulting targets temperature is its thermal conductivity κ (copper $\kappa(300\ \text{K}) =$ 394 W/mK). With continuing radiation the thermal conductivity of copper decreases and hence the temperature increases. If it drops below 373 W/mK the temperature might rise above the long-term stability limit of FKM of 150 °C, defining a lifetime limit of the copper part due to radiation damage.

Besides this simplified example, many other parameters are influenced in different fashions, see Table 7.4. The knowledge on the physics of the influence of radiation on materials, especially the engineering properties is still scarce, as is also the literature. One of the few interesting readings is definitely (Was 2007). An older yet interesting scientific review describes the impact of radiation damage on optical properties and connects the detrimental and desired nature of radiation damage (Ascheron 1991). For understanding radiation damage, as a synergistic topic of solid-state physics, nuclear physics, and engineering, also general books on solid-state physics and materials engineering will help the reader to gain a basic understanding of the mechanisms possible trends and limitations.

In fission reactors this problem was of minor importance, since mostly only the fuel rods suffer from intense irradiation damage. Only few specially adapted materials were developed and hence most reactors run mostly with standard power plant steels. The game changes for accelerator applications, but also future nuclear fusion power plants. In these contexts comparable or even higher neutron flux densities are experienced as in fission reactors. The major difference and complication arises from the significantly higher neutron energies. Higher neutron energies open up additional nuclear reaction pathways, but more importantly induce much more displacement damage. With typical displacement thresholds of some ten eV in most materials,

Table. 7.4 General trends of material properties under increasing irradiation damage. The nomenclature "damage" derives from these trends

Property	Conductivity (heat, elec.)	Hardness	Ductility	Optical absorption	Heat capacity	Yield strength	Volume	Roughness
Trend	↓	↑	↓	↑	→	↑	↑	↑

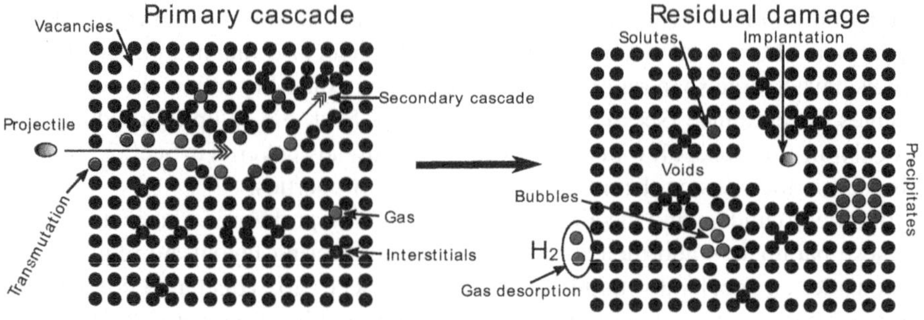

Fig. 7.45 Radiation damage interaction during the particle interaction of a few picoseconds and after a certain relaxation time. The projectile introduces displacements by hitting atoms (PKA, SKA) and undergoing nuclear reactions. Via reaction–diffusion processes these primary damages arrange in energetically more favourable configurations, annihilating a significant aspect of the primary displacements

thermal neutrons cannot induce any direct displacement damage with their spectrum centred around 0.025 eV. In nuclear fusion, neutron energies up to 14.1 MeV and in accelerators even above 30 MeV (plus high energy charged particles) are experienced. Besides lifetime also increased awareness of nuclear waste and safety generates interest in longer lasting and reduced activation materials.

The irradiation damage primarily arises from a number of processes we already know from this book: Collision cascades (a.k.a. nuclear stopping) and nuclear reactions as depicted in Fig. 7.45. Collision cascades primarily alter the lattice structure of solid by introducing displacements of atoms, hence the name displacement damage. Transmutation primarily changes the elemental and isotopic composition via direct reactions and decay. This either generates volatile species (gas damage), mainly as hydrogen and helium, or non-volatiles such as metals (transmutation damage). These damages have similar origin, but behave differently on the long-term as we will see later. Furthermore they can interact with each other leading to synergistic amplification of damage as indicated for example by bubbles and precipitates in Fig. 7.45.

In radiation protection, Sect. 2.7, we came to know the unit Sievert (Sv), describing the amount of damage introduced into living organisms based on the ionising energy deposited by the radiation. Radiation damage quantification also connects with the deposition of energy in the materials, but the response of materials, or at least the interesting part of this response, is very different to living organisms and also among different materials and applications. Therefore, a different standard measure has established in nuclear physics, the so-called displacements per atom (DPA). This number attempts to condense the whole damage into one number, which is in fact an improper generalisation. In spite of the weaknesses of DPA, its easy determination and its connection to irradiation and material properties make it a valid zeroth order damage-estimator.

In the simplest model assumption, the so-called Kinchin-Pease model, DPAs originate from the amount of kinetic energy deposited by the impacting radiation projectiles E_{kin} over a displacement threshold energy E_{disp} (which is basically $= 0$ in living organisms) originating from the binding of atoms in the solids lattice:

$$\frac{\text{Displacements}}{\text{Projectile}} = \frac{E_{\text{kin}}}{2E_{\text{disp}}} \text{for} E_{\text{kin}} > E_{\text{disp}} \tag{7.10}$$

Equation (7.10) allows calculating the DPA from the fluence density of irradiating projectiles, the irradiated depth, and the number of target atoms in this volume. In order to start a displacement, a minimum projectile energy E_{disp} is required or no damage will occur. E_{disp} derives either from other simulations or displacement experiments, in particular with electron beams. Physically, the displacement threshold features a weak temperature and a strong crystal orientation dependence with values in the order of about 10–90 eV. Above this threshold, (7.10) says, the damage per projectile increases proportional to its energy. Implicitly also the damage after a certain irradiation time is proportional to the irradiating flux and depends on its energy spectrum. In other words this implies a tracer assumption (no coupling between individual events), since it includes no flux, time, or fluence dependence. So what will happen if we hit an atom which was already hit?

To understand this we take a deeper look into the physics than just DPA. The displacement collisions are best described by the analogy to the billiard table. Imagine shooting the white ball (the projectile) onto the initial, ordered assembly of the coloured balls. The first coloured ball hit is called the primary knock-on atom (PKA). The PKA itself then hits several other balls, the secondary knock-on atoms, which again hit others forming a displacement cascade as depicted in Fig. 7.46. The projectile energy distributes in the cascade and the amount of energy actually invested into displacements reduces, considering stopping and the displacement threshold energy invested into each individual knock-on atom. The details strongly depend on the involved masses of projectile (electrons, ions, neutrons) and targets due to the collision kinematics as discussed in Sect. 3.3.2 and equations 3.16 and (3.17). For example, calculating the projectile energy threshold of protons or neutrons (1 amu) for displacing a tungsten atom (184 amu) with a displacement threshold of 50 eV yields 2325 eV due to the large mass difference. Typically only 1/3 of the damage estimated by (7.10) actually takes place, due to these dissipation effects. Simulations with Binary collision approximations (BCA) such as SRIM (Ziegler et al. 2008) discussed already in Sect. 3.5, mould these ideas into a tool for calculating displacements.

Calculation of the two nuclear reaction based damages (gas and transmutation) was essentially presented in Chap. 3. Gas production mostly relates to proton and helium producing reactions (e.g. (n, p) or (p, α) reactions), but also stopped/implanted hydrogen and helium beam particles contribute to the gas damage. Varying beam energies allow stacking several Bragg-Peaks, leading to homogeneous gas implantation, similar to proton beam therapy. Transmutation reactions directly connect to these reactions since also the heavy product will differ from the initial target. Nuclear reaction damage is proportional to the DPA quantity for a given energy spectrum of the irradiation, since all damage types are basically described by energy dependent

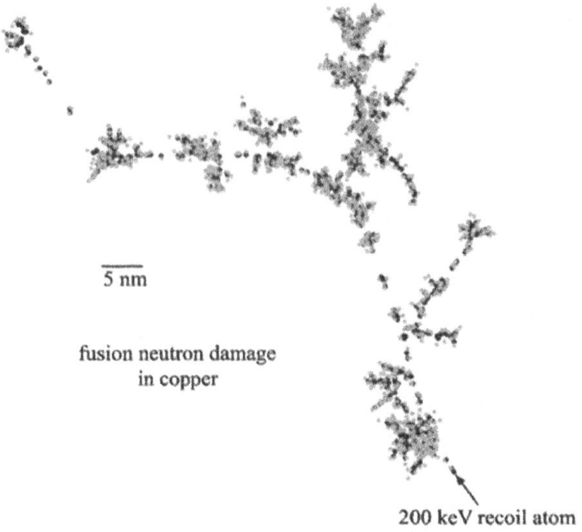

Fig. 7.46 A displacement cascade simulation with hundreds of displaced atoms starting with a 200 keV Cu PKA, produced by a neutron, on the lower right hand side. The PKA hits several other atoms on its way to the upper left hand side, generating secondary cascades in the form of tree branches. Picture reproduced from (Was 2007) Fig. 3.8 with permission by Springer

cross-sections and stopping (Sect. 3.4). As we learned in 3.3.2 nuclear reactions have a *Q*-value and require surpassing the coulomb barrier of projectile and target. Therefore, the transmutation damage features significantly higher projectile energy thresholds than the displacement damage.

Simulation of the nuclear reaction damages with computer codes allows considering the larger number of involved reactions. Nuclear inventory codes such as FISPACT (UK Atomic Energy Authority 2018) or MCNP (Los Alamos National Laboratory 2019) demonstrate the complexity of elemental changes present even in a pure material. Figure 7.47 demonstrates fusion neutron irradiation of iron leading to predominantly production of H, Mn, and Cr, besides radioactive Fe isotopes (which do not represent transmutation damage). The proportionality is valid only in the tracer regime, but it breaks down as soon as the concentrations reach significant levels. Quantities in the order of 100 parts per million (ppm) are generated every year in the fusion reactor example. In comparison to the displacements, the quantity of transmutation damage is smaller by about a factor 10^5 (100 ppm vs. 10 DPA in Fig. 7.47). Yet the different damage mechanism allows for a similar relevance.

The models discussed so far yield quantities of directly produced damages on an atomic scale with a high and proven accuracy (as long as the cross-sections are correct). These primary quantities not necessarily survive the initial steps (some picoseconds) of the reaction or the following equilibration of the material; hence we have to understand their propagation to the next scale of the time domain. This will define residual quantities, the amount of damage surviving long enough to be relevant

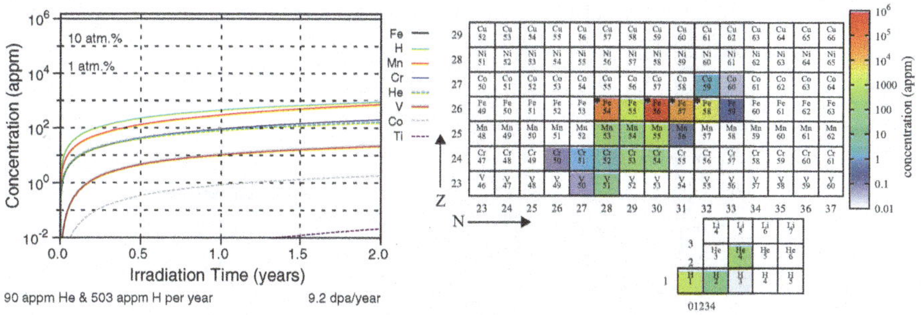

Fig. 7.47 FISPACT calculation of pure iron transmutation in a nuclear fusion reaction. Left: Time evolution of the elemental composition. Right: Isotopic composition after two years; stars mark the initial isotopes. Extracted from (Culham Centre for Fusion Energy 2018)

on the material or component lifetime scale. This survival is a prerequisite for the accumulation of damage. The transfer of primary (modelled) damage quantities to residual quantities is the domain of physical chemistry and its reaction–diffusion processes. Besides this also nuclear decay plays a role in some situations. We will discuss solid materials in the following, since they form the most difficult application case. In liquids, damage effects such as displacement naturally cannot occur and in gases not even percolation and gas damage are possible, reducing the complexity of the problem.

Reaction–diffusion processes in general follow a differential equation in the form of (7.11). The concentration of a given species u evolves over space x and time t via a reaction function f, a diffusion process with coefficient D, and a source-sink term A. The quantity behind u could be for example the concentration of hydrogen atoms in a metal target. As the equation shows we can only change the concentration u by inducing more (residual) damage via A and by changing the target temperature T, the rest is based on predefined natural mechanisms.

$$\frac{\delta u(x,t)}{\delta t} = f(u,T) + D(u,T,x)\frac{\delta^2 u}{\delta x^2} + A(u,x,t) \tag{7.11}$$

An interesting mystery to be solved by this function was found in the context of nuclear fusion materials. Scientists considered tungsten (W) as a material for the plasma facing components due to its high resistance to sputtering and good thermal properties. Unfortunately, it is quite brittle around room-temperature, but it is known that W–Re alloys show significantly improved ductility. The neutron irradiation in a reactor environment will convert about 1% of W into Re every two years. Experimental studies of this situation lead to a confusing result: The Re introduced by irradiation further embrittled the W in contrast to the situation of regular W–Re alloys. Atom resolved measurements (Fig. 7.48) revealed the reasons: Instead of the solute random distribution of Re in W–Re alloys, the Re had precipitated by a

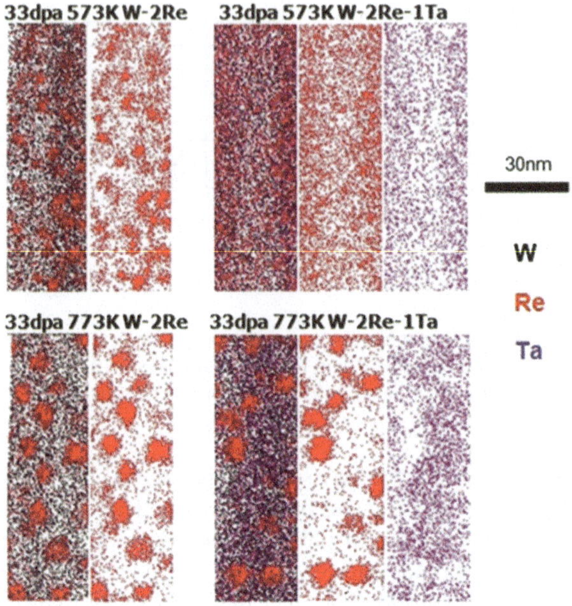

Fig. 7.48 Atom-Probe tomography study of W–Re and W–Re–Ta alloys irradiated by W ions. The originally homogeneously mixed material starts forming Re precipitates after irradiation. Ta on the other hand stays in solution under the same circumstances due to its higher solubility. Reproduced from (Xu et al. 2017) licensed under CC-BY 4.0

synergistic effect of radiation damage. These precipitates block the plastic deformation. Possibly the displacement damage mobilizes the Re in the form of a radiation enhanced diffusion and generates defects providing nucleation sites for the mobile Re. Whether the Re originates from transmutation or alloying makes no difference. Figure 7.48 further shows the problem is specific to Re, as Ta stays solute and even hinders the Re precipitation to a certain extent. Finally only with higher temperatures f and D grow large enough to activate the precipitation process.

Similar temperature dependent effects cluster also displacement damage in the form of vacancy clusters and voids, leading to detrimental effects for material properties and swelling. Gas atoms diffuse through the material and stabilize vacancies or agglomerate in voids forming gas bubbles. Temperature not only induces detrimental effects. If temperatures increase (typically in the order of 50% of the melting temperature) damage anneals and the material recovers its original properties. Gas particles move towards the surfaces and, volatile as they are, leave the material. Transmutation damage will not disappear, but can go into the solute state. The temperature effect usually resembles a 3-act dramaturgy: First, at low temperatures, damage acquires but only little interaction/synergy takes place leading to isolated damage entities. In the second act, at intermediate temperature, the plot densifies and the particles start interacting, leading to major dramatic damage effects. On the other hand also interactions lead to first annealing of damage. Only in the last act, at high temperatures, the situation resolves and damage anneals leading to finally low levels of damage. In conclusion, temperature represents the most important parameter and degree of freedom in radiation damage due to its impact on reaction–diffusion processes.

The physics connecting microscopic/physical mechanisms to macroscopic (engineering) properties are, so far, not well understood. Why is it better to have Re solute than precipitates and is this always the case or only for particular parameters? We approach the problem by further deepening our understanding of the origin of the macroscopic quantities. Let us consider the conductivity. For most metals the Wiedemann–Franz law connects thermal conductivity κ and electrical resistivity ρ via temperature T and the proportionality constant L to a single conductivity via (7.12):

$$\kappa\rho = LT \tag{7.12}$$

This connection of thermal and electrical conduction arises from the dominating conduction via free electrons in both cases. Matthiesen's rule, (7.13), tells us the electrical resistance R (which is one divided by the conductivity) originates from three independent scattering processes of the electrons on their way through the material:

$$R = R_{\text{defects}} + R_{e-e} + R_{e-\text{phonon}} \tag{7.13}$$

The scattering at lattice defects (voids, interstitials, impurities ...) dominates the electrical resistance at temperatures close to 0 K. At intermediate temperatures, electron-electron scattering dominates due to the increasing thermal energy of the electrons. At high temperatures (here some 100 K) the electron-phonon scattering dominates with its typical proportional increase of resistance with temperature. The thermal conductivity on the other hand has the electronic but also a phonon (lattice vibrations) contribution. In metals, with their typically good electronic conduction, the phonon contribution is small, but for insulators the picture looks differently. Therefore L is not literally a constant, but lies within 2.2 to 3.2 × 10^{-8} WΩK^{-2} for most metals. Already this shows us the electronic aspect is far from overwhelmingly dominating the conduction.

Introducing defects via radiation damage will affect each contribution to the total resistance differently. R_{defect} will grow by adding up the residual displacements and impurities introduced by the irradiation. $R_{e\text{-}e}$ on the other hand remains mostly unaffected, since the nuclear transmutation changes the atomic number of the elements (and with that the amount of electrons) usually only insignificantly. Changes in the phonon contributions are more difficult to predict. While the electrical resistance in (7.13) is dominated by $R_{e\text{-phonon}}$ at room temperature for an undamaged material, the situation can change for a damaged material, leading e.g. to a domination of R_{defect}.

The intriguing pitfall of the tracer assumption of radiation damage masks the actual onset of relevant material changes connected with leaving the tracer regime. Tracers, by definition, should not alter the host material, hence if we alter the material properties we are not in the tracer regime! This purely logical consideration tells us more complex models are required for describing the evolution of relevant radiation damage impacts. With the knowledge of primary radiation damage, the amount of damage surviving the first moments of the interaction brings us closer to the actual situation. This residual damage survives long enough to accumulate and hence leave the tracer regime.

Upon leaving the tracer regime via increased accumulated damage (increasing u in (7.11)), composite material effects set in. Figure 7.48 demonstrates the idea of a composite material combining a W matrix with Re precipitates. The primary damage is naturally distributed statistically throughout the material, but at higher temperatures starts to agglomerate forming in this case separate phases in the material. Displacements form voids, transmutation forms precipicates and gas forms bubbles. The original material becomes a composite. Several models for effective parameters of composite materials exist, but they require further knowledge of the form of the inclusions and interfacial effects. Equation (7.14) gives the so-called Bruggemann model, relating the volume fraction and the properties of the individual composite contributions to the effective value of the composite. In the above example Θ represents the volume fraction of the Re precipitates, κ_{Matrix} the conductivity of W, κ_I the conductivity of Re, a the radius of the Re precipitates, R_{int} the interface resistance of W versus Re, and κ_{eff} the overall conductivity. The left side of (7.14) tells us a high fraction of inclusions is required for relevant changes. Furthermore, the damage depends on the difference in conductivity between matrix and inclusion. Finally yet importantly, the effective conductivity can only be smaller than the individual component values if a high interface resistance between both exists.

$$(1-\theta)^3 = \left(\frac{\kappa_{\text{Matrix}}}{\kappa_{\text{eff}}}\right)^{\left(\frac{1+2A}{1-A}\right)} \left(\frac{\kappa_{\text{eff}} - \kappa_{\text{I}}(1-A)}{\kappa_{\text{Matrix}} - \kappa_{\text{I}}(1-A)}\right)^{\frac{3}{1-A}} \text{ with} A = R_{\text{int}}\kappa_{\text{Matrix}}/a \tag{7.14}$$

Most of these effective medium models break down when percolation effects start to induce connected zones of the new phases throughout the material. As these effects require volume fractions theoretically in excess of 18%, the material will anyways hardly survive until this last stage of damage. In a real world situation displacement, transmutation, and gas damage occur in parallel, leading to synergistic effects. This synergistic effect represents the actual crux of radiation damage. The simple combination of swelling and embrittlement already is an engineer's nightmare. The material fixed to a certain structure expands but becomes so brittle that it cannot cope with these dimensional changes without breaking. The result is clear: mechanical failure. In practical situations even more effects add up. Thermo-mechanical loads and regular fatigue related to the application further complicate the situation. These loads can even interact with the damage for example by accumulation of vacancies at stressed parts of a component.

Understanding of the discussed effects profits strongly from experimental studies at accelerators. The flexibility of accelerators in providing different projectile types throughout a vast range of projectile energies allows specific investigation of different effects as depicted in Table 7.5. For the lower energies and heavy ions, DC accelerators come into application due to their flexibility in beam energy. Irradiating with heavy ions leads to extreme displacement damage rates, but nuclear reaction damage is in general not accessible in the considered range up to 250 MeV. Light ions (H, D, He), on the other hand, induce nuclear reactions already above some MeV, but deliver about a factor 1000 less specific displacement damage (remember the Z^2 dependence of elastic scattering cross-section discussed in Chap. 3). For inducing practically homogeneous damage throughout sample with a macroscopic sample thickness, typically over 10 MeV are required. Light ions above 16 MeV require AC accelerators, e.g. cyclotrons also used for medical isotope production, reaching energies up to about 250 MeV with relevant beam currents. Comparing ions with neutrons clarifies the differences in flexibility of irradiation conditions. Fission reactor spectra are fixed and induce completely different ratios of the damage types with a dominant aspect of transmutation due to (n, γ) reactions. In future D-T fusion reactors, dedicated accelerator neutron sources (see Sect. 4.1) other energy spectra will become available with stronger MeV neutron fluxes. Accelerator beam irradiations on the other hand allow for different ratios of gas, transmutation, and displacement damage and variable depth profiles by changing beam species and energy.

Technical limits of direct accelerator irradiation originating from the envisaged testing methods are in contrast to their positive properties. For investigating the discussed connection between primary damage and macroscopic material properties, a macroscopic material testing method has to be applied. These methods require certain sample dimensions to yield reproducible values. Up to now, a consistent set of testing instruments yielding comparable material properties and result quality as for industrial material tests has not been identified for accelerator irradiated samples. The significantly smaller sample dimensions ($\leq$1 mm) compared to regular industrial or fission irradiation testing (Table 7.5) prevent a direct methodological transfer. Thin-films, micro-wires or surface-near testing demonstrated quick solutions, but offer only limited comparability to established engineering methods and databases. Instrumented indentation is one of the common methods for testing accelerator irradiated samples. The method requires the sample to be at least 10 times thicker than

Table. 7.5 Comparison of irradiation damage of different sources in tungsten. Damage calculated via SRIM2013 (Ziegler et al. 2008) and FISPACT-II

Primary damage	30 MeV W	3 MeV p	16 MeV p	30 MeV p	Thermal n	Fusion n
Displ./projectile	110,000	10	17	13	0	
DPA/day	70,000	10	2	1–2	0.003	0.017
Rhenium/DPA (ppm)	0	0	90	400	50,000	700
H/DPA (ppm)	0	0	0.1	30	0	1
Range (mm)	0.002	0.018	0.25	0.5	≈40	≈40

Accelerators calculated for typical currents of 10 μA (W) and 100 μA (p). Estimates for the homogeneous range, defines this as the region where all damage parameters vary less than a factor 2. The total range is typically about twice as high as the homogeneous range the tested depth, according to existing non-active testing norms (ISO 14577). Unfortunately, the first few hundred nano-meters of the surface are often not representative for the bulk properties due to damage introduced by cutting and polishing. On the other hand, having the sample 10 times thicker than its surface-near damage layer (e.g. 25 μm for 3 MeV protons in W) and another 10 times thicker than the testing depth already cancels out several irradiation options in Table 7.5. Accelerator irradiation struggles with conflicts of high energies and currents, thermal limits of targets, and the extreme activation (up to TBq/DPA) connected with transmutation damage, requiring analysis in lead shielded hot-cells (see Sect. 2.7.2.).

This chain of modelling and experimental steps presented in this section aims at developing a complete lifetime modelling of materials including temperature, time, and irradiation parameters. It is unrealistic assuming a regime can be found where all damage effect are cancelled by thermal annealing, outgassing, and formation of suitable alloys under bombardment of high energy particles occurring in the accelerator context. The nature of reaction–diffusion mechanism underlying the material physics hardly allows switching off distinct irreversible damage pathways, but we can still hope understanding and influencing the ratios of processes to significantly prolong material lifetime, reducing cost and improving nuclear safety. Accelerators with their flexibility to induce the three primary damage types in a fast, efficient, and selective fashion provide an important tool for experimental investigation, if new models, technical solutions, and particle sources become available.

References

C. Ascheron, Proton beam modification of selected AIII-BV compounds. Physica Status Solidi a **124**, 11, S. 11–55 (1991)

Brookhaven National Laboratory, *GISAXS Community Website*. Von (2018) https://gisaxs.com abgerufen

I. Bykov, H. Bergsåker, G. Possnert, Y. Zhou, K. Heinola, J. Pettersson, A. Widdowson, Studies of Be migration in the JET tokamak using AMS with 10Be marker. Nucl. Instrum. Methods Phys. Res. Sect. B: Beam Interact. Mater. Atoms **371**, 370–375 (2016)

Culham Centre for Fusion Energy, *FISPACT-II material handbooks*. Abgerufen am 05 2018 von (2018). https://www.ccfe.ac.uk/fispact_handbooks.aspx

B. Chaudhuri, I.G. Muñoz, S. Qian, V.S. Urban, *Biological Small Angle Scattering: Techniques Strategies and Tips* (Springer Nature Singapore Pte Ltd, 2017)

C. Chung, F.Y. Lin, Prompt gamma activation analysis using mobile reactor neutron beam, in *Proceedings of IUPAC International Congress on Analytical Sciences*, S (2001). doi.org/10.14891/analscisp.17icas.0.i633.0

F. De Corte, A. Simonits, Recommended nuclear data for use in the k0 standardization of neutron activation analysis. at. Data Nucl. Data Tables **85**, 47–67 (2003)

C. Dellen, H.-G. Gehrke, S. Möller, C.-L. Tsai, U. Breuer, S. Uhlenbruck, O. Guillion, M. Finsterbusch, M. Bram, Time-of-flight secondary ion mass spectrometry study of lithium intercalation process in $LiCoO_2$ thin film. J. Power Sourc. **321**, S. 241–247 (2016)

W.L. Friedrich, B. Kromer, M. Friedrich, J. Heinemeier, T. Pfeiffer, S. Talamo, Santorini eruption radiocarbon dated to 1627–1600 B.C. Science 28 **312**(5773), 548 (2006). https://doi.org/10.1126/science.1125087

J.H. Gross, *Mass Spectrometry: A Textbook* (Springer, 2006)
P.W. Hawkes, J.C, Spence, *Handbook of Microscopy* (Springer Nature Switzerland AG, 2019)
A. Hilger, I. Manke, N. Kardjilov, M. Osenberg, H. Markötter, J. Banhart, Tensorial neutron tomography of threedimensional dimensional magnetic vector fields in bulk materials. Nat. Commun. **9**, 4023 (2018). https://doi.org/10.1038/s41467-018-06593-4
G.L. Hura, A.L. Menon, M. Hammel, R.P. Rambo II, F.L. Pooli Li, S.E. Tsutakawa, F.E. Jenney, S. Classen, F.A. Frankel, R.C. Hopkins, J.A. Tainer, Robust, high-throughput solution structural analyses by small angle X-ray scattering (SAXS). Nat. Methods **6**, 606–612 (2009)
A. Jablonski, F. Salvat, C.J. Powell, A.Y. Lee, *NIST Electron Elastic-Scattering Cross-Section Database Version 4.0, NIST Standard Reference Database Number 64*. Abgerufen am retrieved 2019 von (2016) https://srdata.nist.gov/srd64/ or https://doi.org/10.18434/T4/1502642
C. Jeynes, M. Bailey, N. Bright, M. Christopher, G. Grime, B. Jones, R. V.V. Palistin, R. Webb, "Total IBA"—where are we? Nucl. Instrum. Methods Phys. Res. Section B: Beam Interact. Mater. Atoms **271**, 107–118 (2012). doi.org/10.1016/j.nimb.2011.09.020
S. Kapishnikov, T. Staalsø, Y. Yang, J. Lee, A.J. Pérez-Berná, E. Pereiro, S. Werner, P. Guttman, L. Leiserowitz, J. Als-Nielsen, Mode of action of quinoline antimalarial drugs in red blood cells infected by *Plasmodium falciparum* revealed in vivo. PNAS **116**(46), S. 22946–22952 (2019). https://doi.org/10.1073/pnas.1910123116
Los Alamos National Laboratory, *MCNP Homepage*. Von (2019). https://laws.lanl.gov/vhosts/mcnp.lanl.gov/index.shtml abgerufen
E. Maire, P.J. Withers, Quantitative x-ray tomography. Int. Mater. Rev. **59**(1), S. 1–43 (2014). https://doi.org/10.1179/1743280413Y.0000000023
M. Mayer, M. Balden, S. Brezinsek, V. Burwitz et al., Material erosion and deposition on the divertor of W7-X. Phys. Scripta, S (2020). https://doi.org/10.1088/1402–4896/ab4b8c
M. Mayer, S. Möller, M. Rubel, A. Widdowson, et al., Nuclear Fusion. **60**(2) 025001 (2019) https://doi.org/10.1088/1741-4326/ab5817
M. Nastasi, J.W. Mayer, Y. Wang,. *Ion Beam Analysis: Fundamentals and Applications* (CRC Press, 2014)
M.T. Postek, A.E. Vladar, J. Kramar, B. Ward, L.A. Stern, J. Notte, S. McVey, Helium ion microscopy: a new technique for semiconductor metrology and nanotechnology, in *AIP Conference Proceedings 931, 161*, S (2007). https://doi.org/10.1063/1.2799363
R. Rieder, R. Gellert, J. Brückner, G. Klingelhöfer, G. Dreibus, A. Yen, S.W. Squyres, The new Athena alpha particle X-ray spectrometer for the Mars Exploration Rovers. J. Geophys. Res., **108**(E12), 8066, S (2003). https://doi.org/10.1029/2003JE002150
M. Salmeron, From surfaces to interfaces: ambient pressure XPS and beyond. Top. Catal. **61**, 2044–2051 (2018). https://doi.org/10.1007/s11244-018-1069-0
H. Schnablegger, Y. Singh, *The SAXS Guide* (Anton Paar GmbH, Austria, 2017)
Z. Siketić, I. Bogdanović Radović, I. Sudić, M. Jakšić, Surface analysis and depth profiling using time-of-flight elastic recoil detection analysis with argon sputtering. Sci. Rep. **8**, S. 10392 (2018).https://doi.org/10.1038/s41598-018-28726-x
M. Stuiver, P.J. Reimer, T.F. Braziunas, High-precision radiocarbon age calibration for terrestrial and marine samples. Radiocarbon **40**, 1127–1151 (1998). https://doi.org/10.1017/S0033822200060161
V.N. Tondare, Quest for high brightness, monochromatic noble gas ion sources. J. Vac. Sci. Technol. A **23**(6), 1498–1508 (2005)
C. Tötzke, N. Kardjilov, I. Manke, S.E. Oswald, Capturing 3D water flow in rooted soil by ultra-fast neutron tomography. Sci. Rep. **7**. Article number: 6192 (2017)
C. Tötzke, N. Kardjilov, N. Lenoir, I. Manke, S.E. Oswald, A. Tengattini, What comes NeXT? High-speed neutron tomography at ILL. Opt. Exp. **27**(20), 28640–28648 (2019)
C. Tuniz, *Accelerator Mass Spectrometry: Ultrasensitive Analysis for Global Science* (CRC Press, 1998)
UK Atomic Energy Authority, *FISPACT-II*. Von (2018) https://fispact.ukaea.uk/abgerufen

E von De Hoffmann, *Mass Spectrometry Third Edition Interscience Principles and Applications* (Wiley, 2007)
P. van der Heide, Secondary ion mass spectrometry, an introduction to principles and practices (Wiley, 2014)
K. Vogel-Mikuš, P. Pongrac, P. Pelicon, P. Vavpetič, B. Povh, H. Bothe, M. Regvar, *Micro-PIXE Analysis for Localization and Quantification of Elements in Roots of Mycorrhizal Metal-Tolerant Plants*. Symbiotic Fungi. Soil Biology, vol. 18 (Springer, Berlin, 2009)
G.S. Was, *Fundamentals of radiation materials science: Metals and alloys* (Springer, Berlin Heidelberg, 2007). ISBN 978–3–540–49471–3
D.B. Williams, C.B. Carter, *Transmission Electron Microscopy: A Textbook for Materials Science* (Springer, 2009)
A.B. Wittkower, H.D. Betz, Equilibrium-charge-state distributions of energetic ions ($z > 1$) in gaseous and solid media. Atomic Data **5**(2) (1973)
A. Xu, D.E. Armstrong, C. Beck, M.P. Moody, G.D. Smith, P.A. Bagot, S.G. Roberts, Ion-irradiation induced clustering in W-Re-Ta, W-Re and W-Ta alloys: an atom probe tomography and nanoindentation study. Acta Mater. **124**, 71–78 (2017)
J.F. Ziegler, J.P. Biersack, M.D. Ziegler, *SRIM—The Stopping and Range of Ions in Matter.* (Chester, 2008)

Chapter 8
Energy Production and Storage

Abstract Nuclear fusion with a magnetically confined plasma is a candidate method to produce vast amounts of energy, neutrons and rare isotopes such as T and ^{3}He. Accelerator based realisations can be simpler compared to plasma devices, but they are considered to be unsuitable, due to the dominance of elastic scattering to the relevant inelastic reactions. This chapter discusses three aspects where accelerators potentially contribute to the energy source landscape: spallation, nuclear batteries, and nuclear fusion. All of them use depth-integrated nuclear reaction with hydrogen isotope ions. Their basic working and development pathways will be derived from this.

Nature is fair, for every good you also receive a bad

All energy applications of accelerators are, so far, theoretical or in early concept stages, but the ideas form an interesting testing ground for the understanding of the technical and physical knowledge presented in this book. The energy production concepts presented here rely on efficiency, optimising the ratios of different cross-sections, stopping power, radiation protection, and nuclide selection. In Chaps. 4 and 5 we learned the physics of nuclear reactions, the Q-value (which is the essence of Fig. 8.1) and the decay of instable nuclides. The three aspects are the essence of the nuclear energy applications of accelerators. In fact, stopping power represents a fundamental limit, but technical concepts for mitigating it exist, as the reader will learn in the following sections.

The interesting aspect of nuclear power in general, not limited to accelerator reactors, lies in the large amounts of energy connected with nuclear bonds, compared to chemical bonds. Figure 8.1 depicts the situation along the nuclide table. By fusion of light nuclei or fission of heavy nuclei, the binding energy increases up to the natural maximum of 8.79 MeV/nucleon in ^{56}Fe. For comparison, the formation of CO_2 from C and O_2 releases about 4.1 eV, more than 10^6 times less energy per reaction. This enormous difference in energy content of nuclear and fossil fuels makes nuclear power attractive. The masses of fuel required decrease accordingly, reducing environmental impact and costs. This high specific energy content makes nuclear energy in particular interesting for space applications, but even more importantly

S. Möller, *Accelerator Technology*, Particle Acceleration and Detection,
https://doi.org/10.1007/978-3-030-62308-1_8

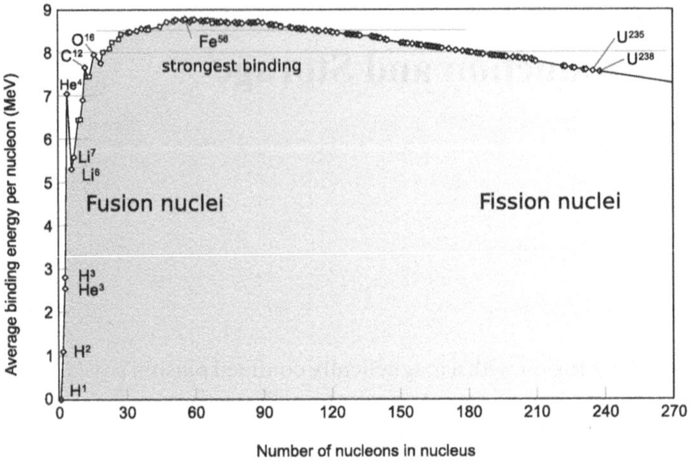

Fig. 8.1 Average binding energy of nucleons. A maximum binding occurs at 56 neutrons and protons (^{56}Fe). The energy released upon separation of heavy nuclei or combination/fusion of light nuclei is in the ten to hundred MeV range, making nuclear fuels million times denser than chemical fuels with their eV binding energies

nuclear power promises also to be an ideal energy source with the perspective of saving humanity from one of its worst threats, the climate change.

Climate change threatens our planets nature and our society's economy. Fatalities related to fossil fuel consumption and the climate change were estimated to be already in the order of 315,000 per year (Annan et al. 2009), with numbers increasing steadily. Nuclear power can provide a carbon-free alternative, but nature is fair and this significant advantage comes with a price. The main risks associated with nuclear power relate to nuclear proliferation (the use of products for weapon or terror purposes) and the risks of nuclear accidents releasing radioactivity. The efforts invested into nuclear safety constitute a major, if not the main, part of the costs of nuclear devices and reactors with significant progress in operational safety until now. The perception of the risks drives nuclear industry as the biggest threat for its commercial success, yet certainly mistakes were made and are always possible in the future.

The only nuclear power source successful until now is the fission of ^{235}U. With a binding energy of 7.59 MeV/nucleon it offers a potential of $235 \times (8.79 - 7.59) = 236$ MeV per ^{235}U nucleus for a complete fission down to the strongest binding (^{56}Fe). Typical reactions split the ^{235}U into two fragments around mass 90 and 140, releasing about 200 MeV and a few neutrons. These neutrons continue the reaction by splitting the next ^{235}U they find. If one or more neutrons per reaction survive the environment to induce the next fission reaction the reactor is called critical, the reaction is self-sustaining. Unfortunately, this depends on the composition of the fission fuel, which changes over time with progressing burn-up. In the end, in all current reactor designs (the so-called generations 1–3) the reactor fuel loses its criticality long before consuming the fissile fuel and we end up with tons of radioactive waste. Future so-called generation 4 fission reactors aim at using a larger part of the fuel, improving economy and safety at the same time. The concepts are interesting, but technically more challenging compared to current designs.

The social and economic acceptance is particularly challenging for nuclear power plants. This contradiction of the actual risks and the publicly perceived risks of fission power compared to the widely accepted fossil power is astonishing for an expert. It has to be admitted, though, this misbalance strongly influenced the head-to-head race of renewables, fossil, and fission power in favour of the renewables. Regarding the costs per kWh renewables probably wouldn't have reached their economic breakthrough without this public support. Consequently, the fundamental ideas of the next generation nuclear power plants must combine improved economics, low waste, proliferation safety, and public safety to bring nuclear power ahead in this competition. Currently operated fission reactor technology compares in its technological complexity to other nuclear technologies such as nuclear fusion or accelerator based technologies as cuneiform inscription to computer writing. In this sense not only uranium and thorium fission reactors, but also fusion reactors (burning mostly deuterium) follow the generation 4 concepts to offer the prospect of new acceptance of nuclear technologies by significant technological advances.

Nuclear fusion research marks an outstanding example of the constraints social acceptance puts on the development and structure of technological thinking. Current technology would allow building a fusion power plant with positive output based on the tokamak or stellarator concept, but no one would do so. Fusion sounds like a good thing so why not do it? The answer is simple: Why risk billions of euros without chances for a return of investment? Fusion science keeps this question in mind allowing it to focus down the research on pathways fulfilling this necessary economic condition by focussing research e.g. on reduced activation materials, longer component service life or reactor down-sizing. Several technological options such as the use of carbon plasma-facing components, the p-^{11}B fusion or a primary helium cooling-cycle in a power reactor were already disregarded or at least postponed for this reason. Economic success needs a clear perspective for involving industry. This perspective has to be provided by public research in the form of a so-called DEMO fusion reactor. Scientists have to see these additional constraints as a chance to discover a clear path through the infinite amount of possible solutions for a technological problem. For the following accelerator based power technologies, this path has still to be revealed, in contrast to the established technologies discussed in the last chapters. Keep that in mind.

8.1 Spallation Fission Reactors

Is it possible to bring a subcritical fission reactor into criticality by particle beams? Definitely yes, but can we realise it with positive energy output, considering stopping power (Sect. 3.2) and why should we do it if a fission reactors also runs fine without accelerator?

Pressurized water fission reactors (PWR) produce vast amounts of isotopes throughout the nuclide chart. A particular nasty subset of isotopes comes by the name minor actinides. Minor actinides is actually an unphysical term, because these actinides are in no way minor to the major (U and Pu) or the normal actinides. The term originates from the fission industry and the fact that these actinides are present in spent fuel from uranium fuelled and water moderated reactors in smaller quantities than U and Pu. Physically some are even major to U and Pu in terms of energy content, see Fig. 8.1. The minor and even the major actinides (Pu and U) will not be completely burnt by the PWR reactor type, since it has certain design limits for its criticality and power output. The criticality represents the number of neutrons produced from the fuel per neutron and has to stay >1 in order to keep the chain reaction running (Fig. 8.2). At the same time, these unused fissile nuclides contribute significantly to the radioactivity of the PWR waste. Consequently, a lot of radioactive and energy-rich waste and furthermore uranium fuel remains to be dumped or chemically separated.

Here the idea of a spallation assisted fission reactor offers a solution. Figure 8.2 shows the decrease of criticality over operational time of a PWR. After time step 26 in Fig. 8.2 the criticality reduces below required value of 1. In other words, the reaction produces less than 1 neutron per consumed neutron (=criticality < 1). An accelerator could fill this criticality gap of the fissile fuel by adding neutrons from proton induced reactions to the equation (red area). The accelerator, not control rods, then defines the reaction rate. The combined fission device is often called Accelerator Driven System (ADS). The ADS makes use of ion induced fission and neutron emitting

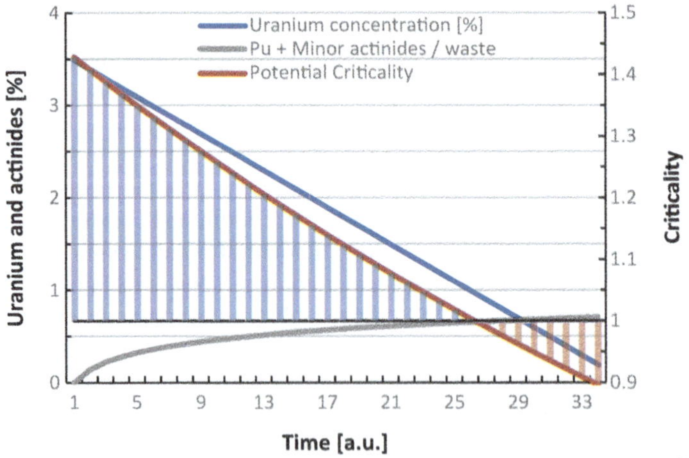

Fig. 8.2 Sketch of the fundamental technological problem of PWRs. The criticality decreases with decreasing fissile U concentration and increasing fission products. Due to the production of non-fissile neutron absorbing isotopes criticality even decreases faster than U content. The control rods keep the criticality slightly below 1 until this becomes impossible due to insufficient fission neutrons; the fuel is spent for the PWR. A PWR has to stop operation at latest at time step 26

reactions summarized under the term spallation reactions. By becoming independent of requiring critical fission fuels (criticality > 1), the technological limit of PWRs, which limits to fuelling by enriched ^{235}U, vanishes and substantial amounts of new isotopes become available as nuclear fuel. A particular nice aspect of these fuels lies in the fact that PWRs and fission explosives rely on fission of the same isotopes with criticality > 1. Consequently, spallation fuels avoid the risk of nuclear proliferation by using fuels with criticality < 1. Technically, the ADS will be a different reactor for which the suitable fuel will be extracted from PWRs and optimised/enriched for the spallation process, but the principle of using the waste of PWRs remains untouched as depicted in Fig. 8.3. Similar to existing nuclear reprocessing, the minor actinides (elements 93–100, except for 94, which is "major" in the PWR speak) will be separated chemically from the spent fission fuel with up to 99.99% separation efficiency, significantly reducing the nuclear waste. These nuclides contribute most of the long-term activity of PWR waste with half-lives of up to 16 million years. In contrast to existing nuclear reprocessing in combination with PWRs, the ADS enables using a significantly larger portion of the reprocessed material.

The spallation technology also works with uranium or thorium fission reactors, but the expensive accelerator technology would negate the economic condition for this approach compared to a critical reactor. Therefore, considerations of spallation currently involve a connection with nuclear fission power by using fission waste as spallation fuel. In addition to power production it also offers the potential for an intense scientific neutron source as discussed in Sect. 4.1. This connection generates additional synergistic value by providing an attractive option for waste treatment in addition to the energy production. Partitioning and transmutation technology of

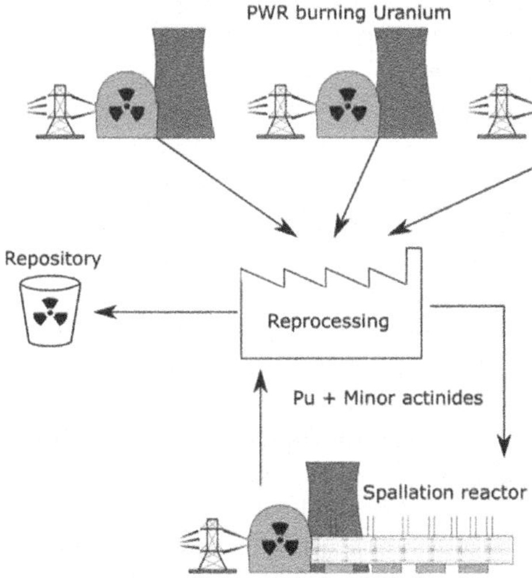

Fig. 8.3 Fuel cycle and nuclear waste with fission and spallation. Fission reactors produce sub-critical nuclear waste. A reprocessing and separation plant makes spallation fuel from this. Only a small amount has to be stored and additional energy is generated from the existing nuclear fuel

the fission waste therefore represents a major technological ingredient to spallation reactors.

High energy ion beams, usually some hundred MeV protons, induce (p, xn) and fission reactions in the ADS as discussed in Sect. 4.1. Fission reactions of the released neutrons act as a multiplicator, increasing the energy release per incident proton. The criticality remains < 1, for example 0.95. The spallation reaction itself requires a different "criticality" limit: For a positive energy output, the energy multiplication factor of the overall system needs to be large enough to compensate the energy invested by the accelerator. In other words, the amount of induced fission reactions per projectile times the released energy per fission event times the energy conversion efficiency has to be larger than the beam energy times the accelerator efficiency. For spallation based neutron sources for scientific or isotope production, the multiplication only defines the beam power of the accelerator part required for a specific neutron flux, since a higher multiplication factor reduces the amount of protons required for emitting one neutron.

The nuclear reactions induced by the beam itself cannot provide a positive energy output, since we learned earlier (Chap. 3) only a few percent of the projectiles induce nuclear reactions by beam-matter interactions and not all of them feature $Q > 0$. The higher the beam energy, the more energy lost per projectile, but also the better the ratio of stopping power to nuclear reaction cross-sections, similar to what was shown in Figs. 3.17 and 3.18. This results in a higher reaction probability or in other words more neutrons per projectile (see Sect. 3.4). An optimum needs to be found. Many of the numerous reaction cross-sections of the target's heavy elements are not experimentally known. The reaction p+^{209}Bi already lists about 90 reactions possibly contributing to the spallation process in (Koning et al. 2015). All of these reactions can hardly be measured for all the involved target isotopes. The uncertainties of theoretical models demand experimental verification for a solid device layout, complicating finding this optimum.

Three major demonstration projects for this technology are currently being implemented. The MYRRHA (Multi-purpose hYbrid Research Reactor for High-tech Applications) project with about 30 MW_{th} aims at providing a research neutron source located at SCK-CEN in Belgium (SCK · CEN Studiecentrum voor Kernenergie 2017). The project China Initial Accelerator Driven System (CiADS) aims at a 10 MW_{th} reactor in a first stage with a potential upgrade to 1 GW by a later addition of acceleration structures. The larger 800 MW_{th} Japanese facility named J-PARC Transmutation Experimental Facility (TEF) intends to demonstrate the power reactor and waste reduction aspect. More details can be found for example in (Nakajima 2014). All facilities rely on linear AC accelerators delivering 500–1000 MeV protons with mA up to several 10 mA beam currents onto Lead-Bismuth eutectic liquid spallation targets. Figure 8.4 depicts the complex accelerator design of the MYRRHA facility.

The TEF intends to demonstrate nuclear fission waste transmutation covering the output of up to 10 running pressurized water reactors. CiADS aims at a technology demonstrator for later commercialisation. MYRRHA will potentially substitute the BR-II research reactor as a new type of research neutron source not relying on the problematic highly enriched uranium.

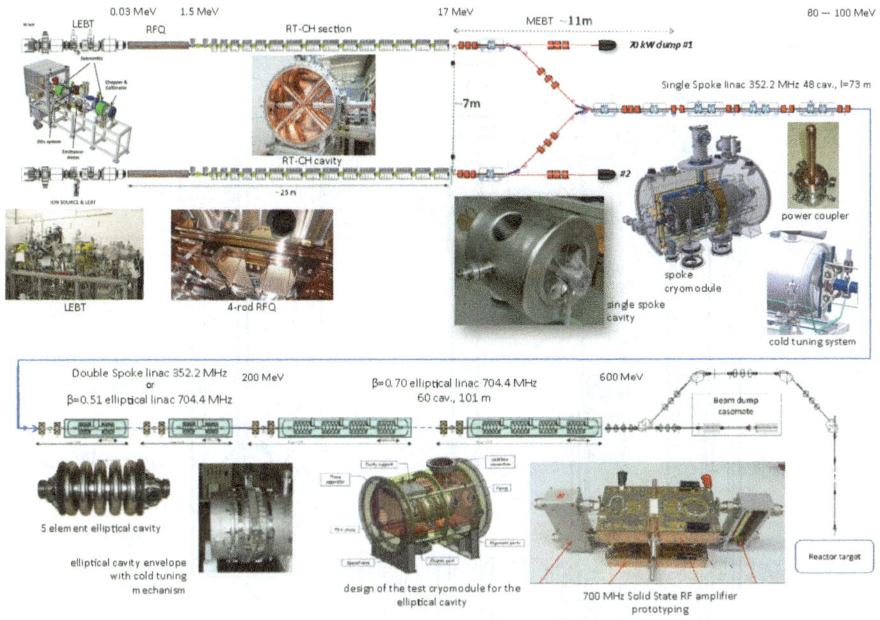

Fig. 8.4 Schematic of the MYRRHA accelerator. The system features two alternative ion sources and about 240 m of RF acceleration cavities reaching 600 MeV with 3.5 mA. The required reliability necessitates redundancy and the use of semiconductor generated AC input power. *Source* SCK · CEN Studiecentrum voor Kernenergie (2017). Reprinted with permission by SCK

Besides the nuclear cross-sections for the different transmutation elements, numerous technical developments are necessary on the way to commercial ADS. In principle all technical components, accelerator, target, and reactor core differ significantly from existing designs. The application of accelerators in the production and safety oriented technological concepts induce additional constraints on the availability of the accelerator. In typical high-energy accelerators irregular malfunctions such as high voltage trips and regular events such as ion source maintenance easily take up $\leq 10\ \%$ of the possible yearly operational time. The MYRRHA accelerator is designed to have on average one incident with more than 3 s of downtime per 250 operational hours. This still sounds a lot considering for example the availability of internet connections or the power grid, but for an accelerator complex of this size it represents a technical milestone, which has yet to be proven possible. The followed strategy uses moderate specific acceleration (2.4 MeV/m), redundancy in the ion source and the acceleration part, and the application of semiconductor generators for the AC driving power instead of the typical Klystrons (Fig. 8.4).

A beam window separates this accelerator part from the target and reactor core as depicted in Fig. 8.5. Technical difficulties of this ansatz were discussed in Sect. 7.4 and limit the ion beam power density and the beam window radiation damage. The target itself consists of a liquid lead-bismuth neutron source located directly behind the window. The heavy metals generate neutrons from the beam and allow for an

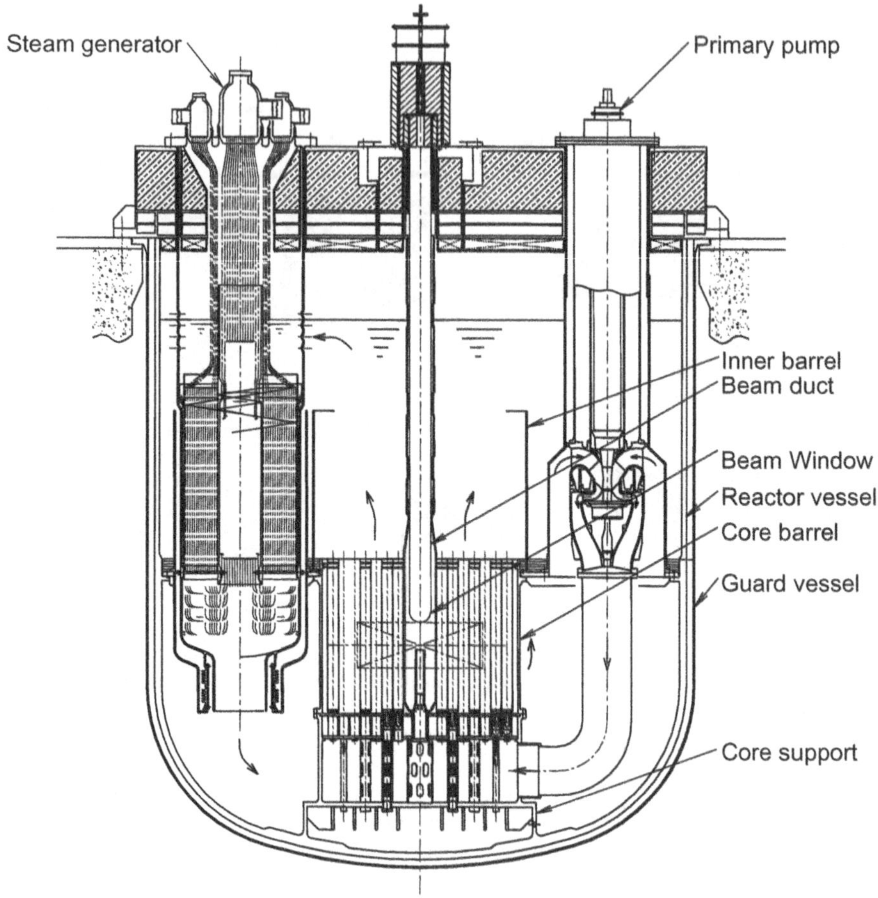

Fig. 8.5 Proposed JAEA ADS design with 800 MW_{th}. The beam enters the reactor from the top centre. A volume of liquid lead-bismuth generates neutrons through spallation. The fissile (MA) fuel in the core barrel around this central spallation target reacts with these neutrons. The heat generated in the core is dissipated via a flow of the liquid lead-bismuth indicated by the arrows. Reprinted with permission from JAEA-Review 2017-003

efficient cooling via pumping. An assembly of fuel rods similar to a PWR will surround this central target. For the TEF design, mixed minor actinides and Pu nitride pellets are envisaged. Finally, the hot liquid metal will power a steam generator. Development of the chemistry and corrosion features of the many different materials and the hot and reactive liquid metal represent major material challenges. The fuel also loses criticality during use, requiring compensation with higher beam current. A reduction of criticality from say 0.97 (33 neutrons per spallation neutron) to 0.93 requires increasing the beam current by a factor 2.3 to reach the same power output. Consequently, the flexibility in beam current limits the fuel cycle time.

The energy production by neutron production reactions is insufficient for a positive outcome, due to the high energy invested into the projectiles. For ADS intending to produce energy, the energy efficiency and energy balance of the spallation process are critical parameters. The neutrons emitted in the spallation reaction react with the fissile actinides, generating additional fission reactions in an independent process chain. Neutron sources with heavy element targets producing about 10 neutrons per ion via spallation (MYRRHA = 6 n/p), see Sect. 4.2. With a criticality of the fissile materials of for example 0.95, these neutrons will be multiplied by a factor 20. Each neutron induces fission reactions with outputs in the 100 MeV per neutron range, releasing about 12 GeV per projectile in the MYRRHA example (similar for JAEA design). Assuming an overall proton acceleration efficiency of 1/3, the 600 MeV projectiles will cost 1.8 GeV of electricity. From the 12 GeV only 4 GeV will become available as electrical power with a typical steam cycle. This results in a multiplication factor of 2.2 or in other words, the facility consumes 45% of its produced electricity only for the accelerator. Considering the above mentioned decay of criticality, 2.2 represents an alarmingly small margin. Coal or PWR plants consume about 10% of the generated power, indicating the technological challenge of an economical ADS power plant. The actual power output of the ADS will be defined by the beam current, but high currents are required for relevant power output (here 4 GW per 1 A beam current). The spallation and neutron reactions themselves produce a vast range of daughter nuclides, but the integrated decay time to natural uranium toxicity levels of the ADS waste potentially lies around only 300 years, 3 orders of magnitude less than PWR waste (Nakajima 2014).

8.2 Nuclear Batteries

Radioactive isotopes contain vast amounts of energy, which constantly releases during their radioactive decay by emission of energetic particles. This physical fact allows seeing them as batteries. Nuclear batteries are primary batteries, meaning they cannot be recharged by an inversion of the discharging process. In contrast to electro-chemical batteries they feature a major disadvantage of having a fixed or load independent power output, respectively. An electro-chemical battery on the other hand adapts to the power requirement and delivers for example also zero output power, equivalent to a maximum usage factor. The big advantage of nuclear batteries on the other hand is their extreme specific energy density. The currently most common nuclear battery isotope, ^{238}Pu, has a theoretical energy density of 632 kWh/g (5.593 MeV α's and $2.53 * 10^{21}$ atoms/g) while an electro-chemical cell based on $LiCoO_2$ (also known as Li-Ion battery) can theoretically store 0.55 Wh/g. These distinct differences open up the application space for nuclear batteries.

In contrast to electro-chemical batteries, nuclear batteries produce fast particles instead of electrical potentials. Technical applications require electrical potentials up to some hundred volts, hence the nuclear battery requires a particle-to-voltage converter. In addition most nuclear decay products feature energies of at least a

Fig. 8.6 A thermally glowing pellet of about 150 g of ^{238}plutonium-oxide emitting 62 W of α-particle power. The oxide form stabilizes the material at the high temperatures required for efficient energy conversion. These types of pellets were used in NASA's Cassini and Galileo missions. Unless otherwise indicated, this information has been authored by an employee or employees of the Los Alamos National Security, LLC (LANS), operator of the Los Alamos National Laboratory under Contract No. DE-AC52-06NA25396 with the U.S. Department of Energy. The U.S. Government has rights to use, reproduce, and distribute this information. The public may copy and use this information without charge, provided that this Notice and any statement of authorship are reproduced on all copies. Neither the Government nor LANS makes any warranty, express or implied, or assumes any liability or responsibility for the use of this information

few keV up to the above mentioned about 5.6 MeV, or in other words they provide extremely low currents at extremely high voltages. A direct conversion by using the particles to charge a capacitor becomes technically infeasible. First converting the particle energy via stopping and absorption into heat enables the usage of thermal energy conversion technology. The Carnot limit defines the maximum efficiency of the thermal approach, requiring high temperatures. ^{238}Pu easily reaches > 800 K as depicted in Fig. 8.6 (Draper point of incandescence). Thermoelectric generators such as BiTe offer a reliable and compact option for energy conversion at the drawback of typical efficiencies < 10%. Heat engines, in particular the Stirling type, offer higher efficiency close to the Carnot limit at the expense of limited lifetime and larger setups with moving parts, but an application ready setup was so far not developed. The thermal conversion route comes into application for isotopes emitting most power via α-decay, such as ^{238}Pu. The short range of a few μm hardly allows these decay products to leave the emitting material. On the other hand the small outside dose rates enable a use near humans without excessive shielding.

For β- and γ-emitters the range is long enough for additionally considering non-thermal conversion techniques. Just like the solar radiation does, the emitted particles also separate charges in solar cells, producing electrical power. These so-called beta-voltaic cells consist of μm thin films of radioactive material evaporated on solar cells (or p-n junctions). The limited amount of radioactive material present in the thin films limits the output to the μW range. This is sufficient to power electronics and MEMS, but hardly scalable for larger applications. While thermal conversion allows for a certain distance, the unavoidable vicinity of emitter and generator introduces problems of radiation damage of the generator for betavoltaics.

The most prominent use of nuclear batteries are NASA's deep space missions. The satellite generators usually feature an output of some hundred watts. Due to the exponential nature of the decay of radioactivity, the generator first provides excess power, which decreases over the years. In consequence, systems of the satellite/spacecraft have to be deactivated to reduce its power consumption. At some point in time, the generator output defines the end of the design lifetime in connection with the minimal power requirement of the spacecraft. A few niche applications for powering pacemakers or remote lighthouses (yes…) existed before the development of powerful electro-chemical batteries.

Traditionally, that means for the main applicant NASA, nuclear batteries are based on ^{238}Pu extracted from fission reactors. Its breeding requires special reactors, which were typically also required for nuclear weapon production. Its properties as nuclear battery are tempting with low external dose rates, high specific power output, and a technically relevant half-life of 87.7 years. Figure 8.7 displays its exponential decay of power output together with the radiation protection aspect. With the decay of the nuclear weapon activities also Pu-238 became rare (Witze 2014), due to the technical connection of both, requiring alternatives.

One of the main approaches of this book is: Things that can be done using a fission reactor are also possible by using an accelerator. Now having the basic knowledge of the functioning of a nuclear battery, and (hopefully) also of accelerators and beam-matter interaction, we can design an accelerator based version. Possible nuclides have to be selected with respect to radiation hazards and power output. Additionally input vs. output energy efficiency gains increased importance in accelerators, since it determines the economy and output power for a given accelerator. Decay details and the nuclear production cross-sections define the energy efficiency. In particular (p, xn) reactions are most relevant, due to the good performance of proton accelerators. Deuterons and Helium ions only slightly extend the set of reachable isotopes. In any

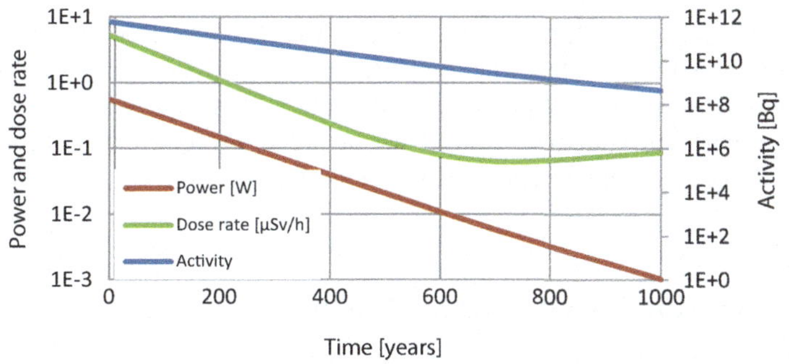

Fig. 8.7 Output of 1 g of ^{238}Pu. The dose rate is given for 1 m distance. The initial output of 0.57 W/g falls by 10% within the first 14 years, but the decay is exponential (line in this logarithmic graph). Only after 800 years, the decay products (mostly ^{234}U) start contributing to the overall activity

case, stable isotopes (including Thorium and Uranium isotopes) will be the basis of the concept.

Accelerators can produce α, β, and γ emitters. In particular the α emitters are interesting for nuclear batteries, but α emitting isotopes sitting in the upmost edge of the nuclide chart can only be reached by reactions with Bismuth, Thorium, or Uranium. If we consider the requirement of about 100 years of half-life, since electrochemical cells cover a few years and longer half-lives just reduce the activity of a given amount of produced isotopes, we end up with ^{209}Po, ^{232}U, and ^{238}Pu as reachable and interesting candidates. The latter ones are reached indirectly via a short-lived β-decay of an intermediate isotope.

Table 8.1 lists these and a few other examples of possible nuclides. Below 1% of the projectiles actually produce a target nuclide in all cases. Figure 8.8 explains the problem of the production. At low energies the production probability strongly increases, but above a few 10 MeV the cross-sections again decrease and the slightly stronger decrease of stopping power can hardly compensate this. Dividing the production probability by the input power in order to consider the production per invested energy leaves us optima in the region < 100 MeV. Moreover, physics limits the energy released per decay to a few MeV. Only the decay of ^{232}U grants a higher value due to a chain of 8 relatively short-lived decays following the primary decay, adding up the released energy.

The production of a quantity of ^{232}U releasing initially 1 W of radiation output power requires the impact of $0.149\ \text{MC} = 9.3 * 10^{23}$ protons on a pure ^{232}Th target. This enormous amount of protons equals 420 h of irradiation with 100 mA beam current. These calculations intend demonstrating the fundamental considerations of energy efficiency, optimisation of beam properties, and isotope selection required for an early assessment of the chances of such a technology. Nobody so far developed accelerator based nuclear batteries, but this case study demonstrates a topic requiring combining all the fundamentals discussed in this book.

P.S.: A charming, but physically questionable variant of nuclear batteries are beam triggered de-excitation isotopes. Long-lived excited state isotopes, most importantly the long lived ^{180m}Ta from which every year potentially 10 GWh are mined from 600 ton of natural Ta, shall be de-excited by a LASER like process started e.g. by collisions with beam particles. The long lifetime of these excited nuclei originates from forbidden transitions of the excited to the ground state due to a mismatch in spin quantum number. The state inversion exists, but how to fulfil the other requirements of a LASER such as low absorption of the emitted photons?

8.3 Accelerator Based Nuclear Fusion

The development of fusion power to give an unlimited supply of clean energy, and a switch to electric cars (Stephen Hawking)

Rising energy consumption, energy costs, and climate change are the most urgent problems of current energy production, endangering the industrialisation itself.

Table 8.1 Estimated properties of a few selected nuclear battery products generated by proton irradiation of bulk primary nuclide materials

Primary nuclide	Product nuclide	Half-life (Years)	Mean γ-energy (MeV)	Energy per decay (MeV)	Specific output power (Atoms/W)	Production probability $\int \sigma/S\, dE$ (%)	Proton dose per output (C/W)	Efficiency at 30, 20* or 6.8** (MeV)
Ne-22	Na-22	2.60	2.20	2.80 (β)	2.64E+20	0.33	1.28E+04	3.08E−04
Cr-54	Mn-54	0.85	0.84	1.38 (β)	1.75E+20	0.33	8.49E+03	1.52E−04
Cu-65	Zn-65	0.67	0.58	1.35 (β)	1.41E+20	0.23	9.82E+03	1.04E−04
Zr-91.92	Nb-91	680.00	0.01	1.25 (β)	1.55E+23	0.20	1.24E+07	8.33E−05
Bi-209	Po-209	102.00	0.005	5.00 (α)	5.80E+21	0.05	1.86E+06	1.25E−04*
Bi-209	Po-208	2.93	2E-5	5.20 (α)	1.59E+20	0.18	1.41E+04	4.68E−04*
Th-232	U-232	69.8	Chain	$\Sigma = 41.5$ (α)	≤1.3E+21	0.14	1.49E+05	8.54E−03**
U-238	Pu-238	87.7	Chain	$\Sigma = 50.5$ (α)	4.94E+20	0.12	6.58E+04	2.02E−03

The evaluation considers (p, n) and (p, 2n) reactions producing ground state nuclei, since these offer the highest cross-sections in the 30 MeV range of protons. In case several primary isotopes can be used for production the one with the highest cross-section is taken and highlighted. Nuclear data are extracted from ENDF/B-VII.1, cross-sections are taken from (Koning et al. 2015). All isotopes are assumed to be produced instantly (with infinite beam current). Calculations for Bi-209 based on 20 MeV and Th-232 on 6.8 MeV protons

Note The high TENDL-2015 cross-sections of the Th-232 reaction are questionable and not verified

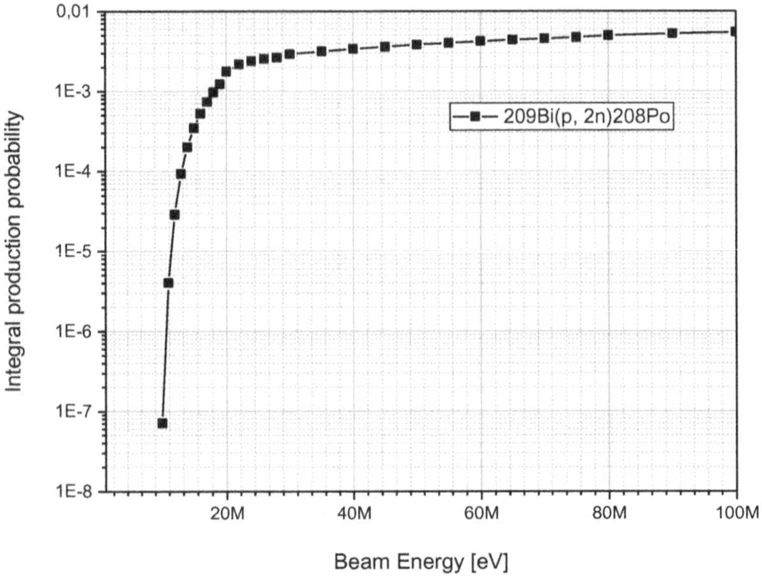

Fig. 8.8 The production probability of Po-208 from irradiation of Bi-209 by protons and its relation to the proton beam energy. At around 20 MeV the production efficiency starts to saturate at some 0.1%

Nuclear fusion has become the metaphor for a clean and sustainable, yet prospering energy future. Fusion would enable us to use water as a primary source of energy by "burning" in particular the deuterium isotope present in water (and other hydrogen containing molecules) in a nuclear fusion reaction. This opens up deuterium as a new environmentally and politically uncritical primary energy source. Making a cheap and easily available resource such as water the main input for our economy would end conflicts and solve pollution and climate change problems. Unfortunately, the technology required for an economical and safe usage of nuclear fusion is as far away from current energy technologies as an abacus is from a super computer.

On the journey for finding the easiest technology for nuclear fusion power plants only a few options have remained realistic. The most promising technologies are the stellarators and the tokamaks (Ongena et al. 2016). These devices confine hot plasmas through the use of complex closed magnetic fields. The fundamental concept of these reactors works similar to the stars, which confine their plasma through gravity. The technical concepts feature the technological disadvantage of requiring a wall and a vacuum vessel around the plasma and mechanisms other than light emission to remove the released fusion energy from the vessel. The plasma production and heating requires an initial energy input (gravity for stars and electricity in fusion reactors), but once the temperature reaches a few ten keV (1 keV = 11,600,000 K) also the technical fusion can be self-sustaining.

These most promising reaction is the D-T reaction D(T, n)^{4}He as it offers a high energy output together with a low reaction barrier (=required plasma temperature).

The D-D reaction would remove the problem of breeding tritium, which is not naturally available, but on the other hand it features about 7 times smaller energy release per reaction. The D-D fusion actually consists of two reactions, namely $D(d,n)^3He$ and D(d, p)T. In D-D and D-T reactors problems arise from the significant transport of energy and particles perpendicular to the magnetic field lines leading to heat loads in the order of MW/m^2 and plasma particle fluxes $> 10^{24}/m^2s$ in the most loaded areas together with about 10^{19} n/m^2s. The necessity for fragile technology such as superconductors together with the materials suffering from the intense neutron radiation from the fusion reactions further complicates the situation.

Accelerators are nowadays applied in a few different roles in nuclear fusion. Direct applications involve the heating of the fusion plasma via neutralized accelerated beams using large DC accelerators as depicted in Fig. 8.9. The process of stopping transfers the beam energy to the plasma. Besides heating via high-frequency absorption in the plasma, this neutral beam heating represents a fundamental technology of reaching the required plasma temperatures. Indirect applications include the application of the analysis methods discussed in Chap. 7 for quantifying the plasma-surface interaction and studying radiation damage and the behaviour of materials under the involved loading conditions.

Fig. 8.9 The ELISE neutral beam injector source. It delivers 23 A of negatively charged protons with a beam area of 1 m^2 which will be neutralised by a stripping gas before injection into the ITER tokamak. Courtesy of Robert Haas, Max-Planck-Institut für Plasmaphysik, Germany

In contrast to the plasma reactors, we have already seen nuclear fusion to be easily induced with particle accelerators. Unfortunately, in accelerators elastic nuclear reactions, which do not produce energy, impose a problem to the particle confinement. For plasma devices elastic collisions have less importance, since the collision partners and their energy remains in the plasma. These elastic reaction cross-sections dominate over the inelastic fusion reactions by orders of magnitude, therefore prohibiting a direct positive energy output. However, this numeric comparison fails to take the energy and particle confinement properties of accelerators, the so-called acceptance (Sect. 2.3), into account.

This section will apply the knowledge gained throughout this book in order to make back-of-an-envelope level calculations for the feasibility of accelerator fusion. Different approaches and possible reactions for accelerator reactors will be considered in this case study. The production of rare isotopes, neutrons and energy are addressed as possible alternative applications. Finally, physical parameters and ratios are developed for a reactor approach to quantify the qualitative understanding of energy confinement and check whether current technology could provide an accelerator reactor operating with positive energy output and if not which technology might be sufficient.

The potpourri of complex technologies involved in the confinement plasma reactor represents the major drawback of these technologies. In the accelerator language, these devices could be described as multi-species storage rings with zero beam energy and an extremely high emittance. If nuclear fusion works on this end of the accelerator parameters, why should it not work on the other end, namely a storage ring with low emittance and finite beam energy? Well, let us assume this as the basis of a technological concept. The technology of an accelerator reactor features some advantages and disadvantages over the plasma reactor. The main problem induced by the finite beam energy is the relevance of elastic scattering events trying to sabotage it. As we learned earlier, elastic collisions transfer energy from the directed (beam energy) to the directionless (emittance) component, forming a major difference between both concepts. While for zero beam energy the confined energy lies solely in the emittance, for finite beam energy the confined energy mostly lies in the beam energy.

So far so good, but this technological disadvantage of accelerators is countered by a technological advantage of an easier confinement of the directed beam energy compared to the chaotic energy of the emittance. Physically it is impossible to separate elastic and inelastic (fusion) reactions of the beam particles, hence we have to find a workaround for the elastic part. The main problem of accelerator fusion therefore reduces to the competition of elastic and inelastic reactions and their cross-sections. In a zeroth order, this is the ratio of fusion to Rutherford cross-section.

In reality this simplifies the physics too extensively. As we learned in the ion-beam analysis Sect. 7.1.4, the elastic cross-sections starts to deviate from the Rutherford value at about 50% of the coulomb barrier of 389 keV for H to H nuclei. In this case it apparently deviates at even lower energy, otherwise the D-T fusion would be impossible with its cross-section maximum at 64 keV. In order to optimize a setup and investigating its potential, the basic nuclear reactions, their cross-section for the energy producing fusion reactions, and the counteracting elastic energy-loss reactions are required. The energy loss reactions are, depending on the reactor concept, electronic stopping and elastic scattering. Therefore, reactions of two light ions are in principle best suited. They require lower initial energy (coulomb barrier) and suffer lower energy-loss cross-sections (mostly proportional to nuclear charge) compared to heavier reaction partners.

An accelerator reactor can be used to produce neutrons, isotopes, or energy. All of these products are basically proportional in production rate and rely on the same reactions, but energy is the only parameter with a threshold (energy production minus device upkeep). Therefore, the following calculations focus on energy production, without loss of relevance for the other products.

Table 8.2 investigates a few light particle reactions by comparing their energy release Q with the approximately required beam energy based on the lowest energy cross-section maximum and a ratio of the Q-value to a ratio representing the invested beam energy E and the charges Z_i responsible for elastic loss reactions. Among the feasible reactions depicted in Table 8.2, the D+T fusion promises by far the highest output efficiency, similar to the situation in plasma reactors. The table illustrates the principle limitation of an accelerator reactor with its different confinement compared to a plasma reactor: For each ion, elastic scattering can lead to a total loss of the invested projectile energy by removing the particle from the beam, therefore each fusion reaction needs to yield a multiple of the projectile energy in order to allow for positive output with finite efficiencies. In other words, the projectile energy scales the loss related to the elastic scattering and the Q-value the gain of the fusion cross-section. In a simple estimation, this factor is 275 for D+T, due to this reactions high Q-value and low required beam energy.

The D-T, D-D and D-^{3}He reaction cross-sections, in the centre-of-mass (CM) system, were parametrized in (Bosch and Hale 1992) with a precision of 2% up to 1 MeV to

$$\sigma_F(E) = \frac{A1 + E(A2 + E(A3 + E(A4 + E * A5)))}{1 + E(B1 + E(B2 + E(B3 + E * B4)))} * \frac{1}{E * e^{B_G/\sqrt{E}}} \quad (8.1)$$

Table 8.2 Compilation of feasible reactions for an accelerator fusion reactor

Reaction	D+D	D+^{3}He	D+T	T+T	P+^{11}B
Energy gain Q (keV)	3650	18,353	17,589	11,332	8682
1st Cross-section peak (keV)	3000	270	64	2100	625
$Q/(E(\sigma_{max}) * Z_1 Z_2)$	1.2	34	275	5.4	2.8

The rows depict the reaction energy gain Q, the 1st maximum in fusion reaction cross-section and a ratio demonstrating their principle feasibility based on a simplified view on the competition of fusion and elastic reactions in relation to the invested projectile energy. From this ratio it becomes clear, that only D+T and T+T are feasible from a practical point of view with typical technical energy conversion efficiency of up to 50% and a closed fuel cycle. D+^{3}He offers an attractive option, if ^{3}He supply can be provided from other sources with E in keV and σ in barn and all other variables being fitting parameters for the individual reactions to the cross-section data. These following calculations are based on these semi-empirical cross-sections.

8.3.1 Neutral Target Reactors

Accelerator reactors with neutral targets, i.e. gas, liquid, or solid, bombarded by accelerated ions are the first considered option. These neutral target reactors are state-of-the-art in many industrial, medical, and scientific applications as discussed above. In Sect. 8.2, we assumed exactly this situation for the production of nuclear batteries.

D+T reactors (Sect. 4.2.1) of this type often apply a Ti or TiD_2 target, which is bombarded by 10–120 keV T^+ and D^+ ions yielding outputs of 10^5–10^{12} neutrons/s. Most of the energy input to the ions is lost by the electronic stopping in the target, which can be calculated by SRIM (Ziegler et al. 2008) for different target situations. Along with the stopping, the fusion reaction cross-sections from (8.1) are applied, after transformation to the lab system. It is always more efficient to accelerate the heavy reaction (i.e. T in D+T) partner due to momentum conservation, thus the calculations are based on this consideration.

For the D-T reaction, Fig. 8.10 compares three different target options to investigate the influence of the stopping power on the reactor performance. A composition with relatively high stopping power is TiD_2, a medium case is given by LiD and the optimum, but gaseous, case is D_2 gas. Figure 8.10 demonstrates higher stopping

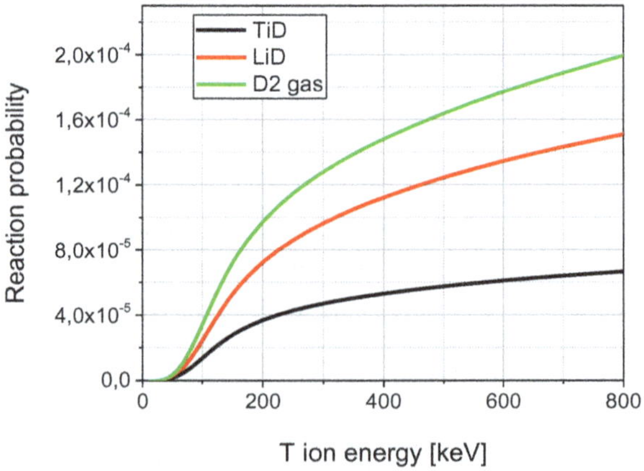

Fig. 8.10 Calculated D+T→^{4}He + n reaction probability for three different neutral targets. The smaller the stopping power, the higher the reaction probability. The optimum energy efficiency is always in the region of 170–180 keV T ion energy, since the stopping decreases slower than the projectile energy rises

powers lead to a lower reaction probability, since fewer nuclei are passed with the same starting energy. For obtaining a high energy-efficiency, a high reaction probability per invested T ion-energy is required, therefore the reaction probability has to be divided by the projectile energy E for obtaining the point of maximum efficiency. This point is 170 keV (TiD_2), 177 keV (LiD), and 180 keV (D_2). The optimal case (D_2) yields a reaction probability of $8.9 * 10^{-5}$. Multiplying this with the D+T Q-value of 17,589 keV, each incident T-ion releases on average an energy of 1.6 keV, while at least 180 keV are invested for acceleration of the T projectile.

The numbers demonstrate clearly: Even with the optimal target, the reactor is not very efficient at neutron production and over a factor 100 away from multiplying the invested energy by a factor 2. Considering typical conversion efficiencies a factor 2 represents the absolute minimum, realistically rather a factor 10 would be required. The presented calculation of reaction efficiency allows an optimisation by optimal choice of projectile energy and target (=giving the maximum number of reactions per input power), but fundamental physics will always prevent positive energy output for this situation. Nevertheless, D+T neutron sources can be optimized in output and efficiency significantly, using the presented considerations.

At 180 keV, the nuclear stopping or elastic collision, respectively, constitute only 0.13% to the total stopping for a D_2 target (Ziegler et al. 2008). Therefore, removing the electrons from the system would provide about 2 orders of magnitude gain in efficiency of the reactor. Another factor $\approx$ 5/3 can be gained by accelerating both particles in a collider instead of a fixed target, avoiding excess centre-of-mass (CM) energy in the laboratory frame. These two factors can only bring us somewhere near the breakeven. For a significant positive output it is not efficient enough to run down the projectile energy from its initial value down to zero as it happens in a neutral target reactor, but we have to keep the projectiles close to the optimal ratio of fusion cross-section to elastic scattering by re-accelerating the particles after non-fusion interactions. With this, the case study of the neutral target reactor quite directly pointed out the technological choices necessary to have a chance for a positive output.

8.3.2 Ion-Beam Collider Reactors

Concluding from the last section, accelerator reactors require removing the electrons from the equation and have both projectile and target moving. Ion-beam colliders, cyclotrons, or Electrostatic-Ion-Beam-Traps (EIBT) constitute technological example implementing these requirements.

To provide a constant energy, the accelerator has to be designed in a way that particle energies are refilled after a certain amount of non-fusion interactions. These features are usually called phase-focussing when the ion-beam is operated in AC mode (bunches) confining it in a certain phase relation to the driving wave. Because of this, the targeted energy and technical details differ from the concepts above. The targeted energy resembles the maximum ratio of fusion to elastic scattering cross-section, but not exactly. For keeping the longitudinal energy constant, a DC beam is not possible, but a bunched beam with AC drivers has to be applied. As the longitudinal phase coherence of the ion-beam adds additional complexity, the whole accelerator design becomes more complex.

In the following, the elastic scattering is calculated, based on Rutherford scattering cross-sections, as elastic scattering imposes the main scientific challenge for accelerator reactors. This time, the calculations will go beyond simply comparing Rutherford to fusion reaction cross-section for analysing accelerator based fusion. The mistake connected with this view is, that the projectile is not necessarily lost after a Rutherford scattering event, since small angular deviations are still confined by the accelerator, certain emittance values are uncritical depending on accelerator acceptance. Most Rutherford collisions are small angle scattering collisions, which induce small momentum and energy transfer, see the scaling with scattering angle φ in the Rutherford cross-section (3.1). Therefore, the projectile still has sufficient energy for a fusion reaction after most Rutherford scattering events and it can stay confined in the beam. In conclusion, similar to a tokamak or stellarator, also an accelerator reactor provides an energy and particle confinement which is in the following quantitatively compared to the requirements of positive energy production.

This confinement depends on the ion optical system acceptance in real and, for AC systems, phase space. For a maximum efficiency of an accelerator reactor, maximal energy and particle confinement/acceptance are required, allowing tolerating plural-scattering and larger φ. The confinement is defined by the acceptance A (Hinterberger, Physik der Teilchenbeschleuniger und Ionenoptik 2008):

$$A = \pi \epsilon_{\max} = \pi * y_{\max}^2 / \beta_{\max} \tag{8.2}$$

The acceptance defines a maximum emittance in the reactor, which is determined by the minimum in extent of the beam tube diameter $y_{\max}$ over the betatron function $\beta_{\max}$ in that position. The equation implies the reactor design should feature a minimal betatron function at a maximal tube diameter throughout the beamline in order to achieve minimal losses. Obvious, but now we have it quantified. An accelerator reactor can be build short and with large diameter, compared to typical particle physics accelerators, since the required kinetic energy is small (64 keV for D+T). In an example of $y_{\max}$= 3 m and $\beta_{\max}$= 10 m the 1σ acceptance is about 10^6 π mm mrad, but which acceptance is required to reach a positive energy output?

Acceptance and Confinement of Accelerator Reactors

The increase in emittance or transversal particle energy, respectively, challenges the acceptance. Conserving the small angle part of the scattering events increases the beam emittance, as this transfers energy from the longitudinal to the transversal movement. The transversal energy is in fact lost for the system as it cannot be recovered to the longitudinal direction, but it has to be removed by beam cooling or particle losses to avoid de-confining the beam. The beam emittance increase per elastic scattering event is given by (Hinterberger, Physik der Teilchenbeschleuniger und Ionenoptik 2008):

$$\Delta\epsilon = \frac{1}{2}\beta\Theta^2_{\text{rms}} \tag{8.3}$$

Here β is the betatron function in the position of the event and θ^2_{rms} is the variance of the root-mean-square (Rutherford) scattering angle. Equation (8.3) illustrates that large angle events contribute over-proportionally to the emittance growth compared to small angle events. For calculating $\Delta\varepsilon$ we have to integrate the Rutherford cross-section, which requires a lower integration limit due to its divergent nature towards small angles. Here a minimum angle of 1π mrad is chosen since this is in the order of the divergence of a beam coming from an ion source. Particles within this divergence are thus assumed to have a negligible influence on the overall beam divergence. Increasing this minimum angle leads to higher $\Delta\varepsilon$ with roughly a square-root dependence.

Assuming that all ions with a Rutherford scattering angle $< \varphi_{\max}$ are confined by the acceptance (thus neglecting the Gaussian shape of the beam), we can determine the smallest $\varphi_{\max}$ with a potential positive energy output (Yield > 0). Using the fusion σ_F, Rutherford loss (scattering angle $> \varphi_{\max}$) cross-sections σ_{RL} and the invested projectile centre-of-mass energy E, this average excess energy yield normalised to the elastic reactions is given by

$$\text{Yield} = (E + Q) * \eta - E * \sigma_{RL}/\sigma_F \tag{8.4}$$

In this calculation, the energy conversion-efficiency η of fusion heat to electricity becomes relevant. A conservative assumption is $\eta = 1/3$ and an optimistic estimate is $\eta = 2/3$. To keep the dimensional freedom in 2D limits, the energy efficiency in producing the ion beams is assumed to be 1, as it can also be absorbed in η. The Rutherford loss cross-section is given by the integral of equation (3.1) from the given $\varphi_{\max}$ to $360° - \varphi_{\max}$. Figure 8.11 plots the resulting yield for 3 different η. For $\eta = 0.33$ the turnover from negative to positive energy output happens at $\varphi_{\max} = 7.7°$ and levels off at about 6 MeV per reaction at 20° acceptance. The energy multiplication, given by the yield divided by the input energy, reaches a value >2 at angles >7.8°. The higher the η, the less large angle scattering event have to be confined. We can use this acceptance angle to determine the average increase in emittance per elastic reaction $\Delta\varepsilon$ using (8.5):

$$\begin{aligned}\Delta\epsilon &= \frac{1}{2}\beta\Theta^2_{\text{rms}} = \frac{\beta}{2}\left(\left(\int_{1\pi\ \text{mrad}}^{\phi_{\max}} \frac{\phi}{\sin^8\left(\frac{\phi}{2}\right)} d\phi\right)^{0.5} / \int_{1\pi\ \text{mrad}}^{\phi_{\max}} \frac{1}{\sin^4\left(\frac{\phi}{2}\right)} d\phi\right)^2 \\ &= \frac{\beta}{2}(75\ \text{mrad})^2\end{aligned} \tag{8.5}$$

In this calculation, the beam energy and the particle charges drop out, as these are only relevant for the probability scaling not its distribution. We obtain a RMS scattering angle of 75 mrad = 4.3°. As an example, $\beta = 10$ m produces an emittance

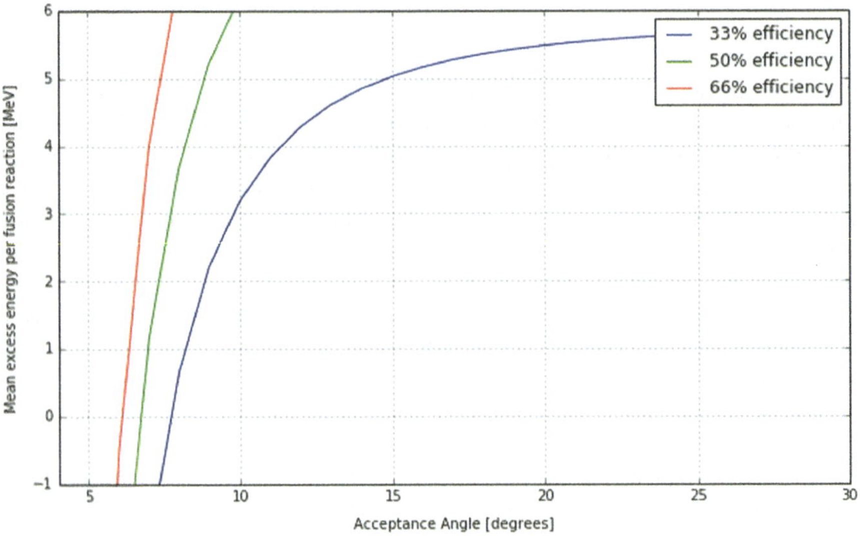

Fig. 8.11 Mean energy yield per D+T reaction versus maximum confined scattering angle $\varphi_{\max}$ at optimal CM energy for 3 different energy efficiencies. At least 6.2° are required for a positive energy output. Higher angles lead to higher mean energy gain, but the benefit levels off at ≈15°. For D+^{3}He at least 17° are required with 470 keV CM energy. Higher accepted angles lead to increased beam emittance, requiring increased beam tube diameters or beam cooling for compensation

of 9 * 10^3 π mm mrad on average per elastic scattering of the beam particles. For example having an acceptance of 10^6 π mm mrad (see example above) and a negligible initial emittance, which is typical for ion sources (Lejeune, THEORETICAL AND EXPERIMENTAL STUDY OF THE DUOPLASMATRON ION SOURCE Part 2: Emissive properties of the source 1974), would allow for $A/\Delta\varepsilon = 110$ scattering events on average, before the beam emittance is larger than the acceptance and a relevant portion of ions gets lost to the walls. The ratio of the confined Rutherford scattering (scattering angle $\leq \varphi_{\max}$) to the fusion cross-section, σ_R/σ_F, on the other hand indicates how many times particles elastically scatter on average before they undergo a fusion reaction.

As the first ratio is independent of particle energy and the second is dependent on it, we now optimize the ion energy in order to determine the relative importance of these two associated particle loss mechanisms. We can maximize the yield by choosing an optimal centre-of-mass energy at the given $\varphi_{\max}$, see Fig. 8.12. A minimum ratio of $\sigma_R/\sigma_F \leq 20$ can only be reached with CM energies >500 keV, but at these energies the energy multiplication becomes too small, see Fig. 8.12. At 81 keV a maximum in the energy multiplication is located. In between the D+T reaction cross-section peak and 81 keV, the Rutherford cross-section decays faster than the fusion cross-section, while the CM energy multiplied with the reacting beam fraction $\sigma_F/\sigma_F+\sigma_R$ is still small compared to the Q-value. In a small window from 64 to 120 keV this leads to a positive energy output. This small energy window and the sharp energy

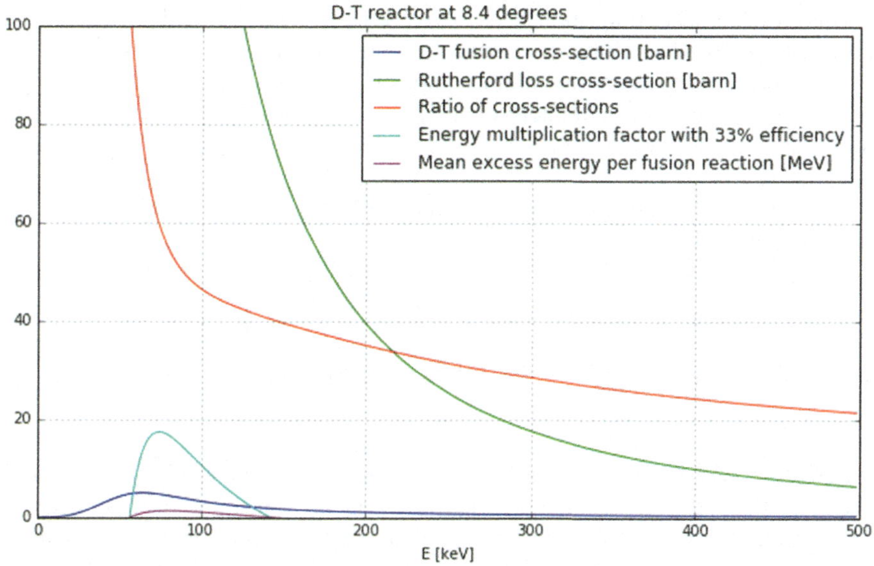

Fig. 8.12 The parameters relevant for the energy production in a D-T reactor with an acceptance of 8.4° and $\eta = 1/3$ for centre-of-mass energies up to 500 keV. After reaching the D-T cross-section maximum at 64 keV also the excess energy per reaction (Yield/E) reaches a maximum of 1.4 MeV at 81 keV, with an only 56 keV wide window of positive output. At energies above this, too much energy is invested for acceleration, as still >95% of the particles are lost before reacting. At energies below the maximum, too many particles are lost before reaction due to low fusion cross-section

multiplication maximum of Fig. 8.12 demonstrate the importance of a fixed CM energy for an efficient accelerator reactor, otherwise the projectiles quickly leave this window towards region of insufficient multiplication factors.

The energy of both beams can be calculated by assuming equal momenta in the CM system. For D+T, the optimum centre-of-mass energy was determined to $W = 81$ keV. Hence we can calculate the deuteron energy E_D and the triton energy E_T with

$$\begin{aligned} m_T * v_T &= m_D * v_D \\ &=> m_T\sqrt{\frac{2E_T}{m_T}} = m_D\sqrt{\frac{2E_D}{m_D}} \\ &=> m_T E_T = m_D E_D \text{ with } E_T + E_D = W \\ &=> \frac{W - E_D}{E_D} = \frac{m_D}{m_T} => \frac{W}{\left(\frac{m_D}{m_T} + 1\right)} = E_D \\ E_D &= 48.6\,\text{keV};\ E_T = 32.4\,\text{keV} \end{aligned} \tag{8.6}$$

This multivariate optimisation leads to a working space in y_{max} and E. Also D+^{3}He and T+T have a small working space, but they require much larger φ_{max}. A $\varphi_{max} > 7.7°$ already represents a substantial number imposing a significant technical challenge, rendering other reactions completely unrealistic. The real number of φ_{max} will be slightly lower, since we assumed every elastically scattered particle to be lost by using σ_R/σ_F.

An approximate wall-loss term $\Delta\varepsilon/A$ can now be added as a negative term to the energy balance in (8.4):

$$\begin{aligned}\text{Yield} &= (E+Q)*\eta - E*\left(\frac{\sigma_{RL}}{\sigma_F}+\frac{\sigma_R\Delta\epsilon}{\sigma_F A}\right)\\ &= (E+Q)*\eta - \frac{E\sigma_{RL}}{\sigma_F} - \frac{\sigma_R E*\beta\beta_{max}*\Theta_{rms}^2}{2\pi\sigma_F y_{max}^2}\end{aligned} \tag{8.7}$$

Equation (8.7) defines the minimum device size y_{max}, for a positive energy yield. Devices with a size $< y_{max}$ cannot produce power but e.g. generate neutrons. Insertion of φ_{max} from (8.5) into (8.7) yields a convergent solution and the modified minimum angle. These parameters strongly depend on the overall reactor properties. The betatron function now comes in with a square relation, making the reactors beam optical design a key point for a technical realisation. Reducing β_{max} with better beam optical technology brings the situation closer to the fundamental physical limit discussed before. Adding emittance cooling could substantially improve the situation since it adds a counteracting effect to the emittance increase by elastic collisions.

Luminosity Runaway

Above we verified that an accelerator can provide sufficient confinement of the elastic losses, but it still needs to be clarified if also significant power output can be generated under those boundary conditions. This question of power output is equivalent to calculating which fraction of the injected ions actually is involved in a fusion reaction before it is lost to de-confining events other than elastic beam scattering. We term this issue luminosity runaway, because a lower luminosity directly leads to a longer dwell time (lower beam density = lower interaction probability) for the beam particles which leads to a higher probability of undergoing these other loss mechanisms. Therefore, it is required to calculate the number of reactions per time dN_R/dt:

$$\frac{dN_R}{dt} = L*\sigma \tag{8.8}$$

With the reaction cross-section σ and the luminosity of the accelerator reactor L for AC and DC accelerators:

$$\begin{aligned}L_{AC} &= \frac{N_B*f*N_D*N_T}{4\pi*r_x*r_y}\\ L_{DC} &= \frac{I_D*I_T}{e^2*\pi r_x r_y}\end{aligned} \tag{8.9}$$

With the number of bunches in the accelerator N_B, the particle circulation frequency f, the average number of particles in the D and T bunches N_D and N_T (D+T reactor), the intensities I of particles per time, and the 1σ values (beam dimension) in the x and y direction r_x, r_y assuming a Gaussian beam profile. The dimensions of the beam are connected to the beam emittance ε and betatron function β

$$r_{x,y} = \sqrt{\epsilon_{x,y} * \beta_{x,y}} \tag{8.10}$$

The luminosity allows us to calculate the extinction length of the ion-beam by combining the integral of equation (8.8) with (8.9) and (8.10):

$$N(x) = N_0 * e^{-\frac{\sigma_F * x * I}{v\pi r_x r_y}}$$
$$=> x = \ln\left(\frac{N_0}{N(x)}\right) * \frac{\pi v r_x r_x}{\sigma_F I} \approx \ln\left(\frac{N_0}{N(x)}\right) * \frac{\pi v \beta(\epsilon_0 + \Delta\epsilon)}{\sigma_F I} \tag{8.11}$$

with the D+T cross-section σ_F(80 keV) = 4.49 barn, the passed distance x and the total velocity v(80 keV) = 2164 km/s (D ions) + 1442 km/s (T ions) = 3606 km/s. Assuming an emittance of 10^6 π mm mrad (acceptance limit discussed above), an intensity equal to 1 A in each beam and β = 1 m, (8.11) yields an extinction length for 90% of the beam to react of $8 * 10^{-18}$ particles/m equal to a range for 90% loss of $1.3 * 10^{16}$ m, corresponding to about one year of necessary confinement time with the given velocities. The emittance increase thus prevents a reaction of the injected ions within their confinement time, because the elastic collisions thin out the beam/reaction partner density too strongly. A higher acceptance can confine them, but they will not induce fusion reactions anymore. On the other end of the technically possible emissivity/beam density we have the initial beam produced by the ion source. Developments of neutral beam injectors for tokamaks report emissivities of $\varepsilon_0 = 0.2\ \pi$ mm mrad and current densities in excess of 1 A/cm^2. With this number a 90% reaction probability is reached after 160 km, keeping the other input parameters as stated above and neglecting the increase of emittance. Therefore a feasible confinement time of <0.1 s would be required, if we could switch off emittance increase. Again, the betatron function is a main factor for an accelerator reactor, as it has a quadratic impact on the beam density (linear for emittance increase and linear for beam dimension).

The number of reactions N_R and the mean energy production per reaction Q define the output power P_0 of the reactor:

$$P_0 = Q * L * \sigma = Q * \sigma * \frac{N_B * f * N_D * N_T}{4\pi * r_x * r_y} \tag{8.12}$$

Since σ and Q are physically fixed, the luminosity is the only technical mean to adjust the power production. The right hand part assumes an AC accelerator. Technically, a continuous DC beam can offer a higher luminosity while offering less complexity as no longitudinal phase coherence is required (smaller emittance), but the refuelling of energy to beam particles can only work in AC accelerators, see

Sect. 2.2.2, ruling out the DC accelerator option. The power generation is proportional to the accelerator mean current $eN_Bf(N_D + N_T)$ and inversely proportional to the square-root of the emittance or the beam size, respectively. The average current can be increased by making the circumference of the machine bigger, if f remains constant but N_B increases. The luminosity can also be increased by increasing the beam diameter at a constant beam density as $N_D * N_T$ depends quadratically on the beam radius $r_{x,\,y}$ leaving a linear gain in P_0. In other words, the technology has a linear upscaling capability.

Technical Conception

The optimal device operation requires a collider setup equal to zero momentum in the centre-of-mass system. Consequently, both ion beams have the same momentum and thus the curvature radii ρ of both beams (charge $q = +1$) in a given magnetic field B are equal (2.33). In principle, both beams can run in the same magnets as a starting inspiration.

The design should follow a mixture of the following goals:

1. Minimize betatron and dispersion functions in the fusion reaction zone
2. Maximize acceptance to improve energy confinement
3. Avoid having delicate systems in the reaction zone (neutron irradiation).

The design differs in the following major points from typical particle physics accelerators:

1. A low, fixed beam energy (81 keV for D+T), but a high beam current (injection of >1 A) are required
2. The beam interaction zone can be a large part up to the whole circumference, since no detectors or the like are required in this reactor type
3. A maximum luminosity is not of prime importance, as the circumference can be small (<100 m) and the beam tube diameter can be large (>1 m) without risking unfeasible device cost.

The beam confinement can be realised in a ring structure, a mirror setup, or a combination of both. The collider ring reactor could be realized by two accelerator storage rings in counter-current operation, similar to the LHC design with its two proton beams. The two rings could be connected, providing the central reaction chamber where the fusion reactions take place surrounded by breeding modules for tritium generation similar to a tokamak reactor. The fusion reaction consumes tritium, therefore it has to be produced via nuclear reactions of the emitted fusion neutron with e.g. lithium. The remaining parts of the two rings offer space for acceleration and emittance cooling. A single magnetic dipole element can combine and separate the counter-propagating beams from their respective rings, since it deflects both beams in the same direction as seen along the direction of motion. Also beam focussing using e.g. multiple quadrupole magnets acts similarly, regardless of the direction the beam enters. For electro-static beam optics the different beam energies prevent a dual use of optical elements. Acceleration facilities, e.g. RF-cavities, required to maintain the design energy rely on a fixed frequency. With two rings the same rotation frequency can be achieved for both species in spite of their different velocity by adjusting the individual ring length accordingly.

A physical feasibility of a positive energy output is not necessarily a technical feasibility, due to conversion efficiencies being $\ll$100% and the aggregates running the reactor also consume power. Typically, aggregates consume $\approx$10% of the power output, but as we saw in Sect. 8.1 this very much depends on the overall conception. Four main aggregates are required to run the accelerator reactor: The ion sources, the gas/vacuum system, beam optics, and energy conversion and T-breeding. This section provides a short overview of the associated details to point out that these aggregates do not prohibit an efficient technical realisation but raise the required energy multiplication factor. For simplicity, the numbers required for this consideration are based on a 1 A ion current system. At many points technical details are based on specifications of state-of-the-art industrial products at others it's just estimations, therefore the numbers must be handled with care.

The D and T ions sources are required to provide maximum beam density at minimum emittance. Currently several options as hot-cathode multi-cusp, duoplasmatron ion sources, or ECR sources provide feasible options. A single ion source can be used to produce D and T ions together with a mass separating beam optical element in order to avoid the need for D-T gas separation in the gas system. According to (Hinterberger, Physik der Teilchenbeschleuniger und Ionenoptik 2008) a duoplasmatron source providing 1 A of positive ion current consumes about 1 kW of electrical power. The initial acceleration can be realized by ion source biasing with the required DC potential, since in the D+T case only <50 kV are required. In the power class of $\geq$100 kW this is considered to provide $\eta_{\text{Acc}} \geq 87\%$ efficiency, according to current European standards for power supplies. In total, the ion source will consume about 100 kW for the two beams.

Besides the vacuum leakage, the ion source determines the requirements for the vacuum system as it imposes the main gas load. For duoplasmatron sources, a ratio of charged to charged + neutral particles leaving the source of 0.1 to 0.8 was observed (Lejeune, THEORETICAL AND EXPERIMENTAL STUDY OF THE DUOPLASMATRON ION SOURCE Part 2: Emissive properties of the source 1974). A construction optimized for minimum emittance showed a maximum of 0.5, but with 0.25 a more conservative number will be assumed here. In conclusion for every ampere of ion current supplied by the ion source, $F = 1.8 * 10^{19} = 7 * 10^{-7}$ bar * m^3/s neutral particles are released into the vacuum system. It has to be noted that 1 A of D and 1 A of T ions are required, but also that hydrogen isotopes will be present as molecules in the vacuum system.

The whole reactor is under vacuum to avoid beam energy loss and emittance increase of the ion beams. the low ion energy is <50 keV requires the residual vacuum pressure to be very low. This is counteracted by gas losses from the ion sources and gas production (helium and neutron induced reactions) from the D+T reaction. Vacuum leakage is considered to be negligible, as this is not physically unavoidable. For the maximum tolerable vacuum pressure, SRIM calculations of the stopping of 32.4 keV T and 48.6 keV D ions in hydrogen enriched air (products + typical residual gas spectrum of vacuum vessels, see Sect. 2.1) are conducted. The required vacuum pressure results from the typically travelled distance corresponding to the mean revelations per D-T reaction times the circle length at the given RF frequency with the requirement of a total energy loss below say 10 keV. The SRIM calculation yields a stopping power of $S = 25$ MeV/m (T) and $S = 31$ MeV/m (D) at 200 g/m^3 ($\approx$1 bar pressure) of 90% hydrogen and 10% oxygen. With a mean travelled distance of $d = 160$ km (see above) and (8.13) this leads to a vacuum pressure requirement of $p_{\max} \leq 2 * 10^{-9}$ bar as the maximum allowed pressure defined by the D ion energy-loss.

$$p_{\max} = \frac{1\ \text{keV}}{d * S} * 1\ \text{bar} \tag{8.13}$$

The gas influx rate from the ion source of $F = 7 * 10^{-7}$ bar $*$ m^3/s given by the ratio of losses of neutral gas to ion current in a duoplasmatron source of 1 A (Lejeune, THEORETICAL AND EXPERIMENTAL STUDY OF THE DUOPLASMATRON ION SOURCE Part 2: Emissive properties of the source 1974) and the He production in the D-T reaction of a 17.6 MW reactor (1 A ion current, $Q = 17.6$ MeV) of 1.25 $* 10^{-7}$ bar $*$ m^3/s with the requirement of $p_{\max}$ lead to a minimum pumping speed K of

$$K = \frac{F}{p_{\max}} = \frac{8.25 * 10^{-7}\,\text{bar}\frac{\text{m}^3}{\text{s}}}{2 * 10^{-9}\,\text{bar}} = 413\,\text{m}^3/\text{s} \tag{8.14}$$

The results of equations (8.13) and (8.14) are also no deal breakers. Vacuum pumps easily reach the desired pressure and the required pumping speeds are high but not orders of magnitude away from technology available on the market. Turbo pumps consume say 1.5 kW/m^{3}s^{-1}, resulting in a total consumption of 720 kW. The compression of the turbo pumps reduces the required pumping speed in the subsequent stages. We assume another 2 stages to reach the atmospheric pressure level with 100 kW power demand in each stage.

The tritium breeding and neutron and α-energy conversion can be made similar to breeder and plasma-facing materials and components developed for tokamaks and stellarators. The complexity is even slightly reduced since only 3.5 MeV α-particles and 14.1 MeV neutrons interact with the materials, not complex plasmas. The main power consumption originates from the cooling water pumps. Designs for tokamak reactors predict a value of 70 MW for a 1.5 GW reactor (Miyoshi et al. 2018), scaling this value down yields 82 kW of power consumption for the 1 A accelerator reactor.

Beam optics for the low energy beams require only little power, since electrostatic or even permanent magnetic systems could be applied. In both cases, the power consumption will be negligible. A certain power consumption will be connected to the re-acceleration or longitudinal and transversal emittance cooling, respectively. This value can only be guessed. Let us estimate a number of say 100 kW.

In total we estimated a power consumption of the aggregates of 1202 kW. With an electrical conversion efficiency of $\eta_{\text{Th}} = 1/3$ this results in an excess power of 4661 kW for the case of the continuous total fusion of a 1 A beam current. The upscaling of the output power requires scaling the ion current. The size of the device has a lower limit as it defines the losses via the acceptance. For higher output power, the reactor vessels scale in size.

This case study investigated the concept of accelerator based fusion using data of ion-stopping and elastic and fusion cross-sections together with some fundamental physical considerations. Re-considering the concept of accelerator fusion with what we learned in this book and the confinement of accelerators in mind lead to several conclusions. Several light ion reactions are compared, as scaling laws point out only light ions are feasible. From these, D+T offers the best chances, since it provides the highest value of Q over the energy of the first fusion cross-section maximum. A concept was developed based on considerations of neutral targets and the kinetic energy of the involved ions. Based on Rutherford elastic scattering cross-sections, an optimum centre-of-mass energy of 81 keV is calculated. Collider reactors were identified as the only possible option for power generation. It was revealed that the reactors have to be large in tube diameter and need extremely small betatron functions (=good confinement), compared to particle physics devices, if excess energy should be produced. These collider reactors also have their difficulties with emittance generation, luminosity runaway, and the betatron function, but with the presented considerations the core problem shifted from the ratio of elastic to fusion cross-sections to a more complex problem.

The understanding accelerator fusion presented here is based on many simplifications and could be called a zeroth order approximation for an academic case study. The idea of accelerator and colliding beam fusion is not new to this book. The electrostatic repulsion of the beams will be very strong resulting in a plasma pressure limiting the achievable beam density. Possible non-Rutherford elastic scattering cross-sections impose a potential risk and chance. The statistical nature of particle beams was only slightly implemented. The detrimental ion-beam neutralisation and betatron stability at high beam currents impose further risks for the technical feasibility. Chances and problems of beam cooling were not discussed. Nevertheless, this section provides a more profound discussion of the technology than just comparing Rutherford and fusion cross-sections. In conclusion, a physically sound concept of accelerator based nuclear fusion with positive energy output might exist, but several designs like "Migma" and intersecting storage rings were already seen to fail. So far, the odds of anything but Tokamaks and Stellarators to work are low. Anything beyond this can only be answered by detailed modelling and experiments. The attractive promises of nuclear fusion power should just keep us open minded for ideas to finally enable the peaceful exploitation of fusion in the twenty-first century.

References

A. Koning, et al., *TENDL-2015: TALYS-based Evaluated Nuclear Data Library* (Von, 2015). https://tendl.web.psi.ch/tendl_2015/tendl2015.html

H.-S. Bosch, G. Hale, Improved formulas for fusion cross-sections and thermal reactivities. Nucl. Fusion **32**, 611–631 (1992)

F. Hinterberger, *Physik der Teilchenbeschleuniger und Ionenoptik* (Springer, Berlin, 2008)

J.F. Ziegler, J.P. Biersack, M.D. Ziegler, *SRIM—The Stopping and Range of Ions in Matter* (Chester, 2008)

K. Nakajima, *Nuclear Back-end and Transmutation Technology for Waste Disposal* (Springer, Tokyo, 2014)

C. Lejeune, Theoretical and experimental study of the duoplasmatron ion source Part 2: emissive properties of the source. Nuclear Inst. Methods **116**, 429–443 (1974)

J. Ongena, R. Koch, R. Wolf, H. Zohm, Magnetic-confinement fusion. Nat. Phys. **12** (2016). https://doi.org/10.1038/nphys3745

Y. Miyoshi, A. Aoki, R. Hiwatari, Y. Someya, Y. Sakamoto, K. Tobita, Cooling water system design of Japan's DEMO for fusion power production. Fusion Eng. Des. **126**, 110–115 (2018). https://doi.org/10.1016/j.fusengdes.2017.11.013

SCK · CEN Studiecentrum voor Kernenergie. (2017). *SCK CEN/20862373.* Mol: Belgian Nuclear Research Centre (SCK · CEN)

A. Witze, Desperately seeking plutonium. Nature **515**, 484–486 (2014). https://doi.org/10.1038/515484a

K. Annan, et al., *Human Impact Report: Climate Change—The Anatomy of a Silent Crisis.* Global Humanitarian Forum, Geneva. ISBN: 978-2-8399-0553-4

Index

S. Möller, *Accelerator Technology*, Particle Acceleration and Detection,
https://doi.org/10.1007/978-3-030-62308-1

The manufacturer's authorised representative in the EU is Springer Nature Customer Service Centre GmbH, Europaplatz 3, 69115 Heidelberg, Germany. If you have any concerns regarding our products, please contact ProductSafety@springernature.com

Printed and bound by CPI Group (UK) Ltd, Croydon, CR0 4YY
30/09/2026
02225254-0015